AAP Research Notes on
Polymer Engineering Science and Technology

HIGH-PERFORMANCE POLYMERS FOR ENGINEERING-BASED COMPOSITES

Edited by

Omari V. Mukbaniani, DSc

Marc J. M. Abadie, DSc

Tamara Tatrishvili , PhD

AAP | APPLE ACADEMIC PRESS

Apple Academic Press Inc.
3333 Mistwell Crescent
Oakville, ON L6L 0A2
Canada

Apple Academic Press Inc.
9 Spinnaker Way
Waretown, NJ 08758
USA

Exclusive worldwide distribution by CRC Press, a member of Taylor & Francis Group

No claim to original U.S. Government works

Printed in the United States of America on acid-free paper

International Standard Book Number-13: 978-1-77188-119-7 (Hardcover)

International Standard Book Number-13: 978-1-4987-2737-2 (eBook)

Typeset by Accent Premedia Services (www.accentpremedia.com)

Library and Archives Canada Cataloguing in Publication

High-performance polymers for engineering-based composites / edited by
Omari V. Mukbaniani, Marc J. M. Abadie, Tamara Tatrishvili.

(AAP research notes on polymer engineering science and technology series)
Includes bibliographical references and index.
Issued in print and electronic formats.
ISBN 978-1-77188-119-7 (hardcover).--ISBN 978-1-4987-2737-2 (pdf)
1. Polymer engineering. 2. Polymers. 3. Composite materials. I. Abadie, Marc J. M., author, editor II. Mukbaniani, O. V. (Omar V.), author, editor III. Tatrishvili, Tamara, author, editor
IV. Series: AAP research notes on polymer engineering science and technology series
TA455.P58H53 2015 620.1'92 C2015-905817-1 C2015-905818-X

Library of Congress Cataloging-in-Publication Data

High-performance polymers for engineering-based composites / Omari V. Mukbaniani,
Marc J. M. Abadie, Tamara Tatrishvili, editors.

pages cm
Includes index.
ISBN 978-1-77188-119-7 (alk. paper)
1. Polymers. 2. Polymerization. 3. Composite materials. 4. Chemical engineering. I. Mukbaniani, O. V. (Omar V.), editor. II. Abadie, Marc J. M., editor. III. Tatrishvili, Tamara, editor.

TA455.P58H5455 2015 620.1'92--dc23 2015031578

Apple Academic Press also publishes its books in a variety of electronic formats. Some content that appears in print may not be available in electronic format. For information about Apple Academic Press products, visit our website at **www.appleacademicpress.com** and the CRC Press website at **www.crcpress.com**

ABOUT AAP RESEARCH NOTES ON POLYMER ENGINEERING SCIENCE AND TECHNOLOGY

The AAP Research Notes on Polymer Engineering Science and Technology reports on research development in different fields for academic institutes and industrial sectors interested in polymer engineering science and technology. The main objective of this series is to report research progress in this rapidly growing field.

BOOKS IN THE AAP RESEARCH NOTES ON POLYMER ENGINEERING SCIENCE AND TECHNOLOGY SERIES

- **Functional Polymer Blends and Nanocomposites: A Practical Engineering Approach**
  Editors: Gennady E. Zaikov, DSc, Liliya I. Bazylak, PhD, and A. K. Haghi, PhD

- **Polymer Surfaces and Interfaces: Acid-Base Interactions and Adhesion in Polymer-Metal Systems**
  Irina A. Starostina, DSc, Oleg V. Stoyanov, DSc, and Rustam Ya. Deberdeev, DSc

- **Key Technologies in Polymer Chemistry**
  Editors: Nikolay D. Morozkin, DSc, Vadim P. Zakharov, DSc, and Gennady E. Zaikov, DSc

- **Polymers and Polymeric Composites: Properties, Optimization, and Applications**
  Editors: Liliya I. Bazylak, PhD, Gennady E. Zaikov, DSc, and A. K. Haghi, PhD

- **Applied Research on Polymer Composites**
  Editors: Pooria Pasbakhsh, PhD, A. K. Haghi, PhD, and Gennady E. Zaikov, DSc

- **High-Performance Polymers for Engineering-Based Composites**
  Editors: Omari V. Mukbaniani, DSc, Marc J. M. Abadie, DSc, and Tamara Tatrishvili, PhD

- **Compositional Analysis of Polymers**
  Editors: Aleksandr M. Kochnev, PhD, Oleg V. Stoyanov, DSc, Gennady E. Zaikov, DSc, and Renat M. Akhmetkhanov, DSc

- **Computational and Experimental Analysis of Functional Materials**
  Editors: Oleksandr Reshetnyak, DSc, and Gennady E. Zaikov, DSc

CONTENTS

LIST OF CONTRIBUTORS

Marc J. M. Abadie
Institut Charles Gerhardt de Montpellier – Agrégats, Interfaces et Matériaux pour l'Energie (IGCM AIME UMR CNRS 5253), Université Montpellier 2, Place Bataillon, 34095 Montpellier Cedex 5, France, E-mail: abadie@univ-montp2.fr

Aladdin Islam Akhmedov
Institute of Chemistry of Additives Named After ACAD. A.M. Quliyev of National Academy of Sciences of Azerbaijan Republic, E-mail: aki05@mail.ru

J. Aneli
Institute of Macromolecular Chemistry and Polymeric Materials, I. Javakhishvili Tbilisi State University, I. Chavchavadze Ave. 13, 0179 Tbilisi, Georgia, E-mail: jimaneli@yahoo.com

R. G. Bakuradze
Georgian Technical University, Department of Cybernetics, 5 S. Euli St. Tbilisi, 0186, Georgia

T. G. Butkhuzi
Physics Department, New York City College of Technology, CUNY, Brooklyn, New York 11201, USA

A. E. Chalykh
Frumkin Institute of Physical Chemistry and Electrochemistry, Russian Academy of Sciences, Leninskii pr. 31, Moscow, 119991, Russia; P. Melikishvili Institute of Physical and Organic Chemistry, Iv. Javakhishvili Tbilisi State University, Georgia

I. A. Chitrekashvili
P.G. Melikishvili Institute of Physical and Organic Chemistry, Iv. Javakhishvili Tbilisi State University, Georgia

D. S. Davtyan
State Engineering University of Armenia, 105 Teryana Str., Yerevan, 0009, Armenia, E-mail: davtyans@seua.am

S. P. Davtyan
State Engineering University of Armenia, 105 Teryana Str., Yerevan, 0009, Armenia, E-mail: davtyans@seua.am

Erol Erbay
Petkim Petrokimya Holding A.S., Izmir, Turkey, E-mail: eerbay@petkim.com.tr

Bakhytzhan Erzhan
JSC "Institute of Chemical Sciences Named After A.B. Bekturov," 050010, Republic of Kazakhstan, Almaty, Sh. Valikhanov st. 106, Kazakhstan

Barnagul Erzhet
JSC "Institute of Chemical Sciences Named After A.B. Bekturov," 050010, Republic of Kazakhstan, Almaty, Sh. Valikhanov st. 106, Kazakhstan

Riyad Fuad oglu Farzaliev
Institute of Petrochemical Processes of National Academy of Sciences of Azerbaijan, Baku, Azerbaijan, E-mail: fnasirov@petkim.com.tr

Vagif Medjid Farzaliyev
Institute of Chemistry of Additives Named After ACAD. A.M. Quliyev of National Academy of Sciences of Azerbaijan Republic

Reza Fayazi
Department of Chemistry, Faculty of Science, Islamic Azad University, Ardebil Branch, Ardebil, Iran

R. A. Gakhokidze
Iv. Javakhishvili Tbilisi State University, Department of Bioorganic Chemistry, Georgia

Alina Galeyeva
Centre of Physical-Chemical Methods of Research and Analysis, Al-Farabi Kazakh National University, 96a, Tole bi str., Almaty, Kazakhstan, E-mail: alinex@bk.ru

V. K. Gerasimov
P. Melikishvili Institute of Physical and Organic Chemistry, Iv. Javakhishvili Tbilisi State University, Georgia; Frumkin Institute of Physical Chemistry and Electrochemistry, Russian Academy of Sciences, Leninskii pr. 31, Moscow, 119991, Russia

Shahriar Ghammamy
Department of Chemistry, Faculty of Science, Imam Khomeini International University, Qazvin, Iran, E-mail: shghamami@yahoo.com

Nazi Goliadze
Ivane Javakhishvili Tbilisi State University Chavchavadze Avenue 1, Tbilisi 0128, Georgia

Abasgulu Guliyev
Institute of Polymer Materials of Azerbaijan National Academy of Sciences, Azerbaijan, E-mail: abasgulu@yandex.ru

M. B. Gurgenishvili
P.G. Melikishvili Institute of Physical and Organic Chemistry, Iv. Javakhishvili Tbilisi State University, Georgia, E-mail: marina.gurgenishvili@yahoo.com

M. G. Hamamchyan
State Engineering University of Armenia, 105 Teryana Str., Yerevan, 0009, Armenia

Gulara Nariman kizi Hasanova
Institute of Petrochemical Processes Azerbaijan National Academy of Sciences, 30, Khodjaly av., AZ 1025, Baku, Azerbaijan

Nazil Fazil oglu Janibayov
Institute of Petrochemical Processes Azerbaijan National Academy of Sciences, 30, Khodjaly av., AZ 1025, Baku, Azerbaijan, E-mail: j.nazil@yahoo.com

Talkybek Jumadilov
JSC "Institute of Chemical Sciences Named After A.B. Bekturov," 050010, Republic of Kazakhstan, Almaty, Sh. Valikhanov st. 106, Kazakhstan, E-mail: jumadilov@mail.ru

Khatuna Kakhiani
Ivane Javakhishvili Tbilisi State University, I. Chavchavadze Ave., 1, Georgia,
E-mail: ktsereteli@gmail.com

Saltanat Kaldayeva
JSC "Institute of Chemical Sciences Named After A.B. Bekturov," 050010, Republic of Kazakhstan, Almaty, Sh. Valikhanov st. 106, Kazakhstan

Mohammad Kanafchian
University of Guilan, Rasht, Iran

Motahareh Kanafchian
University of Guilan, Rasht, Iran

V. Kaulin
Donetsk National Technical University, Donetsk, Ukraine

I. M. Khachatryan
Iv. Javakhishvili Tbilisi State University, Department of Bioorganic Chemistry, Georgia

Anna Khaiauri
Ivane Javakhishvili Tbilisi State University Chavchavadze Avenue 1, Tbilisi 0128, Georgia

Oleg Kholkin
Centre of Physical-Chemical Methods of Research and Analysis, Al-Farabi Kazakh National University, 96a, Tole bi str., Almaty, Kazakhstan

Mahsa Khosbakht
Department of Chemistry, Faculty of Science, Imam Khomeini International University, Qazvin, Iran

N. Z. Khotenashvili
P.G. Melikishvili Institute of Physical and Organic Chemistry, Iv. Javakhishvili Tbilisi State University, Georgia

Saule Kokhmetova
Centre of Physical-Chemical Methods of Research and Analysis, Al-Farabi Kazakh National University, 96a, Tole bi str., Almaty, Kazakhstan

Ruslan Kondaurov
JSC "Institute of Chemical Sciences Named After A.B. Bekturov," 050010, Republic of Kazakhstan, Almaty, Sh. Valikhanov st. 106, Kazakhstan

I. Krutko
Donetsk National Technical University, Donetsk, Ukraine, E-mail: techlab@ukr.net

Andrey Kurbatov
Centre of Physical-Chemical Methods of Research and Analysis, Al-Farabi Kazakh National University, 96a, Tole bi str., Almaty, Kazakhstan

M. K. Kurtanidze
Faculty of Exact and Natural Sciences, Ivane Javakhishvili Tbilisi State University, 3 I. Chavchavadze ave, Tbilisi, 0128, Georgia

Amir Lashgari
Department of Chemistry, Faculty of Science, Imam Khomeini International University, Qazvin, Iran

Giorgi Makharadze
Ivane Javakhishvili Tbilisi State University Chavchavadze Avenue 1, Tbilisi 0128, Georgia, E-mail: giorgi.makharadze@yahoo.com

Giorgi Makharadze
Ivane Javakhishvili Tbilisi State University Chavchavadze Avenue 1, Tbilisi 0128, Georgia, E-mail: giorgi.makharadze@yahoo.com

Tamar Makharadze
Ivane Javakhishvili Tbilisi State University Chavchavadze Avenue 1, Tbilisi 0128, Georgia

B. A. Mamedov
Institute of Polymer Materials of Azerbaijan National Academy of Sciences, Sumgait, S. Vurgun Str.124, Azerbaijan, E-mail: ipoma@science.az

E. Markarashvili
Iv. Javakhishvili Tbilisi State University, Department of Chemistry, I. Chavchavadze Ave. 1, 0179 Tbilisi, Georgia

S. S. Mashaeva
Institute of Polymer Materials of Azerbaijan National Academy of Sciences, Sumgait, S. Vurgun Str.124, Azerbaijan

George Meskhi
Samtskhe-Javakheti State University, Faculty of Engineering, Agrarian and Natural Sciences, Akhaltsikhe, 0800 Rustaveli, 106, Georgia, Email: george.meskhi@yahoo.com

O. Mukbaniani
Iv. Javakhishvili Tbilisi State University, Department of Chemistry, I. Chavchavadze Ave. 1, 0179 Tbilisi, Georgia

L. Nadareishvili
Georgian Technical University, Department of Cybernetics, 5 S. Euli St. Tbilisi, 0186, Georgia, E-mail: levannadar@yahoo.com

O. Nadtoka
Taras Shevchenko National University of Kyiv, Volodymyrs'ka str., 64, 01033 Kyiv, Ukraine, E-mail: oksananadtoka@ukr.net

Fuzuli Akber oglu Nasirov
Institute of Petrochemical Processes of National Academy of Sciences of Azerbaijan, Baku, Azerbaijan; Petkim Petrokimya Holding A.Ş., Izmir, Turkiye, E-mail: fnasirov@petkim.com.tr

Yevgeniya Nikolayeva
Centre of Physical-Chemical Methods of Research and Analysis, Al-Farabi Kazakh National University, 96a, Tole bi str., Almaty, Kazakhstan

G. Sh. Papava
P.G. Melikishvili Institute of Physical and Organic Chemistry, Iv. Javakhishvili Tbilisi State University, Georgia

Sh. R. Papava
P.G. Melikishvili Institute of Physical and Organic Chemistry, Iv. Javakhishvili Tbilisi State University, Georgia

I. S. Pavlenishvili
Georgian Technical University, Department of Cybernetics, 5 S. Euli St. Tbilisi, 0186, Georgia

Lia Wijayanti Pratomo
School of Materials Science and Engineering, Nanyang Technological University, 639798, Singapore

Sevda Rafi kizi Rafiyeva
Institute of Petrochemical Processes Azerbaijan National Academy of Sciences, 30, Khodjaly av., AZ 1025, Baku, Azerbaijan

Gafar Ramazanov
Sumgait State University, Badalbayli Street, Sumgayit AZ5008, Azerbaijan

M. D. Rukhadze
Faculty of Exact and Natural Sciences, Ivane Javakhishvili Tbilisi State University, 3 I. Chavchavadze ave, Tbilisi, 0128, Georgia, Email: marina.rukhadze@tsu.ge

S. N. Rusanova
Kazan National Research Technological University, K.Marx str., 68, Kazan, 420015, Tatarstan, Russia; P. Melikishvili Institute of Physical and Organic Chemistry, Iv. Javakhishvili Tbilisi State University, Georgia

Seymur Salman oglu Salmanov
Institute of Petrochemical Processes of National Academy of Sciences of Azerbaijan, Baku, Azerbaijan

K. Satsyuk
Donetsk National Technical University, Donetsk, Ukraine

Irina Savchenko
National Taras Shevchenko University of Kyiv, 60, Volodymyrska Str., 01033 Kyiv, Ukraine, E-mail: iras@univ.kiev.ua

Sadjad Sedaghat
Department of Chemistry, Faculty of Science, Islamic Azad University, Malard Branch, Malard, Iran

Rita Shahnazarli
Institute of Polymer Materials of Azerbaijan National Academy of Sciences, Azerbaijan

L. K. Sharashidze
Georgian Technical University, Department of Cybernetics, 5 S. Euli St. Tbilisi, 0186, Georgia

V. A. Sherozia
P.G. Melikishvili Institute of Physical and Organic Chemistry, Iv. Javakhishvili Tbilisi State University, Georgia

S. Yu. Sofina
Kazan National Research Technological University, K.Marx str., 68, Kazan, 420015, Tatarstan, Russia; P. Melikishvili Institute of Physical and Organic Chemistry, Iv. Javakhishvili Tbilisi State University, Georgia

O. V. Stoyanov
Kazan National Research Technological University, K.Marx str., 68, Kazan, 420015, Tatarstan, Russia; P. Melikishvili Institute of Physical and Organic Chemistry, Iv. Javakhishvili Tbilisi State University, Georgia, E-mail: ov_stoyanov@mail.ru

Guram Supatashvili
Ivane Javakhishvili Tbilisi State University Chavchavadze Avenue 1, Tbilisi 0128, Georgia

V. Syromyatnikov
Taras Shevchenko National University of Kyiv, Volodymyrs'ka str., 64, 01033 Kyiv, Ukraine

Z. Sh. Tabukashvili
P.G. Melikishvili Institute of Physical and Organic Chemistry, Iv. Javakhishvili Tbilisi State University, Georgia

V. Tarasenko
Taras Shevchenko National University of Kyiv, Volodymyrs'ka str., 64, 01033 Kyiv, Ukraine

T. Tatrishvili
Iv. Javakhishvili Tbilisi State University, Department of Chemistry, I. Chavchavadze Ave. 1, 0179 Tbilisi, Georgia

N. E. Temnikova
Kazan National Research Technological University, K.Marx str., 68, Kazan, 420015, Tatarstan, Russia; P. Melikishvili Institute of Physical and Organic Chemistry, Iv. Javakhishvili Tbilisi State University, Georgia

A. O. Tonoyan
State Engineering University of Armenia, 105 Teryana Str., Yerevan, 0009, Armenia

N. S. Topuridze
Georgian Technical University, Department of Cybernetics, 5 S. Euli St. Tbilisi, 0186, Georgia

Kakha Tsereteli
Independent Expert, P. Melikishvili Institute of Physical and Organic Chemistry, Iv. Javakhishvili Tbilisi State University, Georgia

A. Ya. Valipour
Institute of Polymer Materials of Azerbaijan National Academy of Sciences, Sumgait, S. Vurgun Str.124, Azerbaijan

Anahit Varderesyan
State Engineering University of Armenia, Yerevan, Teryan Str.105, Armenia, E-mail: anahitvarderesyan08@mail.ru

N. S. Vassilieva-Vashakmadze
Iv. Javakhishvili Tbilisi State University, Department of Bioorganic Chemistry, Georgia, E-mail: nonavas@rambler.ru

Eldar B. Zeynalov
Institute of Petrochemical Processes Named After ACAD. Y.G. Mamedaliyev, Azerbaijan National Academy of Sciences, Khojaly Aven 30, AZ1025 Baku, Azerbaijan,
E-mail: zeynalov_2000@yahoo.com

LIST OF ABBREVIATIONS

AA	acetic acid
AAM	acrylamide
APS	ammonium persulfate
ATRP	atom transfer radical polymerization
AIBN	azobisisobutyronitrile
BP	benzoyl peroxide
BMA	buthylmethacrylate
BA	butyl acrylate
CPU	central processing unit
CTA	chain transfer agent
CS	chitosan
CS	coherent state
DMA	decylmethacrylate
DDA	degree of deacetylation
DFT	density functional theory
DCRS	developing drug controlled release systems
DCM	dichloromethane
DCP	dicumylperoxide
DCPC	dicyclohexylperoxydicarbonate
DEAC	diethylaluminum chloride
DTA	differential thermal analysis
DTG	differential thermo-gravimetric
DAA	dinitrileazo (bis) isooilacid
EB	electron beam
EOM	equation of motion
ETS	ethyl silicate
EVA	ethylene with vinyl acetate
EHA	ethylhexylacrylate
FTIR	Fourier-transform infrared spectroscopy
FP	frontal polymerization
FGMs	functionally graded materials'

GPC	gel permeation chromatograph
GOPs	gradually oriented/stretched polymers
GOS	gradually oriented/stretched state
GPU	graphical processing unit
HMA	hexylmethacrylate
HMC	high-molecular compounds
HG	hydrogels
IR	infrared analysis
KS	Kohn-Sham
LC	liquid crystal
LDPE	low-density polyethylene
MFR	melt flow rate
MMA	methyl-methacrylate
MMA-co-MA	methylmethacrylate-co-methacrylic acid
MW	molecular weight
MWCNT	multi-wall carbon nanotubes
NMP	nitroxide mediated polymerization
NMR	nuclear magnetic resonance
NA	nutrient agar
NB	nutrient broth
OMA	octylmethacrylate
OLED	organic light emitting diodes
PMC	paramagnetic centers
PCM	pitch composite materials
gP2M5VP	poly-2-methyl-5-vinylpyridine hydrogel
PSMA	poly(styrene-alt-maleic anhydride)
PAAm	polyacrylamide
gPAA	polyacrylic acid hydrogel
PANI	polyaniline
PE	polyelectrolytes
PEO	polyethylene oxide
PE	polymer electrolytes
PMMA	polymethacrylic acid
PVC	polyvinyl chloride
PES	potential energy surface
RAFT	reversible addition–fragmentation chain transfer

SEM	scanning electron microscope
SE	Schrödinger equation
SIMS	secondary positive and negative ions
SWCNT	single-wall carbon nanotubes
SPV	softening point by Vicka
SEC	static exchange capacity
THF	tetrahydrofuran
TG	thermogravimetric
TGA	thermogravimetric analysis
TEGDM	triethylene-glycol-dimethacrylate
TFA	triflouroacetic acid
UV	ultraviolet
UCPS	upper critical point of solubility
VCP	vinylcyclopropanes
VTF	Vogel-Tamman-Fulcher
XRD	x-ray diffraction analysis

PREFACE

This book provides coverage of new research in polymer science and engineering with applications in chemical engineering, materials science, and chemistry. In addition to synthetic polymer chemistry, it also looks at the properties of polymers in various states (solution, melt, solid). The chapters provide a survey of the important categories of polymers, including commodity thermoplastics and fibers, elastomers and thermosets, and engineering and specialty polymers. Basic polymer processing principles are explained as well as in-depth application of the latest polymer applications in different industrial sectors. This new book systematically reviews the field's current state and emerging advances. With contributions from experts from both the industry and academia, this book presents the latest developments in polymer products and chemical processes. It incorporates appropriate case studies, explanatory notes, and schematics for more clarity and better understanding.

This new book:

- familiarizes readers with new aspects of the techniques used in the examination of polymers, including chemical, physicochemical, and purely physical methods of examination;
- gives an up-to-date and thorough exposition of the present state of the art of polymer chemistry;
- features a collection of articles that highlight some important areas of current interest in polymer products and chemical processes;
- describes the types of techniques now available to the polymer chemist and technician, and discusses their capabilities, limitations, and applications; and
- provides a balance between materials science and mechanics aspects, basic and applied research, and high-technology and high-volume (low-cost) composite development.

ABOUT THE EDITORS

Omari V. Mukbaniani, DSc

Omari Vasilii Mukbaniani, DSc, is Professor and Director of the Macromolecular Chemistry Department of I. Javakhishvili Tbilisi State University, Tbilisi, Georgia. He is also the Director of the Institute of Macromolecular Chemistry and Polymeric Materials. For several years he was a member of the advisory board of the *Journal Proceedings of Iv. Javakhishvili Tbilisi State University* (Chemical Series), and a contributing editor of *Polymer News* and the *Polymers Research Journal*. His research interests include polymer chemistry, polymeric materials, and chemistry of organosilicon compounds. He is an author more than 360 publication, eight books, three monographs, and 10 inventions.

Marc J. M. Abadie, DSc

Professor Marc J. M. Abadie is Emeritus Professor at the University Montpellier, France. He was Head of the Laboratory of Polymer Science and Advanced Organic Materials – LEMP/MAO. He is currently the 'Michael Fam' Visiting Professor at the School of Materials Sciences and Engineering, Nanyang Technological University NTU, Singapore. His present activity concerns high-performance polymers for PEMFCs, composites and nanocomposites, UV/EB coatings, and biomaterials. He has published 11 books and 11 patents. He has advised nearly 95 MS and 52 PhD students with whom he has published over 402 papers. He has more than 40 years of experience in polymer science with 10 years in the industry (IBM, USA – MOD, UK & SNPA/Total, France). He created in the 1980s the 'International Symposium on Polyimides and High Temperature Polymers,' (STEPI), which takes place every three years in Montpellier, France. The next symposium, STEPI 10, will be in June 2016.

Tamara Tatrishvili, DSc

Tamara Tatrishvili, PhD, is Senior Specialist at the Unite of Academic Process Management (Faculty of Exact and Natural Sciences) at Ivane Javakhishvili Tbilisi State University as well as Senior Researcher of the Institute of Macromolecular Chemistry and Polymeric Materials in Tbilisi, Georgia.

PART I

APPLICATION OF POLYMER CHEMISTRY AND PROMISING TECHNOLOGIES

CHAPTER 1

RAFT POLYMERIZATION OF ACRYLIC ESTERS USING NOVEL CHAIN TRANSFER AGENTS ON THE BASIS OF THIOCOMPOUNDS

RIYAD FUAD OGLU FARZALIEV,[1] FUZULI AKBER OGLU NASIROV,[1,2] EROL ERBAY,[2] and NAZIL FAZIL OGLU JANIBAYOV[1]

[1]*Institute of Petrochemical Processes of National Academy of Sciences of Azerbaijan, Baku, Azerbaijan,*
E-mail: fnasirov@petkim.com.tr

[2]*Petkim Petrokimya Holding A.S., Izmir, Turkey,*
E-mail: eerbay@petkim.com.tr

CONTENTS

ABSTRACT

Polyalkylacrylates with a narrow polydispersion were synthesized with RAFT polymerization of butyl acrylate or 2-ethylhexylacrylate using novel S-alkylarylthiophosphate type chain transfer agents. Well-defined polyalkylarylacrylates with a wide range of controlled molecular weight and narrow polydispersity ($1.22 < Mw/Mn < 1.42$) were obtained. Results indicate that the new S-alkylarylthiophosphate type chain transfer agents are effective for controlled/living RAFT polymerization of acrylates with different alkyl groups and polymerization process is well controlled. A linear increase of molecular weight occurred with respect to conversion and the polydispersity was relatively low.

Synthesized polyalkylacrylates with narrow polydispersity were successfully tested as a viscosity additive to base motor oils.

1.1 INTRODUCTION

Controlled free radical polymerization is a convenient and versatile technique to synthesize polymers with a controlled molecular weight, narrow polydispersity and designed architecture. The majority of this work in the literature has been carried out on nitroxide mediated polymerization (NMP) [1, 2], Atom transfer radical polymerization (ATRP) [7] and RAFT polymerization [4–12] to form narrow polydispersity polymers or copolymers from most monomers amenable to radical polymerization. Over the past few years it has been demonstrated that polymerization through reversible addition–fragmentation chain transfer (RAFT) is an extremely versatile process. RAFT polymerization is a relatively simple methodology that can be used to polymerize a wide range of monomers under various conditions. It is possible to obtain high conversion and to achieve commercially acceptable polymerization rates *via* RAFT polymerizations. The chain transfer agent (CTA) plays an essential role to accomplish successful polymerization, because the key feature of RAFT is the sequence of addition-fragmentation equilibration involving chain transfer agents. Reported difficulties with RAFT polymerization (such as, retardation and poorer than expected control) are frequently attributable to inappropriate choice of CTA for the monomer(s) and/or reaction conditions. CTAs that perform well under a given set of circumstances are not necessarily optimal for all circumstances. In order to guide the selection and design of CTAs for RAFT polymerization of various monomers (exactly, alkyl acrylates), relative reactivity's of different functional groups were investigated by our group at the Institute of Petrochemical Processes of Azerbaijan National Academy of Sciences (IPCP of ANAS).

Polyalkylacrylates are important functional materials because of their useful viscosity properties when added to base motor oils.

1.2 EXPERIMENTAL PART

High purity monomers, butyl acrylate (BA) and 2-ethylhexylacrylate (EHA), were purchased from Sigma and used without further purification. Initiators, benzoyl peroxide (BP), dicumylperoxide (DCP), and dinitrileazo (bis) isooilacid (DAA), were also were purchased from Sigma and used after recrystallization from isopropanol.

As chain transfer agents, S-alkylarylthiophosphates were used and synthesized by the reaction of O,O-diphenyl-S-toluildithiophosphate and O-butyl-toluil- or 3,5-di-tert-butyl-4-hydroxyphenyl-xynthogenate, with the following chemical structures:

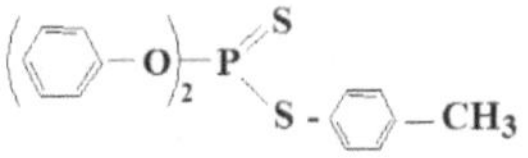

CTA-1

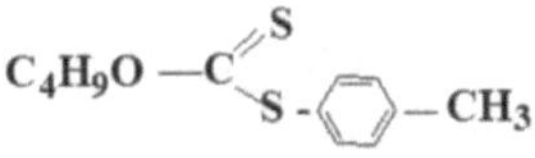

CTA-2

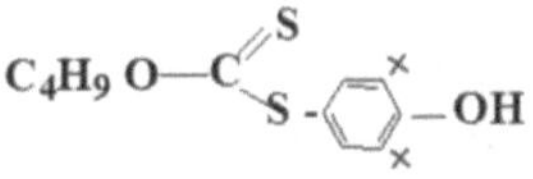

CTA-3

In the polymerization reaction, toluene, benzene and p-xylene were used as solvents. The general procedure was as follows: an ampule tube was filled with RAFT-1 benzoyl peroxide and p-xylene. The tube was purged with argon for 15 min to remove oxygen in the solution, sealed with argon and then placed into an oil bath at the desired polymerization temperature. The polymerization was stopped after a desired period of time by cooling the tube under cold water. Afterwards, the tube was opened. The contents were transferred and dissolved in tetrahydrofuran (THF), and precipitated via the addition of a large amount of methanol. The conversion of monomer was determined gravimetrically.

Molecular weights and the molecular weight distributions were measured using Waters 1515 GPC with THF as the mobile phase and at a column temperature of 30°C. Polystyrene standards were used to calibrate the columns.

IR spectrum was recorded by the means of a Nicholet NEXUS 670 instrument between 400 cm^{-1} to 4000 cm^{-1} wave numbers, using a film prepared with KBr, received from THF or toluene solution.

1.3 RESULTS AND DISCUSSION

We have investigated the effects of different initiators, CTAs and solvents on RAFT polymerization of butyl- and 2-ethylhexyl-acrylates. The results of the experiments are presented in Table 1.1.

As seen in Table 1.1, without CTA, BA-monomer polymerizes with BP (Exp. 1), DCP (Exp. 4), and DAA in p-xylene solvent to a high molecular weight of poly-BA and yield of 85–95%. But the obtained polymers have wide range molecular weight distribution, MWD of 2.8–4.2. This is observed also in 2-EHA polymerization (Exp. 10, 13, 16). Polymer yield was 90–96% and MWD was 2.5–4.5.

In the presence of various CTAs, BA polymerizes to poly-BA with a yield of 65–96% and narrow MWD of 1.35–1.18 both in toluene and p-xylene solutions (Exp. 2, 3, 5, 6, 8, 9). The same results were obtained in the polymerization of 2-EHA. In toluene and p-xylene solution when using BP, DCP, and DAA, the yield of poly-2-EHA was in the range of 82–98% with a narrow MWD of 1.20–1.33. The results showed that synthesized S-alkylarylthiophosphates are effective chain transfer agents (CTAs) for the polymerization of butyl- and 2-ethylhexyl-acrylates in order to obtain narrow polydispersity.

Experimental results of different CTAs types, concentrations, reaction temperatures, and reaction times, which effect RAFT polymerizations of BA and 2-EHA, are presented in Tables 1.2 and 1.3.

Reaction concentrations: [M] = 2.0 mol/L; [I]=1.0×10^{-3} mol/L, [CTA-1] = 5.0×10^{-3} mol/L, reaction temperatures of T = 70°C and reaction times of τ = 6 h CTA-1 allow for the production of polybutylacrylate with the yields of 80%, molecular weights (Mn) of 85,000, and polydispersity (Mw/Mn) of 1.15.

TABLE 1.1 Effects of Initiators, CTAs and Solvents Types on RAFT Polymerization of Butyl- and 2-Ethylhexyl-Acrylates

№	Monomer, M	Initiator, I	CTA	Solvent	[M], mol/L	$[I]\times10^3$, mol/L	$[CTA]\times10^3$, mol/L	T, °C	Reaction time, hours	Polymer yield, %	$Mn\times10^{-3}$	M_w/M_n
1	BA	BP	-	p-xylene	2.0	5	0	80	6	95	85	2.80
2	"_"	BP	CTA-1	toluene	2.0	5	10	60	6	75	66	1.35
3	"_"	BP	CTA-2	p-xylene	3.0	10	50	90	12	96	85	1.23
4	"_"	DCP	-	p-xylene	2.0	5	0	120	6	88	65	3.50
5	"_"	DCP	CTA-3	toluene	3.0	1	5	120	6	80	54	1.22
6	"_"	DCP	CTA-1	p-xylene	2.0	5	10	130	12	93	72	1.24
7	"_"	DAA	-	p-xylene	2.0	5	0	70	6	85	100	4.20
8	"_"	DAA	CTA-2	toluene	1.0	1	10	65	6	65	65	1.18
9	"_"	DAA	CTA-3	p-xylene	3.0	5	10	80	12	90	95	1.20
10	2-EHA	BP	-	p-xylene	2.0	5	0	80	6	96	86	2.50
11	"_"	BP	CTA-3	toluene	3.0	5	50	90	6	88	60	1.20
12	"_"	BP	CTA-2	p-xylene	3.0	10	10	70	12	95	92	1.31
13	"_"	DCP	-	p-xylene	2.0	5	0	120	6	92	115	3.10
14	"_"	DCP	CTA-1	toluene	3.0	10	50	115	6	82	95	1.25
15	"_"	DCP	CTA-2	p-xylene	2.0	5	10	130	12	94	82	1.32
16	"_"	DAA	-	p-xylene	2.0	5	0	70	6	90	118	4.50
17	"_"	DAA	CTA-3	toluene	2.0	1	5	70	6	86	65	1.28
18	"_"	DAA	CTA-1	p-xylene	2.0	5	10	80	12	98	88	1.33

TABLE 1.2 Effects of CTA Type, Concentration, Reaction Temperature and Reaction Time on RAFT Polymerization of Butyl Acrylate (Reaction Conditions: [M]= 2.0 mol/L; initiator (I) – benzoyl peroxide; solvent –p-xylene)

№	$[I]\times10^3$, mol/L	CTA	$[CTA]\times10^3$, mol/L	Temperature, °C	Reaction time, hours	Polymer yield, %	$Mn\times10^{-3}$	Mw/Mn
1	1.0	CTA-1	5.0	70	6	80.0	85.0	1.15
2	5.0	CTA-1	10.0	80	6	85.0	79.0	1.20
3	1.0	CTA-2	5.0	70	6	90.0	81.5	1.25
4	10.0	CTA-2	50.0	90	6	82.0	75.0	1.12
5	1.0	CTA-3	5.0	70	6	93.0	71.0	1.18
6	1.0	CTA-3	10.0	90	1	42.0	1.2	1.65
7	1.0	CTA-3	10.0	90	3	70.0	40.0	1.37
8	1.0	CTA-3	10.0	90	6	85.0	58.0	1.25
9	1.0	CTA-3	10.0	90	12	92.0	66.0	1.24
10	1.0	CTA-3	10.0	90	24	98.0	73.0	1.23
11	1.0	CTA-3	10.0	90	36	99.0	74.0	1.22

TABLE 1.3 Effects of CTA Type, Temperature, and Reaction Time on RAFT Polymerization of 2-Ethylhexylacrylate (Reaction Conditions: [M] = 2.0 mol/L; initiator, I – benzoyl peroxide; solvent –p-xylene

№	$[I]\times10^3$, mol/L	CTA	[CDA]$\times10^3$, mol/L	Temperature, °C	Reaction time, hours	Polymer yield, %	$Mn\times10^{-3}$	Mw/Mn
1	5.0	CTA-1	5.0	70	6	85.0	95.0	1.80
2	5.0	CTA-1	10.0	80	6	75.0	32.5	1.55
3	5.0	CTA-1	50.0	90	6	65.0	25.0	1.32
4	1.0	CTA-2	2.5	70	6	80.0	40.5	1.60
5	10.0	CTA-2	50.0	90	6	72.0	30.0	1.33
6	5.0	CTA-3	10.0	80	6	80.0	35.0	1.35
7	10.0	CTA-3	50.0	90	1	50.0	1.6	1.42
8	10.0	CTA-3	50.0	90	3	65.0	19.0	1.32
9	10.0	CTA-3	50.0	90	6	75.0	30.0	1.28
10	10.0	CTA-3	50.0	90	10	90.0	46.0	1.25
11	10.0	CTA-3	50.0	90	20	95.0	51.0	1.24
12	10.0	CTA-3	50.0	90	36	99.0	55.0	1.23

Under the same conditions, CTA-2 and CTA-3 allow production of polybutadiene with the yields of 90% and 93%, molecular weights (Mn) of 81,500 and 71,000, and polydispersities of 1.25 and 1.18, respectively.

In the presence of CTA-3, an increase in reaction time from 1 to 36 h results an increase of polybutylacrylate yield from 42.0% to 99.0%. This results in an increase in molecular weight of the polymer from 1200 to 74,000 and a decrease of molecular weight distribution (Mw/Mn) from 1.66 to 1.22. As can be seen in Figure 1.1(a), the molecular weight of the polymer increases linearly with monomer conversion.

The same results were received in the polymerization of 2-ethylhexylacrylate (Table 1.3). The molecular weight increases linearly with conversion and the molecular weight distributions remain relatively low after 65% conversion of monomer, indicating that the polymerization remained controlled (Figure 1.1). Results consistent with living-type polymerization with fast initiation and little to no termination were obtained at lower temperatures. The molecular weight distributions obtained were narrower (1.22 < Mw/Mn < 1.42) with a higher initiator efficiency (Tables 1.2 and 1.3).

The collective results shown demonstrate that the new S-alkylarylthiophosphate type chain transfer agents are effective for the controlled/living RAFT polymerization of acrylates with different alkyl groups.

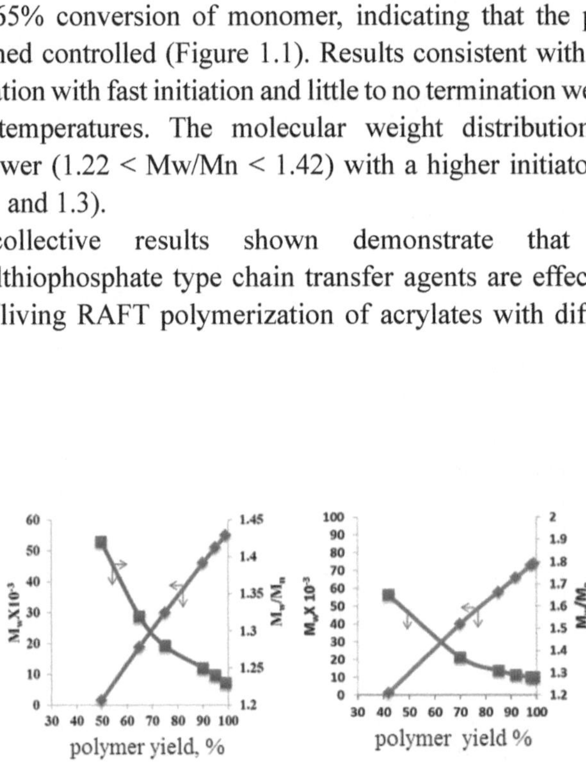

(a) (b)

FIGURE 1.1 Molecular weight/conversion data for the polymerization of butyl- (a) and 2-ethylhexyl-acrylates (b) RAFT polymerization with synthesized novel CTAs.

1.4 CONCLUSION

The newly synthesized S-alkylarylthiophosphate type chain transfer agents were investigated in the RAFT polymerization of butyl- or 2-ethylhexyl-acrylates initiated by benzoyl peroxide in p-xylene. Well-defined polyalkylarylacrylates with a wide range of controlled molecular weights and narrow polydispersities (1.22< Mw/Mn< 1.42) were obtained. The polymers were characterized by FTIR and GPC. The molecular weight increases linearly with conversion and the molecular weight distributions remained relatively low after 65% conversion of monomer, indicating that the polymerization was controlled. Results consistent with living-type polymerization with fast initiation and little or no termination were obtained at lower temperatures. The collective results shown demonstrate that the new S-alkylarylthiophosphate type chain transfer agents are effective for the controlled/"living" RAFT polymerization of acrylates with different alkyl groups.

Various narrow molecular weight distribution polyalkylarylacrylates were successfully tested as a viscosity additive in base motor oils.

KEYWORDS

- **alkyl acrylates**
- **chain transfer agent**
- **narrow polydispersity**
- **RAFT polymerization**
- **S-alkylarylthiophosphates**

REFERENCES

1. Matyjaszewski K., Controlled/Living Radical Polymerization. Progress in ATRP, NMP and RAFT, Proceedings of a Symposium on Controlled Radical Polymerization, 22–24 August 1999, New Orleans, In: ACS Symp. Ser., 768, 2000.
2. Matyjaszewski K., Davis T. P. Handbook of Radical Polymerization. New York: Wiley-Interscience, 920 p, 2002.

3. Percec V., Barboiu B. Macromolecules, 28, 7970, 1995.
4. Coca S., Jasieczek C., Beers K. L., Matyjaszewski K. J. Polym.Sci., Part A: Polym. Chem., 36, 1417, 1998.
5. Moad G., Chong Y. K., Postma A., Rizzardo E., Thang S. H. Polymer, 46, 8458–8468, 2005.
6. Matyjaszewski K., Coca S., Jasieczek C. B. Macromol. Chem. Phys., 198, 4011, 1997.
7. Grmaud T., Matyjaszewski K. Macromolecules, 30, 2216, 1997.
8. Haddleton D. M., Jasieczek C. B., Hannon M. J., Shooter A. J. Macromolecules, 30, 2190, 1997.
9. Wang J. L., Grimaud T., Matyjaszewski K. Macromolecules, 30, 6507, 1997.
10. Percec V., Barboiu B., Kim H. J. J. Am. Chem. Soc., 120, 305, 1998.
11. Zhang X., Xia J., Matyjaszewski K. Macromolecules, 31, 5167, 1998.
12. Qiu J., Gaynor S. G., Matyjaszewski K. Macromolecules, 32, 2827, 1999.
13. Chiefari J., Chong Y. K., Ercole F., Krstina J., Jeffery J., Le T. P. T., Mayadunne R. T. A., Meijs G. F., Moad C. L., Moad G., Rizzardo E., Thang S.H. Macromolecules, 31, 5559–5562, 1998.
14. Moad G., Rizzardo E., Thang S. H. Australian Journal of Chemistry, 59, 669–692, 2006.
15. Moad G., Rizzardo E., Thang S.H. Australian Journal of Chemistry, 58, 379–410, 2005.

CHAPTER 2

RING-OPENING POLYMERIZATION OF VINYLCYCLOPROPANES

ABASGULU GULIYEV,[1] RITA SHAHNAZARLI,[1] and GAFAR RAMAZANOV[2]

[1]*Institute of Polymer Materials of Azerbaijan National Academy of Sciences, Azerbaijan, E-mail: abasgulu@yandex.ru*

[2]*Sumgait State University, Badalbayli Street, Sumgayit AZ5008, Azerbaijan*

CONTENTS

ABSTRACT

The vinylcyclopropanes (VCP) with various polar and non-polar, cyclic and non-cyclic, vicinal and geminate substituents in cyclopropane ring containing methyl group and chlorine atoms in double bond and in three-membered cycle have been synthesized and characterized. Their reactivity in the processes of radical polymerization has been established. The model addition reaction of thiophenol to the synthesized VCP has been carried out and it has been shown that the addition reaction as well as polymerization proceeds on scheme of 1.5-addition, that is, opening of cyclopropane ring. It has been revealed the nature influence of functional substituent on polymerization ability and properties of the prepared polymers. It has been shown that the synthesized VCPs are of interest as the reactive monomers for preparation of polymers with specific properties and the adducts prepared on their basis possess biological activity.

2.1 INTRODUCTION

The chemistry of vinylcyclopropane (VCP) is essentially unstudied field of organic and polymer chemistry, and therefore, is of large interest for theory and practice. On the basis of these compounds one can prepare practically valuable materials for development of economy of the country. They can be used as the biologically active substances (insecticides for agriculture), reactive monomers, especially during synthesis of polymers with technically valuable properties (resistance materials for microelectronics), semiproducts, synthones (as "building blocks") for synthesis of compounds of more complex structure (pheromones, prostaglandins, synthetic analogs of natural pyretroids), as the active additions to polymer materials, etc.

One of the main problems of the polymer chemistry recently is the preparation of functional polymers or they possess unique properties and are used in various fields of science and technique.

In this work the results of the investigations prepared during synthesis and radical polymerization of a series of derivatives of VCP have been presented. The functionally substituted VCPs, which are characterized by high tension of cycle and are easily subjected to the polymerization with ring-opening were chosen as the object [1]. The synthesis of functionally substituted VCPs has been carried out by cyclopropanation one of two double bonds of conjugated diene hydrocarbons by various carbenes generated both in catalytic conditions in the presence copper salts and in interphase conditions in the presence of Makoshi catalyst [2, 3].

2.2 EXPERIMENTAL PART

2.2.1 MATERIALS

All used reagents and solvents have been preliminarily purified according to the standard methods.

2.2.2 TECHNICS

The IR-spectra of the synthesized monomers and polymers and also adducts were taken on apparatus "Specord M-80" in the field of prisms KBr, NaCl, LiF as the thin films. The NMR-spectra were taken on spectrometers BS-487 B of "Tesla" (80 MHz) in various solvents (internal standard – hexamethyldisiloxane). The isomer composition and purity of the synthesized compounds were determined by the methods of thin-layer and gas-liquid chromatography. The characteristic viscosities of the polymer products were determined in Ostwald viscosimeter. The molecular weights were determined by a method of analytical ultracentrifugation according to the data of gel-permeating chromatography. The photosensitive characteristics of polymers were investigated by a method of irradiation of surface of the sample by quartz lamp DPT-220. The biological activity of the synthesized compounds was studied by disk and emulsion-contact methods on strains of some microorganisms.

2.2.3 SYNTHESIS

The synthesis of ethyl esters of vinylcyclopropane carboxylic acids was carried out by interaction of conjugated dienes (divinyl, isoprene and chloroprene) with ethyldiazoacetate in the catalytic conditions in the presence of anhydrous copper salts (sulphuric or chloride) on methodology described in Ref. [3], and acids and alcohols have been prepared by saponification [4] and reduction [5] of the corresponding esters. Allyl, glycidyl and alkoxymethyl derivatives have been prepared by standard methods from corresponding acids and alcohols. The synthesis of 1,1-dichlorsubstituted VCPs was carried out by cyclopropanation of the corresponding diene by dichlorcarbene in the conditions of interphase catalysis. The synthesis of adducts of substituted VCPs with thiophenol was carried out on [6].

The polymerization of VCP was carried out in sealed ampoule in solution in the presence of initiator (AIBN) in the range of 60–70°C.

For determination of photosensitive characteristics of the prepared polymers from each polymer three solution with various concentrations was made. The solutions of resists were applied on glass substrates by centrifugation at 2500 revolutions per minute. The thickness of the prepared films was 0.30–0.45 mcm. The drying of resists was carried out in optimal conditions at 80°C for 20 min. The light intensity was measured by radiation thermoelement RTN-10C. The irradiated films were sustained in developer (mixture of dioxane with isopropyl alcohol at ratio 1:2) for 2 min and were again dried.

The investigations on revealing of biological activity of VCPs synthesized by us and their adducts with thiophenol were carried out by emulsion-contact and disk methods on test-strains: *Staphylococcus aureus, Escherichia coli, Candida albicans, Aspergillus niger, bacillus serratia (anthracoid).*

The emulsion-contact method was carried out in Petri dishes with agar on emulsions of test-cultures with content in 1 mL 500 mln microbial bodies. The content of studied substance – 0.001 in 0.1 mL of 1%-solution (1:2000). The dishes were placed into thermostat for 24 h at temperature 34°C, on the expiry of this time the result was noted.

Disk method. A disk by diameter of 1 cm made from filter paper and impregnated by studied substance was applied on surface of agar. After 24 h incubation in thermostat at 37°C the sizes of zones of delay of bacteria growth around of disks was measured.

The modifying action of some synthesized monomers and oligomer products prepared from them was studied using them in compositions on the basis of polyester and epoxy diane resin in various quantities. The hardening was carried out for 16 h at room temperature, 2 h – at 60°C, 2 h – at 80°C and 2 h – 120°C. The initiating system for curing of polyester composition was the mixture of isopropyl benzene hydroperoxide with cobalt naphthenate and for hardening of epoxide composition was used polyethylene polyamine.

2.3 RESULTS AND DISCUSSION

With the aim of establishment of dependence between structure of VCP and their behavior in free-radical reactions of addition and polymerization and also preparation of polymers on the basis of compounds containing two types of substituents – double bond and functional group, 2-substituted (mono- and di-) VCPs were chosen as the possible initial monomers.

The interaction of ethoxycarbonyl-, diethoxycarbonyl- and dichlorocarbenes with symmetric and asymmetric conjugated diene hydrocarbons was chosen as the reaction for synthesis of intended VCPs:

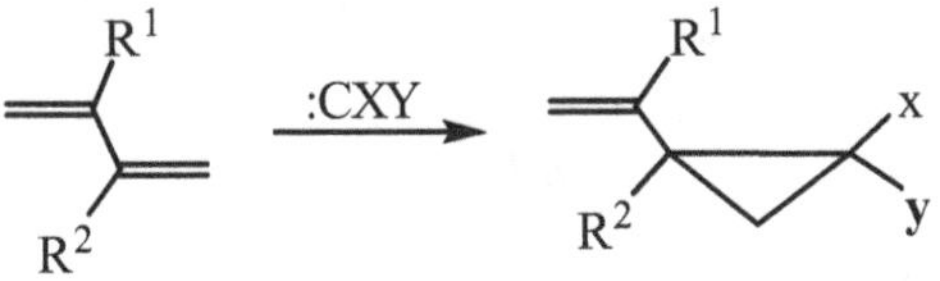

$y = H$; $x = CO_2R$; CH_2OR ($R = Me$; Et; Gly; All)

$x = y = CO_2R$; CH_2OR; $x = y = Cl$ ($R^1 = R^2 = H$; Me; Cl)

It has been shown that cyclopropanation of diene hydrocarbons of symmetric structure of ethoxycarbonylcarbene leads to the formation only mixture of *cis*- and *trans*-isomers while in use of diene hydrocarbons of asymmetric structure along with geometric isomers the position isomers are formed [4]:

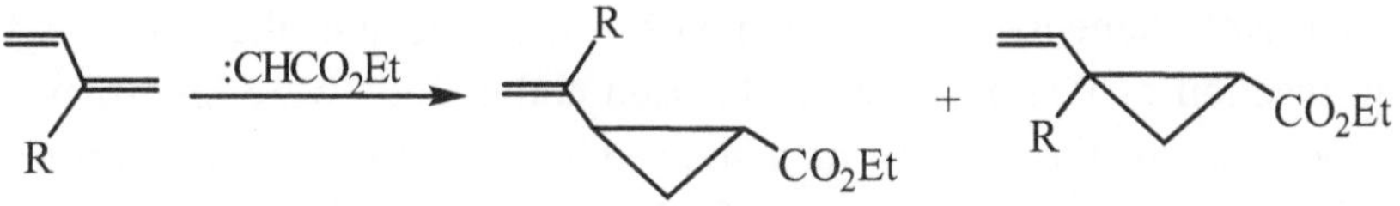

In establishing of structure of the synthesized compounds the chemical shifts of signals of protons of methyl groups connected with double bond and three-membered cycle (in the first case a resonance of protons of methyl groups occurs in more weak fields) have been considered. Depending on location of methyl groups – in double bond or in three-membered cycle – integral intensities of signals of olefin and cyclopropane protons are strongly differed.

In establishing of geometry of individual stereoisomers the chemical shifts of separate protons of cyclopropane ring and character of their spin-spin interaction have been taken into account [7].

The structural attributions made for stereoisomer VCPs on the basis of the spectral data have been completely confirmed by some chemical conversions: favorable location in a space of double bond and carbonyl group in *cis*-isomers allowed in the presence of mineral acids to transfer them to the corresponding lactones (or perlactones in the presence of sodium bicarbonate) [8]:

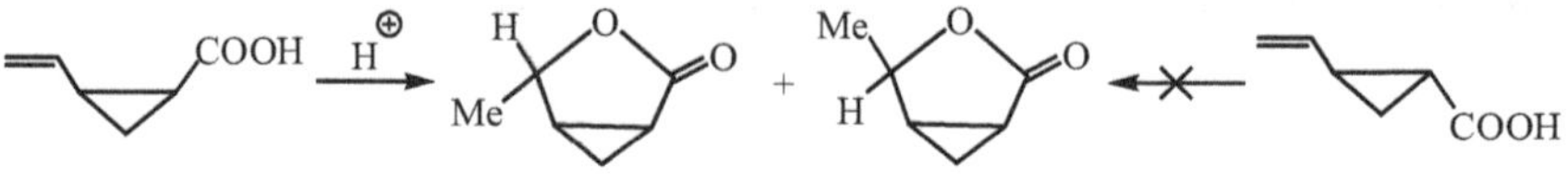

endo- exo-

On the basis of the synthesized VCPs of carboxylic acids and alcohols prepared by establishing of the corresponding esters the allyl and (thio) glycidyl ethers have been then prepared:

R= –CO; $-CH_2$; R′= Me, Et, $CH_2=CH-CH_2-$;

With the aim of study of stereochemical aspects of the processes of stereo- and regioselective addition and study of mechanism of separate stages of polymerization processes and also to make a conclusion about structure of elementary link of chain of macromolecules the model addition reaction of PhSH to the synthesized compounds were investigated. The analysis of the spectral data showed that the addition reaction proceeds with formation of mixture of alicyclic unsaturated sulfides having *cis*- and *trans*-configurations of internal double bond. It means that an addition proceeds regioselectively in 1,5-position:

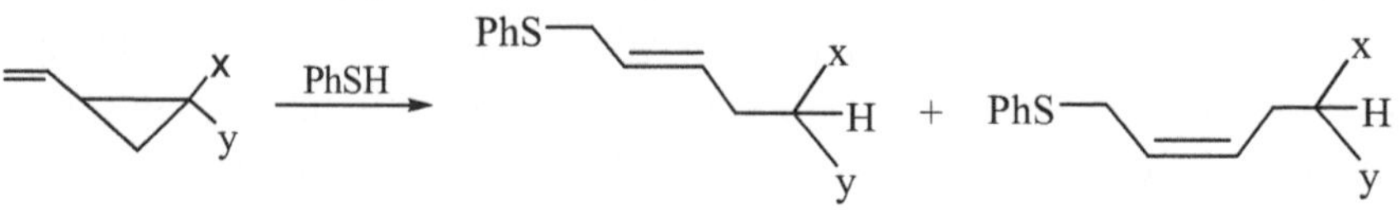

trans- cis-

In Figure 2.1 the spectra of adducts $PhSCH_2CH{=}CH{-}CH_2{-}CH(Cl)_2$ (a) and $PhSCH_2CH{=}CH{-}CH_2{-}CH(CO_2Et)_2$ (b) are presented:

The results of kinetic and spectral investigations and also model reaction showed that VCPs are subjected to the polymerization in the presence of radical initiators with formation of polymer with pentenamer links. In other words, the polymerization of VCP proceeds as a result of simultaneous opening of vinyl group and three-membered cycle. It has been also established that the selective opening of cyclopropane ring in the process of polymerization has been stipulated by stabilizing action of functional substituent on growing radical.

Consequently, a high tension of cyclopropane ring and stabilizing action of functional substituent is the moving force leading to the proceeding of polymerization of VCP with ring-opening [9]:

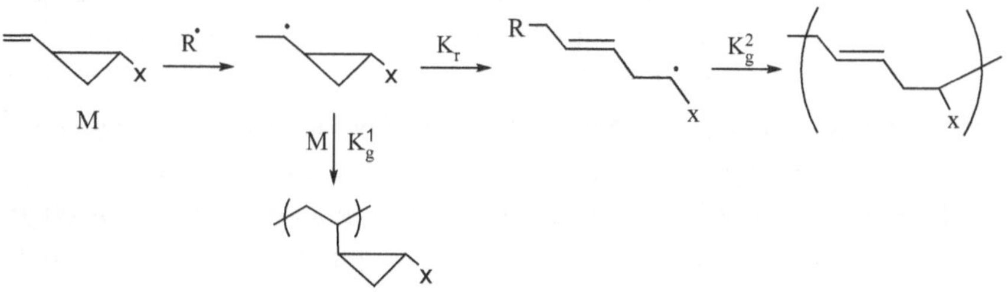

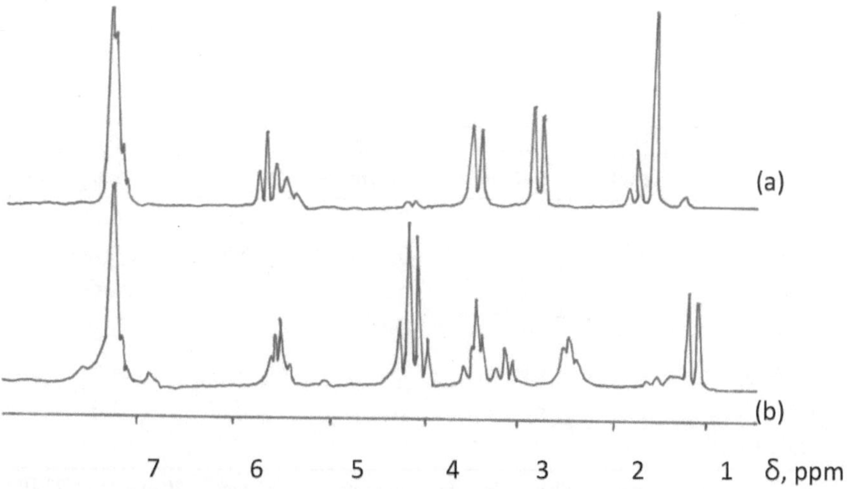

FIGURE 2.1 NMR-spectra of adducts of substituted VCPs with thiophenol.

According to the data of GLC-analysis the investigation of interaction reaction of thiophenol with VCP having – CH_2OR-group as substituent showed that a mixture of two linear and cyclolinear adducts, that is, 1,5- and 1,2-regioisomers is formed. The prepared data showed an availability of elements of both structures in composition of adducts. Consequently, at polymerization of VCP one can expect a formation of both 1,2-, and 1,5- structural links a ratio of which would be depend on conditions of polymerization [11]. In other words, VCPs, as bifunctional monomers could be polymerized both due to opening of double bond and on both group immediately. Indeed, the prepared results showed that a proceeding of radical polymerization of VCP basically takes place on scheme including rearrangement of intermediately forming radicals. A formation of large quantity of rearranged links in this case in comparison with non-rearranged ones means that intermediately forming radicals in greater degree are inclined to rearrangement with ring-opening than addition reaction [10]. This natural consequence results from the fact that a removal of stress of three-membered cycle as a result of its opening has been also conjugated with energy winning in a quantity ~ 106.86 kJ/mol (Table 2.1).

Thus, as a result of competitive reactions – monomolecular rearrangement and intermolecular chain growth the polymers containing both

TABLE 2.1 Polymerization of VCP of Total Formula

(X)	[M], mol/L	Yield of polymer %	[η] of polymers, dl/g	$MM \cdot 10^{-3}$	$V \cdot 10^3$, mol/L·min
$(CO_2Et)_2$	1.5	54	0.46	69.0	6.75
CO_2Et	1.5	47	0.43	12.5	5.10
CO_2H	1.3	42	0.47	6.9	4.55
CO_2Gly	1.0	30	0.32	–	2.50
CH_2OGly	3.0	61	–	0.77	1.53
CH_2OCH_3	3.0	65	0.20	3.56	1.82
CH_2OCOCH_3	3.0	70	0.14	3.32	1.75

Conditions of polymerization: solvent – benzene, temperature – 70°C, time – 120 min, $[J]=5.72 \cdot 10^{-3}$ mol/L.

saturated (with cyclic group in side appendage) and unsaturated linear structures are formed. The found values $r_n = \frac{k_n}{k_p^1} = \sim 261$ allow to conclude that an elementary link with linear structure should prevail over other structural links.

The analysis of spectral data and also data of chemical analysis showed that in the process of polymerization the high-molecular compounds a structure of macromolecules of which corresponds to the simultaneous opening of both polymerization-capable groups – double bond and three-membered cycle are basically formed. As a result a mixture of two *cis*- and *trans*-1,5-structural links with side functional groups is formed:

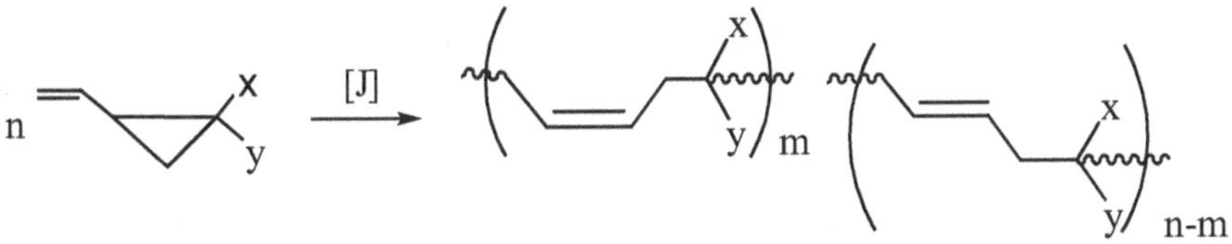

A ratio of a number of olefin protons (or protons of methyl groups in double bond) in polymers prepared from VCP with polar substituent's to total number of protons shows that all samples of polymers basically consist of linear 1,5-structural links. A quantitative determination of unsaturation of the synthesized polymers is completely agreed with those results, which have been prepared from data of the spectral analysis. The good correspondence of the chemical and spectral data allowed to exclude an availability of any other structural units in macromolecular chain of the synthesized polypentenamers.

Recently an interest to the monomers polymerizing with ring-opening sharply grew in connection with revealing of possibility of preparation of polymers with valuable properties on their basis. By this reason a number of the investigations carried out in the field of synthesis and polymerization of cyclic monomer compounds continue to increase. An illustrative example – the works of Japan, German, Korean and other scientist [12–17].

VCPs containing $-CH_2OR$-group as a functional substituent are polymerized with formation along with 1,5-structural links and other links, that is, in the chain of macromolecules along with rearranged links there are also the links not containing double bonds. Such structures, possibly, are formed either as a result of polymerization in position 1,2, or as a result of intramolecular cyclization of growing allyl carbinyl radical (direction B). In addition, it is possible a proceeding of polymerization with expansion of cycle (direction A):

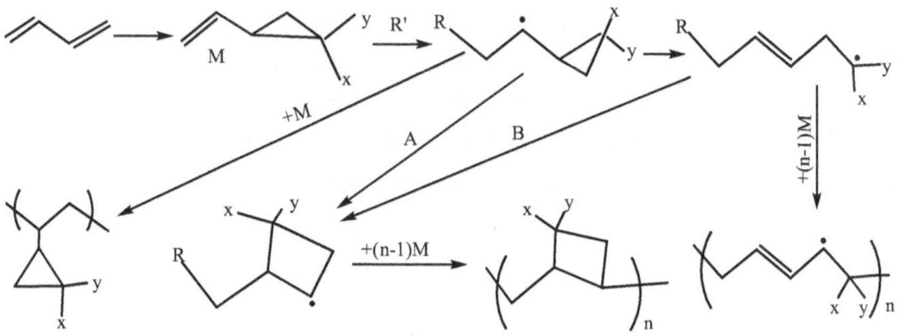

A determination of unsaturation of the prepared polymers testifies to the fact that in macromolecules along with linear structure there are other structural links not containing double bonds. The experiments showed that with increase of concentration of monomers in initial mixture an unsaturation of the forming polymers falls. It means that a monomolecular rearrangement proceeds considerably quickly in comparison with bimolecular chain growth reaction on intermediately forming radical. That is why an increase of concentration of monomer in the initial mixture leads to the increase of fraction of links with cyclopropane ring in the chain of macromolecule. An availability of links with cyclopropane rings as the side appendages in the chain of macromolecules shows that the substituents being in three-membered cycle influence less than polar ethoxycarbonyl groups on stabilization of allylcarbinyl radicals forming in the process of polymerization.

As far as all synthesized VCPs (with the exception of gem-diethoxycarbonyl- and chlorosubstituted) are the mixture of two geometrical isomers, the polymerization should be considered as a copolymerization of stereoisomer's. Therefore it was necessary to consider the relative polymerization activity of these isomers by a method of copolymerization.

The determined values r_1 and r_2 (Table 2.2) indicate to "ideal case" of copolymerization, that is, $r_1 \cdot r_2 = 1$. Consequently, the considered system is characterized by the fact that there are two various monomer and one total

TABLE 2.2 Relative Polymerization Ability of Isomer VCPs

R^1	R^2	R^3	$r_{cis}(K_{11}/K_{12})$	$r_{trans}(K_{22}/K_{21})$
H	H	CO_2H	0.94±0.081	1.06±0.06
H	CH_3	CO_2H	0.97±0.06	1.02±0.05
CH_3	H	CO_2H	0.98±0.031	0.97±0.01
CH_3	CH_3	CO_2H	0.98±0.03	1.02±0.036

radical on which a chain growth takes place in it. Due to equality $K_{11}=K_{12}$ or $K_{22}=K_{21}$ $r_1 \cdot r_2=1$. Consequently, the isomer monomers are differed only by values Q and for them ($e_1=e_2$).

The investigations of polymerization of the synthesized monomers showed also that on polymerization ability the VCPs with methyl group in cycle have a less activity in comparison with VCPs not having methyl group. It has been established by comparison of values of the corresponding rate constants that the largest rate constants is observed in a case of polymerization of monomers having isopropenyl group and chlorine atom in double bond. An availability of methyl group directly in cycle slightly decreases a polymerization activity of monomers which can be connected with destruction of transoid conformation (location of vinyl group) advantageous for these compounds, this directly influences on conjugation of the initial molecules of monomers.

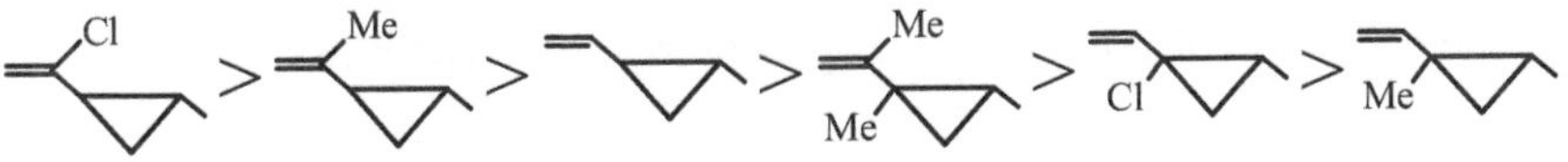

During investigation the influence of nature of functional substituent the following series of activities has been revealed:

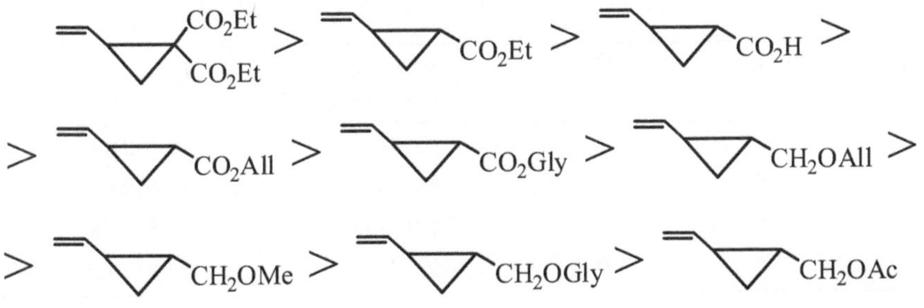

For some of the synthesized polymers the photosensitive properties have been determined. In Table 2.3 the data for polymers prepared from alkyl esters of VCP of carboxylic acids are presented.

The tests showed that the synthesized VCPs and their adducts with thiophenol show a high bactericide and fungicide activity. In comparatively small concentration they cause a death of gram-positive and gram-negative bacteria and also fungi of genus *Candida* for 10–20 min and fungi *Aspergillus niger* die for 20 min., unlike from isomer *cis*-structure, the compounds with *trans*-structure show comparatively high bactericide and fungicide activity. Among studied class of compounds the most effective was 5-phenylthio-3,4-dimethylpenten-3-carboxylic-1 acid which gave the greatest repression zone on fungi and at all concentrations exceeded a standard. All studied compounds possess a bactericide and fungicide property, which, apparently, has been connected with availability of biologically active fragments – carboxyl and hydroxyl groups, heteroatom's of sulphur and chlorine and also three-membered cycle and aromatic ring in their structure.

For improvement of technological, heat-physical and physical-chemical indices of compounds made on the basis of epoxide or polyester resins their modification by oligomer products containing active functional groups in its composition, undergoing the chemical reaction with resin in the process of its hardening was carried out. Epoxy oligomers prepared from VCP with glycidoxymethyl functionality have been tested as the modifiers of polyester and epoxide resin of type ED-20.

TABLE 2.3 Photosensitive Characteristics of Some Polymers With Structure:

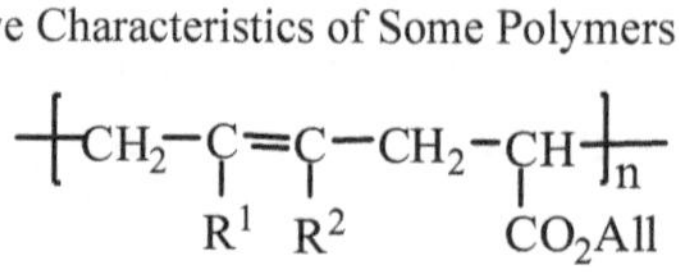

R^1	**R^2**	**MM·10^3**	**Softening temperature, °C**	**Photosensitivity, cm^2/Wt**
H	H	6.2	95–98	18.3–21.4
H	CH_3	5.4	85–90	17.6–20.7
CH_3	H	6.5	101–108	17.5–20.8
CH_3	CH_3	5.8	92–95	16.8–20.2

TABLE 2.4 Composition and properties of polymer compositions

Compositions, mass h.			Properties of hardened composition		
Polyester resin	Epoxide resin ED-20	Epoxy-oligomer	Gel-formation time	Brinell hardness, kgs/cm^2	Vicat heat-resistance, °C
100	–	–	15	16.8	165
100	–	10	24	17.4	170
100	–	30	28	15.0	180
–	100	–	2	13.5	185
–	100	10	3.5	16.8	200
–	100	30	5.0	13.5	190

The prepared compositions showed good heat-resistance and stability at storage and also high strength characteristics (Table 2.4).

2.4 CONCLUSIONS

1. The synthesis of a series of VCPs with various substituents in cyclopropane ring has been carried out and their behavior in the reactions of radical addition and polymerization has been studied. It has been shown that the model reaction of radical addition of thiophenol to VCP proceeds, basically, as 1,5-addition with formation of cyclopropylcarbinyl radicals on intermediate stage the formation of which has been directly proved by EPR-spectroscopy in technique of spin trap.
2. It has been shown on the basis of data of kinetic, stereochemical, model and structural investigations that the mechanism of polymerization of VCP includes the stages of intermediate formation of cyclopropylcarbinyl radicals and their rearrangement to homoallyl radicals accompanying by migration of double bond. It has been established that the substituents being in cyclopropane ring and possessing π-electron system increase a monomolecular rearrangement rate of cyclopropylcarbinyl radicals to allylcarbinyl favoring thereby realization of mechanism of 1,5-addition.
3. The polymerization of VCP has been studied and the following regularities of the process have been established:
   - unlike vinyl monomers, the polymerization VCP in the radical conditions proceeds both on scheme of 1,5-addition with rearrangement

of vinylcyclopropane skeleton and on scheme of 1,2-addition with conservation of cyclopropane ring;
- proceeding of polymerization in that or other direction is controlled by relative stability forming in the process of radicals; polymerization of VCP with polar substituent's in three-membered cycle leads to the formation of macromolecules with linearly building pentenamer links whereas the monomers with substituent's of aliphatic character give "cyclolinear" polymers containing 1,5- and 1,2-structural units.
- polymerization of VCP with polar functional groups proceeds with high degree regioselectively on scheme of 1,5-addition with formation of macromolecular chains containing pentenamer structures.

4. It has been established that VCPs as well as their adducts with thiophenol are the effective biologically active compounds and modifiers for thermoreactive polymers. The polymers from VCP are the high-effective photo- and electrono-resists of negative type and show the high lithographic characteristics. The synthesized polymers with side epoxide groups are thermoreactive polymers and their use in composition of unsaturated polyesters and epoxide resins leads to the formation of materials with high strength characteristics.

KEYWORDS

- **1,5-addition**
- **model reaction**
- **radical polymerization with ring-opening**
- **vinylcyclopropane**

REFERENCES

1. Nefedov, O. M., Ioffe, A.I, Menchikov, L. G. Chemistry of carbenes. M.: Khimiya, 177–197, 1990.
2. Makoshi, M. Reaction of carbanions and halocarbenes in two-phase systems. Usp. khimii, 46 (12), 2124–2202, 1977.
3. Lishanskiy, I. S., Guliyev, A. M., Pomerantsev, V. I., Gurova, A. D. Synthesis and establishment of structure of adducts of carbenes with cis- and trans-pentadiene-1,3. J. Org. Chem., 5, 918, 1970.

4. Modern methods of organic synthesis. Edited by, B. V. Ioffe. Izd-vo LGU. L-d, 53–55, 1980.
5. Landgrebe, J. A., Becker, J. W. Synthesis, stereochemistry and solvolysis of 2-bromobicyclopropyl and closely related structures. J. Org. Chem., 3(3), 1173–1178, 1968.
6. Guliyev, A. M., Lemeshev, A. N., Kasimova, S. P., Lishanskiy, I. S. Alkenylcyclopropanes in the addition reaction. I. Stereochemitry of free radical addition of thiphenol to substituted alkenyl- cyclopropanes. J. Org. Chem., 14(2), 346, 1983.
7. Plemenkov, V. V. Electron and spatial structure of monofunctional cyclopropanes. J. Org. Chem., 7(13), 849–859, 1997.
8. Guliyev, A. M., Guliyev, K. G., Mustafaeva Ts.D., Babakhanov, R. A. Identification of stereoisomers of alkenylcyclopropane carboxylic acid by lactonization. Azerb. khim.zhurn, 1, 103–105, 1979.
9. Guliyev, A. M., Ramazanov, G. A., Guliyev, M. F., Gasanova, S. S. Radical polymerization of 1-vinyl-2-acetoxymethylcyclopropane. Vysokomolek.soyed., B, 29(8), 581–584, 1987.
10. Lishanskiy, I. S., Guliyev, A. M., Zak, A. G. et al. Free radical isomerization polymerization as a method of synthesis of regular-constructed, including optically active polymers. Symposium on macromolecular chemistry. Budapest, 140, 1969.
11. Ioshi, R., Zvolinskiy, B. Heats of polymerization, their physical meaning and bond with structure of monomers. In book. Polymerization of vinyl monomers. Edited by, D.Khem. M.: Khimiya, 250–302, 1978.
12. Bailey, W. J., Sun, R. L., Katsuti, H. et al. Ring-Opening Polymerization with Expansion in Volume. Ring-Opening Polymerization, ACS Symposium Series 59, Robert, F.Gould, Editor, Washington, D.C., 38–59, 1977.
13. Cho Iwan, Kahn Kwang-Dun. Polymerization of substituted cyclopropanes. I. Radical Polymerization of 1,1-disubstituted 2-vinylcyclopropanes. J. Polym. Sci. Polym. Chem. Ed., 17, 3169–3182, 1979.
14. Sanda, F., Takata, T., Endo, T. Radical polymerization behavior et 1,1-disubstituted 2-vinylcyclopropanes. Macromolecules, 26(8), 1818–1824, 1993.
15. Moszner, N., Zeuner, F., Volkel Th, Reinherder, V. Synthesis and Polymerization of Vinylcyclopropanes. Macromol. Chem. Phys., 200, 2173–2187, 1999.
16. Sanda, F., Endo, T. Radical Ring-Opening Polymerization. J. Polym. Sci. Polym. Chem., vol.39, 265–276, 2001.
17. Guliyev, A. M. Radical polymerization of methyl substituted ethyl vinylcyclopropyl carboxylates. Iran, ISPST, 4, 2003.

CHAPTER 3

SYNTHESIS OF METAL DITHIOPHOSPHATES ON HLAY AND HY ZEOLITES AND POLYMERIZATION OF 1,3-BUTADIENE WITH HETEROGENEOUS CATALYTIC DITHIOSYSTEMS

FUZULI AKBER OGLU NASIROV,[1,2] SEVDA RAFI KIZI RAFIYEVA,[1] GULARA NARIMAN KIZI HASANOVA,[1] and NAZIL FAZIL OGLU JANIBAYOV[1]

[1]*Institute of Petrochemical Processes Azerbaijan National Academy of Sciences, 30, Khodjaly av., AZ 1025, Baku, Azerbaijan, E-mail: j.nazil@yahoo.com*

[2]*Petkim Petrokimya Holding A.Ş., Izmir, Turkey, E-mail: fnasirov@petkim.com.tr*

CONTENTS

ABSTRACT

The phosphoro-sulfurization of alcohols and phenols with phosphorus pentasulfide occurs throughout the hydroxyl groups present in their chemical structures. Zeolites also have hydroxyl groups and it is possible to carry out phosphoro-sulfurization, syntheses of O,O-disubstituted dithiophosphoric acids, and obtain metal complexes on zeolite matrices. In this article the problems associated with the synthesis of dithiophosphates of HY and HLaY zeolites and synthesis of Ni-, Co-, Fe-, Cr-containing dithiophosphates are considered. The synthesized nickel- and cobalt-containing zeolite dithiophosphates in combination with the aluminum organic compound, diethylaluminumchloride, are highly active and selective catalysts for preparing low and high molecular weights of 1,4-cis polybutadienes. Their productivities reach 1200–2550 kg polymer/g Me.hr (for cobalt catalysts) and 700–850 kg polymer/g Me.hr (for nickel catalysts), which is much more than those in homogeneous nickel- or cobalt-containing catalytic dithiosystems 270–400 kg polymer/g Me.hour.

3.1 INTRODUCTION

Metal dithiophosphates are widely investigated as various chemical additives such as stabilizers of polymeric materials, engine oils and corrosion inhibitors, etc. [1–5]. In recent years these metal complexes were also investigated as a component of bifunctional catalysts-stabilizers for oligomerization and polymerization reactions of dienes, in particular 1,3-butadiene [6–16].

Homogeneous catalysis has a number of advantages such as higher catalyst activity and selectivity. However, the most important obstacle to their large-scale use in the industry is the separation of homogeneous catalysts from the reaction medium. Consequently this leads researchers to use heterogeneous catalytic systems, but with the equal advantages of those in homogeneous catalytic systems. Proceeding from it; there was a task in view to develop metal dithiophosphates chemically connected with zeolites on a matrix.

It is known that the phosphoro-sulfurization of alcohols and phenols with phosphorus pentasulfide (P_2S_5) occurs with hydroxyl groups. Zeolites have hydroxyl groups in their structure and thus it is possible to carry out phosphoro-sulfurization, synthesize O,O-disubstituted dithiophosphoric acids on zeolite matrices, and then receive metal complexes. The proposed mechanism of the reaction are as follow:

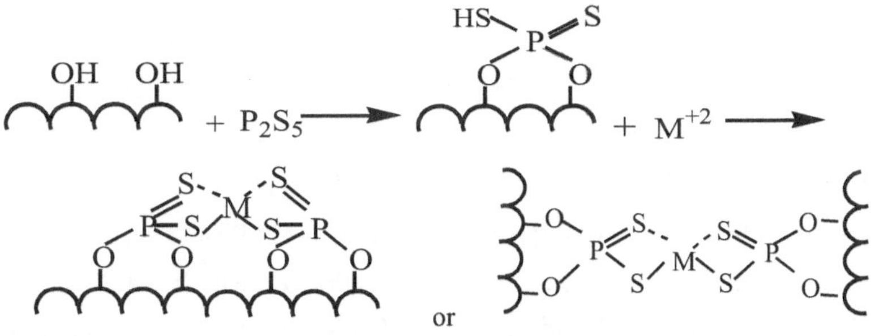

In the case of 3 valence metals there is probably participation of one, two and three dithiophosphorilic zeolite matrixes.

3.2 RESULTS AND DISCUSSION

3.2.1 SYNTHESIS AND CHARACTERIZATIONS

The phosphoro-sulfurization of the draiming zeolite is carried out in a glass reactor, supplied with a mixer, a circulating thermostat, a gas remove tube, and a neutralizer, H_2S.

First zeolite, then p-xylene was added into a reactor at temperatures of 80–90°C. After emulsion was formed by addition of fine-grained P_2S_5 gradually to the reaction medium under intensive mixing, release of H_2S was observed. The temperature then rises to 120–130°C, and the process is conducted before full release of H_2S. After full removal of H_2S from the reaction zone, the process is completed under a small vacuum (650–700 mm Hg). After the end of the reaction, the mass is washed out by ethanol (2–3 times) and dried at 60–70°C under vacuum. Thus dithiophosphorilated zeolites containing phosphorus of HY (3.46% and 3.53%) and HLaY (3.16% and 3.30%) are obtained.

For the obtaining of metal complexes the dithiophosphorilated zeolites were placed into a reactor, ethanol was added, and to the suspension, drops of ethanol solution of chlorides of Ni (II), Co (II), Cr (III) or Fe (III) were added.

The mixing was proceeded for 4–5 hours with heating in a water bath. After cooling, the reaction product was filtered and dried under vacuum at 60–70°C. It has been established that metal complexes give negative charge to chlorine, which somewhat confirms prospective structure. For the obtained metal complexes, phosphorus percentages have been defined as follow:

HУ – DTPh-Ni – 3.36 and 3.52	HLaУ – DTPh-Ni – 4.91 and 5.02
HУ – DTPh-Co – 3.44 and 3.53	HLaУ – DTPh-Co – 4.03 and 4.30
HУ – DTPh-Cr – 3.27 and 3.47	HLaУ – DTPh-Cr – 4.83 and 4.97
HУ – DTPh-Fe – 3.11 and 3.22	HLaУ – DTPh-Fe – 4.73 and 4.80

Obtained dithiophosphorilated metal zeolites were investigated by the means of DTA and IR spectroscopy.

3.2.1.1 Differential Thermal Analysis (DTA)

DTA curves have show an endothermic peak at 175°C, being characteristic of initial zeolite HLaУ is exposed to change after phosphoro-sulfurization and appears at 130°C. For the metal containing zeolite, this peak appears depending on the metal present at the respective temperatures: Ni – 140°C; Co – 130°C; Cr – 135°C; Fe – 135°C, however endothermic peaks for metal complexes are observed at 90–100°C.

The zeolite HУ endothermic peak is also present at 130°C. In the dithiophosphorilared sample, this peak is observed at 120°C and for metal containing samples (Ni, Co and Cr) it is observed at 155°C, 150°C and 135°Crespectively.

Such sharp changes of the temperature characteristics of initial zeolites after dithiophosphorilation and metal introduction into the zeolite matrix can be acknowledged by the formation of dithiophosphorilic metal containing zeolites.

As an example in Figure 3.1, DTA curves of zeolite HLaУ and products of its transformation are depicted.

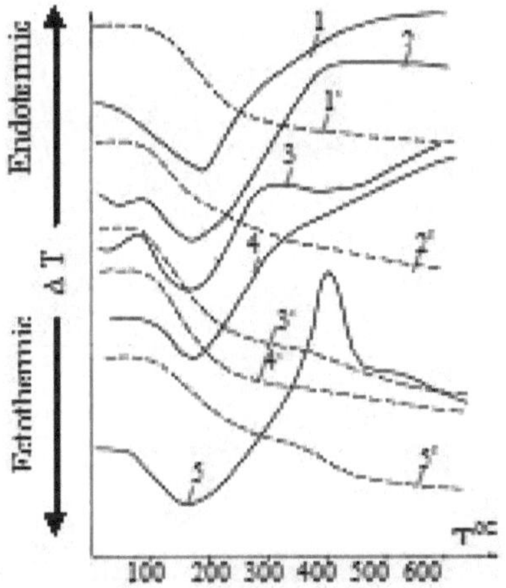

FIGURE 3.1 Derivatograms (– DTA, –TQ): 1 – HlaY zeolite; 2 – HlaY – DTPh; 3 – HLAY – DTPh-Ni; 4 – HLaY – DTPh-Co; 5 – HLaY – DTPh-Cr.

3.2.1.2 Infrared Analysis (IR)

In order to observe changes in a matrix of zeolites, the products of phosphoro-sulfurization and metallic compounds have been investigated using IR spectroscopy. It is established that absorptions at 1395 cm^{-1} and 1655 cm^{-1} are characteristic of initial zeolites of the HLaY type. After phosphoro-sulfurization, these areas disappear in the region of 1400 and 1640 cm^{-1}. For metal containing zeolites, these peaks are as follows for the respective metal complexes: Ni-complexes at 1400 and 1630 cm^{-1}, Co-complexes at 1400 and 1625 cm^{-1}, for Cr-complexes along with peak in the field of 1400 cm^{-1}, absorption is also observed at 1460 cm^{-1} and peak in the field of 1645 cm^{-1} (Figure 3.2).

The same changes are observed in the IR-spectra of zeolite HY and its transformed products. In a spectrum of initial zeolite of bases, areas characterizing full occluding are observed in the field of 1000 and 1600 cm^{-1}. For the dithiophosphorilated zeolites, these fields are displaced in 950 and 1500 cm^{-1}, and for the metal containing products: Ni-complex at about 1560 and 950 cm^{-1}, Co at 1590 and 1030 cm^{-1}, and Fe at 1560 and 1020 cm^{-1}. On Figure 3.2, IR-spectrums of zeolite HLaY and transformed products are shown.

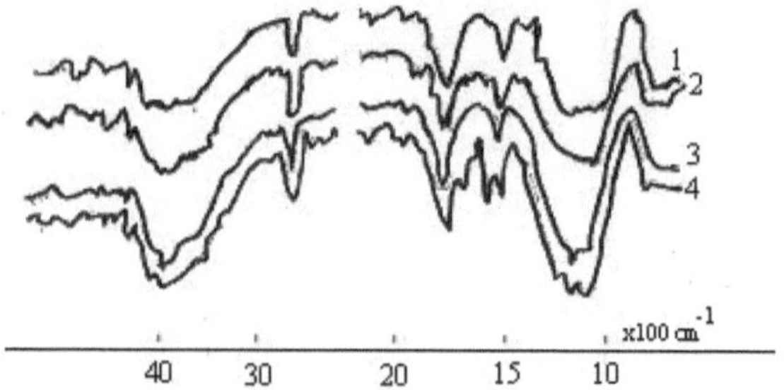

FIGURE 3.2 IR spectrums: 1 – HlaY zeolite; 2 – HlaY – DTPh-Ni; 3 – HlaY-DTPh-Co; 4 –HlaY-DTPh-Cr.

On the basis of analytical data, regarding the structural features, it is possible to conclude that after phosphoro-sulfurization and treatments of metal complexes at zeolites HLaУ and HУ occurs the remarkable structural changes mentioning a skeleton of zeolites that confirms chemical bond formation between a matrix of zeolites and dithiophosphorilic group and metal through it.

The synthesized metal containing dithiophosphorilated zeolites were investigated as heterogenized component of catalytic systems for butadiene polymerization. It has been established that in a combination with an aluminum organic compound, diethylaluminum chloride (DEAC), these catalytic systems possess higher activity and selectivity in the synthesis of 1,4-cis-polybutadiene than their homogeneous analogues (organic O,O-disubstitued dithiophosphates of nickel or cobalt). Table 3.1 summarizes the results, including the polymerization of butadiene in suspension and gas phases, HLaУ-DTPh-Co, and HLaУ-DTPh-Ni which were heterogenized by the same way, and by using homogeneous nickel- or cobalt-containing catalytic dithiosystems of IPCP-O,O-di-4-metylfenildithiophosphate cobalt (DCDTF-Co) + DEAC or O,O-diethyldithiophosphate nickel (DEDTPh-Ni) + DEAC. In these reactions, diethyl aluminum chloride (DEAC) is used as a co-catalyst at a reaction temperature of 25°C and withtoluene as a suspension medium.

The data demonstrate that in the gas phase polymerization of butadiene using the heterogenized cobalt-containing catalytic systems lead to high molecular weight 1,4-cis polybutadienes with yields of 90–95%, and

TABLE 3.1 Results of Butadiene Polymerizations With Nickel- and Cobalt-Containing Catalytic Dithiosystems

Catalyst	Polym. type	Reaction time, min.	Polym. yield, %	Productivity, kq PBD/q Me.hour	Intrinsic viscosity, [η], dL/q	Microstructure, %		
						1,4-cis	1,4-trans	1,2-
HLaY-DTPh-Co	Gas Phase	45	95.0	2500	2.8	96	3	1
HY-DTPh-Co	"—"	60	90.0	1900	2.3	93	6	3
HLaY-DTPh-Co	"—"	60	94.0	2000	2.5	95	4	1
HLaY-DTPh-Ni	Suspension	60	92.0	1100	1.2	93	5	2
HLaY-DTPh-Ni	"—"	90	91.0	850	1.1	92	5	3
HY-DTPh-Ni	"—"	90	88.0	800	0.85	88	8	4
HLaY-DTPh-Co	"—"	90	93.0	2200	2.5	95	4	1
IPCP Co-Cat.*	Homog.	60	98.0	400	2.3	92	6	2
IPCP Ni-Cat*	Homog.	20	80.0	270	0.1	80	17	3

Note: *Homogeneous DCDTPh-Co (or DEDTPh-Ni)+DEAC catalysts of IPCP of ANAS ([Co]=1.0×10^{-4} mol/L; ([Ni]=5.0×10^{-4} mol/L, Al:Co=100:1, T = 25°C; solvent – toluene).

productivities of 1900–2500 kg polybutadiene/g Co.hour. The synthesized polymers were characterized with intrinsic viscosities of 2.0–2.8 dL/g and 1,4-cis contents of 92–96%.

In the case of suspension polymerization these catalysts allow to produce also high molecular weight polybutadiene with intrinsic viscosity of 2.1–2.5 dL/g, 1,4-cis content of 93%, yield of 95%, and productivity of 2200 kg polybutadiene/g Co.hour.

Nickel-containing catalytic dithiosystems provide the preparation of relatively low molecular weight polybutadiene with intrinsic viscosities of 0.85–1.2 dL/g, polymer yields of 80–92%, and productivities of 800–1110 kg polybutadiene/g Ni.hour. Surprisingly, in the case of heterogenized nickel-containing catalytic systems, polybutadienes with high 1,4-cis content of 88–93% were observed, which is not observed in the presence of homogeneous nickel containing catalysts.

From the data, it can be concluded that directly heterogenized nickel- and cobalt-containing catalytic dithiosystems show very high activity and selectivity for butadiene suspension and gas phase polymerization processes. Their productivities reached 1900–2500 kg polymer/g Me.hour (for cobalt catalysts) and 800–1110 kg polymer/g Me.hour (for nickel catalysts), which is much more than that of homogeneous nickel- or cobalt-containing catalytic dithiosystems, 270–400 kg polymer/g Me.hour.

3.3 CONCLUSION

Novel nickel- or cobalt-containing heterogenized dithiosystems were synthesized by direct phosphorylation of surface hydroxyl groups on the zeolite matrix. Using dithiophosphorilated zeolites, metal dithiophosphates (particularly nickel and cobalt) were synthesized by connecting zeolites on a matrix.

These metal dithiophosphates, incombination with the aluminum organic compound, diethylaluminum chloride (DEAC), were used as a component in heterogenized catalytic dithiosystems for gas phase and suspension polymerizations of butadiene. These catalytic systems possess higher activity (yield of polymer – 86–95%) and selectivity (1,4-cis content – 88–96%) in the production of 1,4-cis-polybutadiene than their homogeneous analogues (organic O,O-disubstitued dithiophosphates of nickel or cobalt).

KEYWORDS

- **butadiene**
- **heterogenization**
- **metaldithiophosphates**
- **polymerization**
- **zeolite**

REFERENCES

1. Janibayov, N. F. Prof. Dr. Theses, IPCP of ANAS, Baku, 1987, p. 374 (in Russian).
2. Guliyev, A. M. Chemistry and Technology of additives to oils and fuels; Khimiya, Moscow, 1972, p. 367 (in Russian).
3. Pat. 852930, USSR, 1981.
4. Mamedov, M. Kh., Markova, Y. I., Rafiyeva, S. R., etc. J. Problems of Chemistry, 2011, №1, p.105 (in Russian).
5. Pat. 6610636, USA, 2004.
6. Pat. 2030210, Russia, 1987.
7. Nasirov, F. A. Petrochemistry, 41(6), 403, 2001, (in Russian).
8. Janibayov, N. F., M. Kh. Mamedov; etc. In Proceedings of the Polychar 10. World Forum on Polymer Applications and Theory, Denton, USA, 228, 2002.
9. Janibayov, N. F, Nasirov, F. A., Rafiyeva, S. R. J. Processes of Petrochemstry and Oil Refining, 10, 3–4 (39–40), 279, 2009.
10. Kubasov, A. A. Soros, J., 6 (6), 44, 2006.
11. Pat. 6093674, USA, 2000.
12. Pat. 6255420, USA, 2001.
13. Pat. 20020198335, USA, 2002.
14. Pat. 20030018142, USA, 2003.
15. Houssin J-M. Y. Nanoparticles in Zeolite Synthesis. PhD, University of Eindhoven, Germany, 2007.
16. Lee, H. A. New strategy for Synthesizing Zeolites and Zeolite-like Materials. Ph. D., Pasadena, CA, USA, 2005.

CHAPTER 4

POLYACRYLAMIDE HYDROGELS OBTAINED BY FRONTAL POLYMERIZATION AND THEIR PROPERTIES

ANAHIT VARDERESYAN

State Engineering University of Armenia, Yerevan, Teryan Str. 105, Armenia, E-mail: anahitvarderesyan08@mail.ru

CONTENTS

ABSTRACT

Method of Frontal Polymerization was used for Polyacrylamide (PAAm) Hydrogels (HG) obtainment. The states of frontal waves' stable propagation were established. The impacts of different kinetic parameters on hydrogels' synthesis (initiator concentration, initial temperature, concentration of surfactant active substance and on pH value of reaction media) were investigated.

4.1 INTRODUCTION

Hydrogels based on PAA have attracted increasing scientific and technological interest in recent years. PAAm HG is a cross-linking polymer with sewed macromolecules, which has ability to absorb the high amounts of water and various active substances by increasing its volume.

We managed to synthesize PAAm HG in a high-performance method – Frontal Polymerization (FP). The FP method is a nontraditional mode for polymers and copolymers synthesis. In contrary to the traditional polymer synthesis methods, the FP takes place from the local edge of the reaction ampoule (were having been given local heat) and propagates by the auto-waves mechanism due to the thermal conduction and temperature-dependent reaction rates. In a HG obtainment the method of FP has big advantage in comparison with traditional methods carried out in industry by the polymerization of monomers mixture in two or more steps. In addition to this case a residual monomer still remains in the resulting in PAAm HG [1].

In literature many theoretical and experimental features of the FP [2–7] for polymers, polymer composites and PAAm HG synthesis have been investigated as well as the conditions of stable front waves propagation have been discussed.

An important advantage of the PAAm HG is that they are chemically inert, and stable in a wide range of pH, temperature and ionic force. PAAm HG have suitable mechanical and physical properties and the main important property – swelling capacity. Thanks to this ability PAAm HG are widely used in various fields, especially agriculture, bioengineering, medicine, pharmaceutics (drug delivery devices).

In the presented article we want to draw attention to the FP's aspects for the synthesis of PAAm HG, with required properties of mechanical strength, appropriate swelling-deswelling capacity and most important fact-the absence of acrylamide in the final gel.

4.2 EXPERIMENTAL PART

The hydrogels were synthesized on the basis of acrylamide, in presence of the surface active material – sodium lauryl sulfate ($C_{12}H_{25}SO_4Na$)

(SAM), initiator – potassium persulfate ($K_2S_2O_8$). Resultant hydrogels were characterized by Fourier transform infrared spectroscopy (FTIR), thermogravimetric analysis (TGA) and by investigating their chemical compositions and structure and morphology. Equilibrium swelling ratios, swelling and deswelling kinetics were also carried out at different temperatures and pH to check their pH and thermo-sensitivities.

4.3 RESULTS AND DISCUSSION

The gel's swelling capacity depending on the synthesis conditions was investigated.

On the Figure 4.1 the kinetic curves of water swelling depending on the SAM ($C_{12}H_{25}SO_4Na$) quantities are presented. The swelling capacity of hydrogel is highest in first ~10 hours (SAM quantity = 3%).

On the Figure 4.2 the influence of the SAM quantities (1, 3, 5, 7 mass percent) on the gel's swelling capacity is given. Results show that swelling capacity of the gel depends on the quantities of SAM: the maximum swelling is observed at 3% of SAM and minimum at 1% and 7%.

Thermogravimetric analysis was used to investigate the thermal stability and the crosslink density of obtained hydrogels.

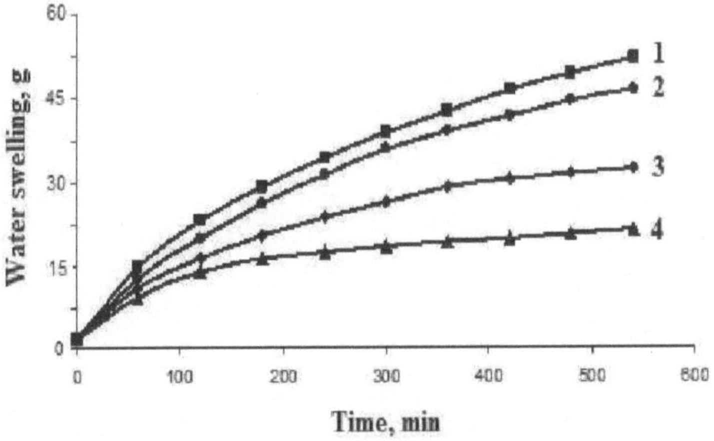

FIGURE 4.1 Kinetic curves of water swelling depending on the SAM quantities, 1–3; 2–2; 3–5; 4–7% mass.

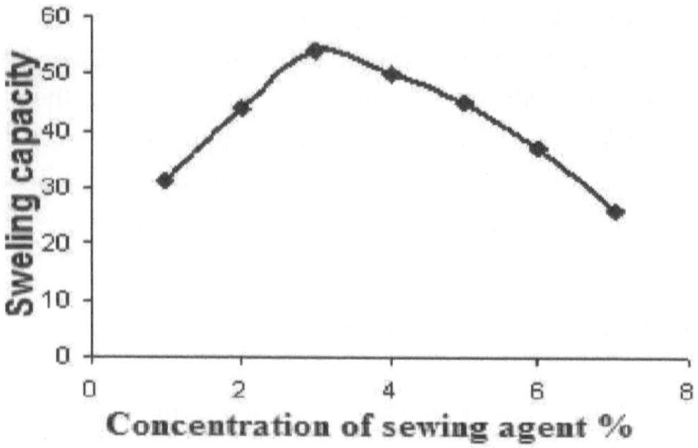

FIGURE 4.2 Swelling capacity depending on the sewing agent SAM concentration.

On the Figure 4.3 the weight loss curve of the dried hydrogel is shown. There is a step on the curve at the 200°C which is because of the sample's weight loss, about 15% by weight.

This may be attributed to the evaporation of residual water restrained by the hydrophilic bonds in the hydrogels. Hidrogels at the temperatures lower than 400°C have a good thermal stability. Thermooxidative destruction occurs at temperatures above 400°C.

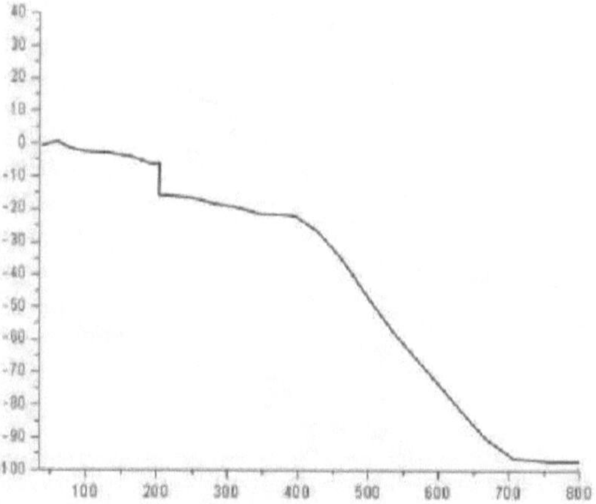

FIGURE 4.3 Thermogravimetric analysis of hydrogel.

4.3.1 PH SENSITIVITIES

The hidrogel's stability depending on the pH value was investigated. It was found out that the hydrogel is stable in the range pH=1–11.

Hydrogel samples were kept in solutions with various pH ranging from 1 to 11 until the equilibrium swelling is reached. On the basis of the experiment results shown on the Figure 4.4, it can be stated that the hydrogel reaches the maximum swelling capacity in the solution where the pH value is 7.

4.3.2 THERMOSENSITIVITIES

Swelling-deswelling of hydrogel in distilled water at different temperatures (25, 50, and 100°C) has been investigated. As it is obvious from Figure 4.5, the maximum swelling capacity is observed in the case where the water temperature is 25°C.

By means of infrared Spectroscopy (FTIR) hydrogels were analyzed for purpose to identify characteristics of functional groups of AAm (Figure 4.6). On the Figure 4.6 infrared spectra for acrylamide

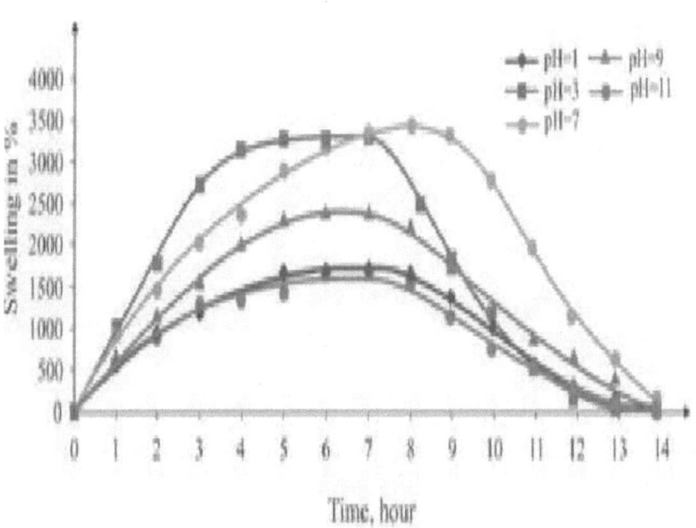

FIGURE 4.4 Swelling-deswelling of hydrogel depending on pH.

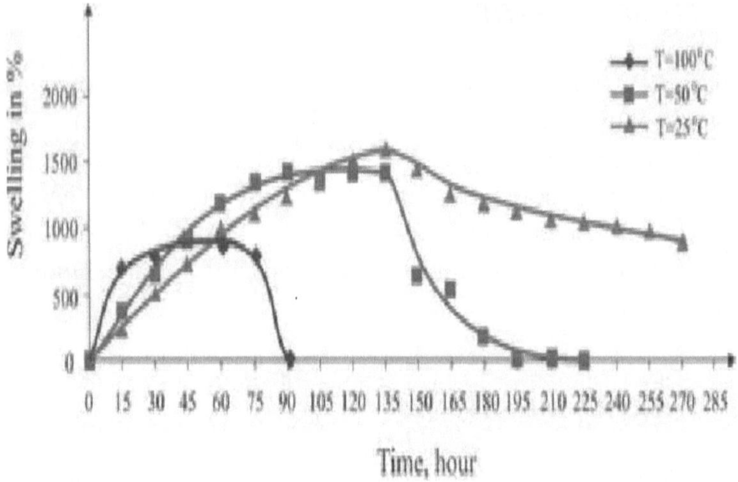

FIGURE 4.5 Dependence of swelling-deswelling ratio of hydrogel on temperature.

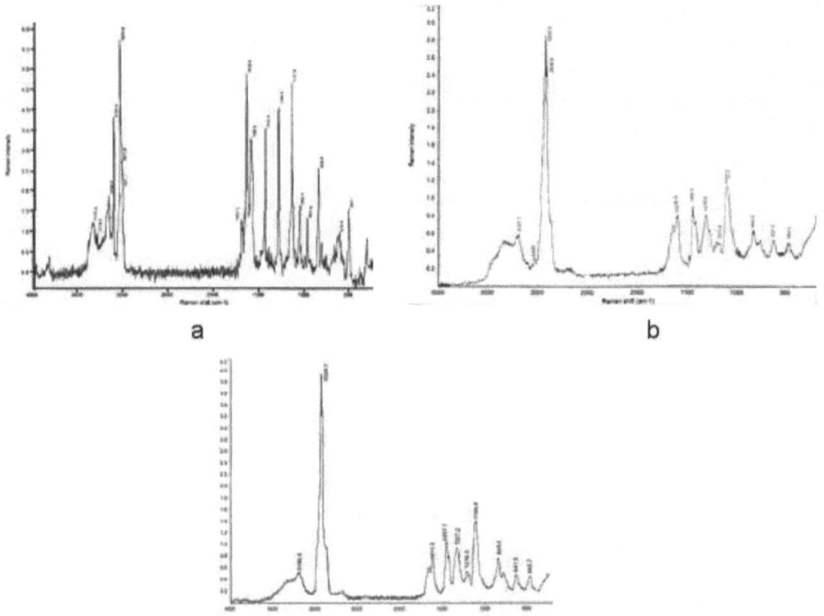

FIGURE 4.6 FT–Raman spectra ($\lambda_{exit.}$= 976 nm) of aacrylamide (a), crystalline polyacrylamide (b) and hydrogel (c).

(a), crystal polyacrylamide (b) and the synthesized hydrogel (c) are shown. From comparison of spectra we can conclude that crystalline polyacrylamide contains acrylamide (an absorption strip acrylamide 3034 cm^{-1}). Figures 4.6a and 4.6c show that synthesized hydrogel does not contain any traces of acrylamide.

4.4 CONCLUSIONS

We propose a novel frontal polymerization method to prepare macroporous, thermo- and pH-responsive release behaviors of hydrogels. Due to the absence of even traces of toxic acrylamide the obtained hydrogels have the potential to be safely used in agriculture, separation membranes, biosensors, bioengineering, medicine, pharmacy (drug delivery devices), cosmetology, etc.

KEYWORDS

- **deswelling**
- **frontal polymerization**
- **swelling**

REFERENCES

1. Gavini, E., Mariani, A., Rassu, G., SimoneBidali; Spada, G., Bonferoni, M. C., Giunchedia, P. "Frontal Polymerization as a New Method for Developing Drug Controlled Release Systems (DCRS) Based on Polyacrylamide," Eur. Polym. J., 45, 690–697, 2009.
2. Tonoyan, A. O., Kuvarina, L. V., Aleksanyan, G. G., Davtyan, S. P., et al. The low of vinyl monomer's radical polymerization in adiabatic regime, Visokomol. Soed., 16, 1005–1011, 1974.
3. Davtyan, S. P., Tonoyan, A. O. Theory and practice of adiabatic and frontal polymerization.
4. Monograph, P. 680, Palmarium Academic Publishing, 2014.
5. Davtyan, S. P., Berlin, A.A, Tonoyan, A. O. On principal approximations in theory of frontal radical polymerization of vinyl monomers, Russian Chemical Reviews, 79(3), 205 – 218, 2010.
6. Davtyan, S. P., Avetisyan, A. S., Berlin, A. A., Tonoyan, A. O. Synthesis and Properties of Particle_Filled and Intercalated Polymer Nanocomposites, Review Journal of Chemistry, 3, (1), 1–51, 2013.

7. Davtyan, S. P., Hambartsumyan, A. F., Davtyan, D. S., Tonoyan, A. O., Hayrapetyan S, Bagyan, S. H, Manukyan, L. S. The structure, rate and stability of autowaves during polymerization of Co metal-complexes with acryl amide, European Polymer, J., 38(12), 2423–2431, 2002.
8. Gevorgyan, L. A., Varderesyan, A. Z., Alaverdyan, G.Sh., Tonoyan, A. O. Synthesis of polyacrylamide superabsorbent hydrogel by the method of photochemical polymerization and investigation of its properties. Bulletin of SEUA. Collection of scientific and methodical papers. 2011, 3(2), 448–452.

CHAPTER 5

NOVEL LANTHANIDE POLYCOMPLEXES FOR ELECTROLUMINESCENT DEVICES

IRINA SAVCHENKO

National Taras Shevchenko University of Kyiv, 60, Volodymyrska Str., 01033 Kyiv, Ukraine, E-mail: iras@univ.kiev.ua

CONTENTS

ABSTRACT

Lanthanide complexes that contain polymerizable groups can be polymerized or copolymerized together with another monomer. This results in a polymer or copolymer in which the lanthanide complex is part of the polymer backbone or of the side chain. One macromolecule contains two active layers of electroluminescent cell – the emission layer (unsaturated lanthanide complex), electronic conduction layer (1,10-phenanthroline).

Copolymers of 2-methyl-5-phenylpentene-1-dione-3,5 with styrene in ratio 5:95, which containing Eu, Yb and Eu, Yb with 1,10-phenanthroline were synthesized at the first time. The luminescence spectra of obtained metal complexes and copolymers in solutions, films and solid state are investigated and analyzed. The solubilization of β-diketonate complexes with 1,10-phenanthroline was shown to change luminescence intensity in such complexes. Obtained copolymers can be used as potential materials for organic light-emitting devices.

5.1 INTRODUCTION

Organic light emitting diodes (OLEDs) are considered the next generation of technology for flexible flat panel displays and low cost solid-state lighting. In particular, organic light emitting diodes that have the potential to achieve an internal quantum efficiency close to 100% have attracted considerable research interest. Lanthanide *β*-diketones have attracted much attention because of their spectroscopic properties. Investigation of luminescence and optical absorption properties of lanthanide *β*-diketones is of great importance since these complexes are widely used as light-converting optical materials, light-emitting diodes, luminescent probes, polymer sol–gel derived glasses, electroluminescent devices [1–4]. Light emitting devices are usually made of solid-state materials that emit lights of various wavelengths upon the stress with an electric field.

The photoluminescence properties of rare-earth (lanthanide) compounds have been fascinating researchers for decades [5–7]. An attractive feature of luminescent lanthanide compounds is their line-like emission, which results in a high color purity of the emitted light. The emission color depends on the lanthanide ion but is largely independent of the environment of a given lanthanide ion. Most of the studies on these compounds have been limited to either inorganic compounds (lanthanide phosphors) or molecular lanthanide compounds (for instance, the *β*-diketonate complexes). On the basis of the unique photo physical properties of lanthanide cations (long luminescence lifetime and very sharp emission band), rare earth metal complexes, especially europium(III) complexes, as luminescent materials have received increasing attention for application such as analytical sensors, imaging techniques, displays and organic light-emitting diodes [8–14]. Recently, europium complexes have attracted more interest in organic light-emitting diodes for their saturated red-emission [9, 10]. Also several europium complexes have been applied as red emitters in electroluminescent devices [15–17].

Using the monomer complex has a number of disadvantages connected with aggregation or crystallization of the film.

Therefore, there is a necessity of the polymeric materials synthesis. It is well known that metal polymers are mainly produced by intercalation of metals in the polymer ligand matrix. This method has a lot of disadvantages such as partial degradation of the polymer chain and low yield of the synthesized polymers as well as low coordination level, which results in composition heterogeneity. All these shortcomings have an influence upon physical characteristics of obtained compounds.

The aim of this work were synthesis of Eu, Yb complexes with 2-methyl-5-phenylpentene-1-dione-3,5 and 1,10-phenanthroline as well as copolymers based on them with styrene and investigations of optical properties of metal-containing polymeric systems depending on influence of 1,10-phenanthroline configured-in the complex coordination sphere on the luminescence properties and concentration of rare earth elements complexes in polymeric materials.

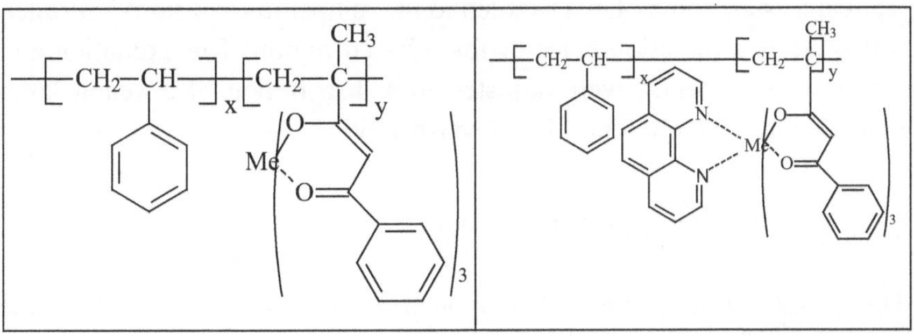

Copolymers of 2-methyl-5-phenylpentene-1-dione-3,5 with styrene in ratio 5:95, which containing Eu, Yb and Eu, Yb with 1,10-phenanthroline were synthesized at the first time.

5.2 EXPERIMENTAL PART

NMR: ^1H (D_2O) δ (ppm): 3.27 (singlet, 3H, CH_3); 3.47(singlet, 1H, =CH-); 5.27 (singlet, 1H, $=CH_2$); 5.58 (singlet, 1H, $=CH_2$); 7.15–7.60 (mul'tiplet, 5H, Ph).

Complexes obtained by an exchange reaction between equimolar amounts of lanthanide acetate and sodium 2- methyl-5-phenylpentene-1-dione-3,5 salt in a water-alcohol solution at pH 9–9.5 with a slight excess of the ligand.

The polymerization was carried out at 80°C in the thermostat in dimethylformamide solution with monomer concentration 0.03 mol/L and initiator 2,2'-azobisisobutyronitrile concentration 0.003 mol/L. Thus, obtained metal polymers precipitate out from propanol-2 solution.

The synthesized compounds have been studied by NMR, IR-, electronic absorption and diffuse reflectance spectroscopy. The infrared spectra were recorded in KBr tablets at a range of 4000–400 cm^{-1} with Spectrum BX II FT – IR manufactured by Perkin Elmer. The electronic absorption spectra were recorded using spectrophotometer Shimadzu "UV-VIS-NIR Shimadzu UV-3600" and the diffuse reflectance spectra were obtained using the Specord M-40 spectrophotometer in the range of 30,000–12,000 $cm^{-1.}$ The excitation and luminescence spectra of solid samples and solutions (10^{-3} M, $CHCl_3$) were recorded on a spectrofluorometer "Fluorolog FL 3–22," "Horiba Jobin Yvon" (Xe-lamp 450 W) with the filter OS11. The InGaAs photoresistor (DSS-IGA020L, Electro-Optical Systems, Inc., USA) cooled to the temperature of liquid nitrogen was used as a radiation detector for infrared region. The excitation and luminescence spectra were adjusted to a distribution of a xenon lamp reflection and the photomultiplier sensitivity.

5.3 RESULTS AND DISCUSSIONS

The IR-spectra of synthesized compounds were registered to establish the type of coordination of the lanthanide ion with a mphpd functional groups. Analysis of IR spectra showed a presence of stretching vibrations of C-O and C-C bonds at 1500–1600 cm^{-1}, that confirms the cyclic bidentate coordination of ligand to metal ions. A slight shift of the main absorption band for the complex in a comparison with β-diketone sodium salt to the long-wave region indicates a weakening of the metal – ligand bond, due to increase of covalent bond. The low intensity band at 1660 cm^{-1} corresponds to the valency vibration of the double bond ν(C = C). Also there is a broad absorption band of coordinated water molecules at 3400–3200 cm^{-1}. Thus, obtained results indicate a cyclic bidentate coordination of mphpd molecules in the complex.

Electronic spectra of the monomer as well as metal polymeric complexes have a set of bands corresponding to europium ion (Table 5.1). Shift of the main absorption bands in the long wavelength region in comparison with the spectra of aqua-ions, and their increase in intensity indicates the formation of metal complexes. Electronic spectra of the Yb(mphpd)$_3$·2H$_2$O have a singular transition band $^2F_{7/2}$→$^2F_{5/2}$ corresponding to ytterbium ion. A slight shift of the maximum which is observed in absorption spectra of the complex Yb(mphpd)$_3$·2H$_2$O in comparison with the spectra of aqua-ions indicates the formation of metal complex and a weakening of the metal – ligand bond (Figure 5.1). The absorption spectrum of Yb(mphpd)$_3$ · 2H$_2$O presents two main bands. The band attributed to the ligand absorption lies in the near-ultraviolet region. The second band (Figure 5.2) extended in the NIR region from 900 to 1015 nm is due to f-f transition from ground $^2F_{7/2}$ to exited $^2F_{5/2}$ states of the Yb^{3+} ion. The absorption maximum (Table 5.2) is shifted to higher wavelengths for about 2 nm compared to ytterbium aqua-ion (i.e., nephelauxetic effect) due to the partially covalence of the coordination bonds [18, 19].

The luminescence of NIR emitting lanthanide ions is efficiently quenched by O–H vibrations in coordinated water molecules. In order to overcome this hindrance synthesis of 1,10-phenanthroline adduct of the studied complex was performed. The adduct was prepared by reaction between YbCl$_3$, sodium salt or 2-methyl-5-phenylpentene-1-dione-3,5 and 1,10-phenanthroline (taken in molar ratio 1:3:1) in alkaline (pH = 8 – 9) aqueous-alcoholic medium:

$$\mathrm{YbCl_3 + 3Na(mphpd) + Phen \rightarrow Yb(mphpd)_3 \cdot Phen + 3NaCl}$$

where Phen stands for 1,10-phenanthroline.

Similarity of electronic absorption and diffuse reflectance spectroscopy (ESA and SDR) show a similar structure of the complexes in solution

TABLE 5.1 Some Distinctive Absorption Bands For Metallic Complexes

Complex	$^7F_0$®$^5H_6$	D$_1$*	$^7F_0$®$^5L_6$	D$_2$*	$^7F_0$®$^5D_2$	D$_3$*	$^7F_0$®$^5D_1$	D$_4$*
Eu(mphpd)$_3$·2H$_2$O	31,420	30	25,180	70	21,280	220	18,550	150
Eu(mphpd)$_3$·Phen	31,400	50	25,190	60	21,270	230	18,580	120

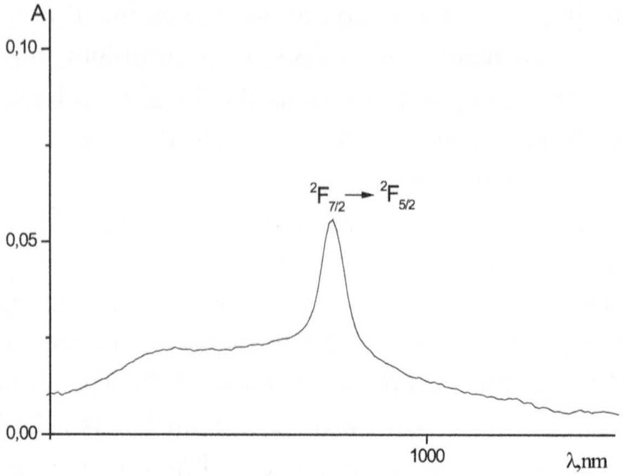

FIGURE 5.1 Electronic absorption spectrum of $Yb(mphpd)_3 \cdot 2H_2O$.

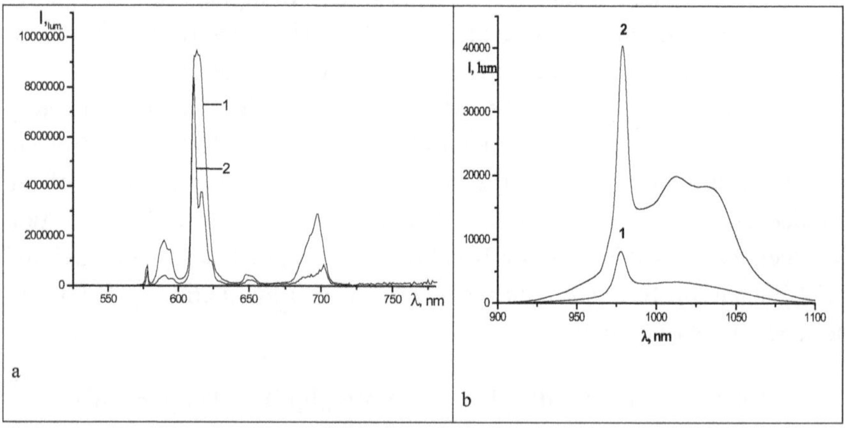

FIGURE 5.2 Luminescence spectra of: (a) 1-$Eu(mphpd)_3 \cdot Phen$; 2-$Eu(mphpd)_3$, in solid state, T=298 K, λ=362 nm; (b) 1 – $Yb(mphpd)_3$, 2 – $Yb(mphpd)_3Phen$.

TABLE 5.2 Absorption Maxima of $^2F_{7/2} \rightarrow {}^2F_{5/2}$ Transition

Compound	λ_{max} (nm)
$YbCl_3$[a]	973
$Yb(mphpd)_3 \cdot 2H_2O$	975
$Yb(mphpd)_3 \cdot Phen$	976

[a]Ref. [18].

and polycrystalline state. A slight shift of the maximum which is observed in absorption spectra of the complex Eu(mphpd)$_3$·Phen indicates a replacement of water molecules in the nearest coordination environment without significant changes in the coordination polyhedron geometry.

The luminescence spectra of europium β-diketonate complexes in all samples are similar to each other and approving the structure similarity of coordination polyhedrons, which are distorted antiprism. All the samples have an equal number of the magnetic and electric dipole transitions as well as the forbidden transitions (see the Table 5.3).

The absorption spectrum of Yb(mphpd)$_3$ · Phen is similar to that of Yb(mphpd)$_3$ · 2H$_2$O. The absorption maximum is insignificantly (~1 nm) shifted to higher wavelengths indicating the substitution of water molecules in the coordination sphere of Yb^{3+} ion without considerable changes of coordination polyhedron.

Under excitation at 410 nm the complex and its adduct both exhibit NIR luminescence corresponding to $^2F_{5/2} \rightarrow {}^2F_{7/2}$ transition (Figure 5.3). The luminescence intensity and its quantum efficiency are higher for adduct (Table 5.3) due to the absence of water molecules in the inner sphere of Yb^{3+} ion.

Europium luminescence spectra (Figure 5.2a) at 77K allow to establish the short-range coordination environment symmetry.

The transition band $^5D_0 \rightarrow {}^7F_0$ in luminescence spectra of all compounds appears as a symmetrical single line and indicates a presence of

TABLE 5.3 Energy Transition in Eu Luminescence Spectrum

Transition	Eu(mphpd)$_3$, cm^{-1}	Eu(mphpd)$_3$(Phen)], cm^{-1}
$^5D_0 \rightarrow {}^7F_0$	17,331	17,331
$^5D_0 \rightarrow {}^7F_1$	16,978	16,977
	16,849	16,921
$^5D_0 \rightarrow {}^7F_2$	16,287	16,340
	16,234	16,233
$^5D_0 \rightarrow {}^7F_3$	15,408	15,384
$^5D_0 \rightarrow {}^7F_4$	14,468	14,490
	14,347	14,347

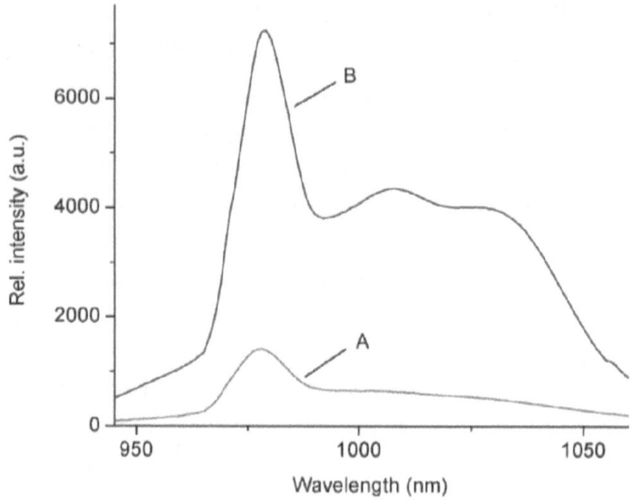

FIGURE 5.3 Luminescence spectra of $Yb(mphpd)_3 \cdot 2H_2O$ (a) and $Yb(mphpd)_3 \cdot$ Phen (b) in $CHCl_3$ under excitation at 410 nm.

one luminescence center. High-intensity lines caused by electric dipole transition $^5D_0 \rightarrow {}^7F_2$ compared with relatively low intensity magnetic dipole transition $^5D_0 \rightarrow {}^7F_1$ suggests not center-symmetric nature of the environment for all investigated compounds.

As for phenanthroline complexes, obviously phenanthroline is a part of the complex and forms an adduct but not a mixed-complex due to spaciousness of the diketonate fragment. Based on the number of the splitting components we can assume a significant rhombic distortion.

The luminescence intensity of complexes based on Eu(III) and Yb(III)-phenanthroline is greater in comparison with complex without additional ligand (Figures 5.2 and 5.3). The phenanthroline displaces the coordinate water out of the coordinate sphere, which is quenching agent.

The luminescence intensity of copolymer of styrene with Eu(III)-phenanthroline complex (Figure 5.4) is identical practically with gomopolymer of $Eu(mphpd)_3 \cdot$Phen [9].

The luminescence intensity of copolymer of styrene with Yb(III)-phenanthroline complex is greater in several times in comparison with copolymer with $Yb(mphpd)_3 \cdot 3H_2O$ (Figure 5.5) and respective monomeric complexes (Figure 5.2b).

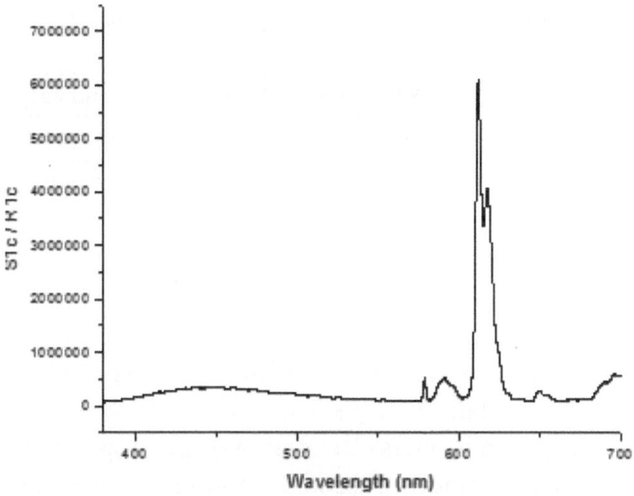

FIGURE 5.4 Luminescence spectrum of Eu(mphpd)$_3$·Phen-co-styrene in solid state, T=298K, λ=358 nm.

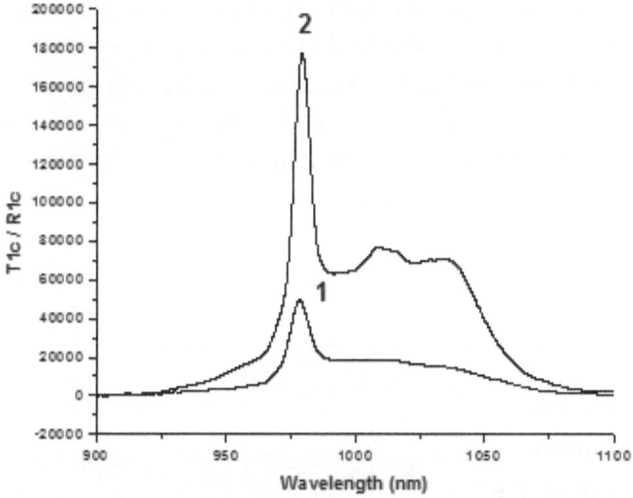

FIGURE 5.5 Luminescence spectra of: (a) Yb(mphpd)$_3$·3H$_2$O-co-styrene; (b) Yb(mphpd)$_3$·Phen-co-styrene in solid state, T=298K, λ=336 nm.

The luminescence intensity and its quantum efficiency are higher for adduct (Table 5.4) due to the absence of water molecules in the inner sphere of Yb^{3+} ion.

TABLE 5.4 Quantum Efficiency of Studied Complex and Its Adduct With 1,10-Phenanthroline

	$Yb(mphpd)_3 \cdot 2H_2O$		$Yb(mphpd)_3 \cdot Phen$	
	Solid	**$CHCl_3$ solution**	**Solid**	**$CHCl_3$ solution**
Quantum efficiency (%)	0.022	0.013	0.032	0.015

5.4 CONCLUSIONS

The investigations conducted in the present wirk-allowed to determine the composition, structure and properties of the complexes and metallopolymers on their basis obtained for the first time.

The similarity of monomers electronic absorption spectra with copolymers spectra confirms of identical coordinative environment of lanthanide ions in both cases.

The solubilization of Eu(III) and Yb(III) β-diketonate complexes with 1,10-phenanthroline, was shown to change luminescence intensity in these complexes. The luminescence intensity of these complexes is greater in comparison with complex without additional ligand.

Thus, obtained copolymers which containing 5% emission components (lanthanide complex) in polymer chain only are comparably with suitable gomopolymers concerning luminescence properties and can be perspective for optical application.

KEYWORDS

- **copolymers**
- **lanthanide**
- **luminescence**
- **metal complexes**
- **organic light emitting diodes**
- **β-diketone**

REFERENCES

1. Jiu, H., Liu, G., Zhang, Z., Fu, Y., Chen, J., Fan, T., Zhang, L. Fluorescence enhancement of Tb(III) complex with a new β-diketone ligand by 1,10-phenanthroline. J. Rare Eart. l(29), 741–745, 2011.
2. Kuo, Y., Chi-Chou Lin. A light emitting device made from thin zirconium-doped hafnium oxide high-k dielectric film with or without an embedded nanocrystal layer. Applied Physics Letters. 102, 031117 (1–7), 2013.
3. Sun, M., Xin, H., Wang, K.-Z., Zhang, Y. A., Jin, L.-P., Huang, C.-H. Bright and monochromic red light-emitting electroluminescence devices based on a new multifunctional europium ternary complex. Chem. Commun. vol. 6, 702–703, 2003.
4. Stathatos, E., Lianos, P., Evgeniou, E., Keramidas, A. D. Electroluminescence by a Sm^{3+}-diketonate-phenanthroline complex. Synth. Met. vol.139 (2), 433–437, 2003.
5. Zheng, Y. X., Liang, J. L., Lin, Q., Yu, Y. N., Meng, Q. G., Zhou, Y. H., Wang, S. B., Wang, H. A., Zhang, H. J. A comparative study on the electroluminescence properties of some terbium (3-diketonate complexes. J. Mater. Chem. vol. 11 (10), 2615–2619, 2001.
6. Sun, J., Zhang, X., Xia, Z., Du, H. Luminescent properties of $LiBaPO_4$:RE (RE=Eu^{3+}, Tb^{3+}, Sm^{3+}) phosphors for white light-emitting diodes. J. of Applied Physics. vol. 11, 013101 (1–7), 2012.
7. Zeng, L., Yang, M., Wu, P., Ye, H., Liu, X. Tb-containing electroluminescent polymer with both electron- and hole-transporting side groups for single layer light emitting diodes. Synth. Met. vol.144 (3), 259–263, 2004.
8. Ling, Q., Yang, M., Zhang, W., Lin, H., Yu, G., Bai, F. PL and EL properties of a novel Eu-containing copolymer. Thin Solid Films. vol. 417 (1–2), 127–131, 2002.
9. Savchenko, I., Bereznitskaya, A., Smola, S., Fedorov Ya., Ivakha, N. Novel electroluminescent materials on polymer metal complexes. Functional mater., vol.19 (4), 541–547, 2012.
10. Neng-Jun Xiang, Louis, M. Leung, Shu-Kong So, Meng-Lian Gong. Preparation and photoluminescence of a novel β-diketone ligand containing electro-transporting group and its europium(III) ternary complex. Spectrochimica Acta Part A: Molecular and Biomolecular Spectroscopy. vol. 65, 907–911, 2006.
11. Zeng, L., Yang, M., Wu, P., Ye, H., Liu, X. Tb-containing electroluminescent polymer with both electron- and hole-transporting side groups for single layer light emitting diodes. Synth. Met. vol. 144 (3), 259–263, 2004.
12. Smith, N. A., Sadler, P. J. Photoactivatable metal complexes: from theory to applications in biotechnology and medicine. Philos Trans. A Math. Phys. Eng. Sci. vol. 371, 20120519, 1–13, 2013.
13. Pekka Hänninen, Harri Härmä. Lanthanide Luminescence: Photophysical, Analytical and Biological Aspects: Springer-Verlag, Berlin, Heidelberg. 385, 2011.
14. Crislene, R. S. Morais, C. G. Gameiro, et al. Thermal decomposition of lanthanide(III) complexes with 4,4,4-trifluoro-1-phenyl-1,3-butanedione. Photoluminescent properties and kinetic study. Journal of Thermal Analysis and Calorimetry. vol. 87 (3), 887–891, 2007.

15. Petrochenkova, N. V., Petukhova, M. V., Mirochnik, A. G., and Karasev, V. E. Europium(III) Acrylatodibenzoylmethanate: Synthesis, Spectroscopy (IR, Luminescence), and Polymerization Properties. Russian Journal of Coordination Chemistry. vol. 27 (9), 676–679, 2001.
16. Semenov, V. V., Zolotareva, N. V., Klapshina, L. G., et al. Synthesis of C-Functionalized Acetylacetone and Its Europium Complex. Preparation and Study of Luminescence of Europium-Containing Sol-Gel Films. Russian Journal of General Chemistry. vol. 79 (9), 1802–1810, 2009.
17. Semenov, V. V., Cherepennikova, N. F., Grigor'ev, I. S., et al. Europium, Terbium, and Ytterbium 3-(3'-Triethoxysilylpropyl)pentane-2,4-Dionates. Synthesis and the Formation of Luminescent Sol–Gel Films. Russian Journal of Coordination Chemistry. vol. 33 (1), 68–78, 2007.
18. Poluektov, N. S. et al. Spectrophotometric and Luminescence Methods for Determination of Lanthanides, Naukova Dumka: Kiev (in Russian), 1989.
19. Sastri, V. S. et al. Modern Aspects of Rare Earths and their Complexes, Elsevier Science, B. V.: Amsterdam, 2003.

CHAPTER 6

NOVEL HETEROGENIZED COBALT CONTAINING CATALYTIC DITHIOSYSTEMS FOR GAS PHASE POLYMERIZATION OF BUTADIENE

SEYMUR SALMAN OGLU SALMANOV,[1] FUZULI AKBER OGLU NASIROV,[1,2] and NAZIL FAZIL OGLU JANIBAYOV[1]

[1]*Institute of Petrochemical Processes of National Academy of Sciences of Azerbaijan, Baku, Azerbaijan,* [2]*Petkim Petrokimya Holding, Izmir, Turkiye, E-mail: fnasirov@petkim.com.tr*

CONTENTS

ABSTRACT

Cobalt-containing bifunctional catalytic dithio systems has been heterogenized on various supports (such as silica gels, aluminum oxides, zeolites, carbon black, etc.) by the methods of direct deposition or pre-alumination. The received data demonstrate that the pre-alumination method shows a very high activity in the gas phase polymerization of butadiene. The productivity of these catalysts reaches 650.0–2200.0 kg PBD/g. Co•h. which is much higher than the output of solution process using the same homogenous catalysts (57.0 kg PBD/g Co•h.) and the known gas phase process using the heterogenized neodymium-catalysts (500.0 kg PBD/g Nd•h.).

High activities of heterogenized bifunctional cobalt-containing catalytic dithiosystems allow developing principally new ecologically favorable and economically benefited technology of butadiene gas phase polymerization process.

6.1 INTRODUCTION

The industrial production of polydienes is carried out with solution polymerization technology by using homogeneous Ti, Co and Ni-containing Ziegler-Natta type catalytic systems. The biggest obstacle preventing their large scale application in the industry is the problem of separating homogeneous catalysts from the reaction medium, inefficient removal and recovery of the solvents and monomers after polymerization, the latter of which often requires more process steps and energy than the actual polymerization.

As known, gas-phase polymerization methods for the production of thermoplastics, such as polyethylene and polypropylene, have proved to be particularly advantageous and have gained acceptance. The gas-phase process has many advantages over the solution process. Solvents are not used in this process, therefore there are no technological, economical or ecological problems concerning cleaning and drying the solvents, washing off catalyst residues from the polymerizate, polymer degassing, polymer solubility, and viscosity of reaction medium. In the gas-phase polymerization process, the complicate procedures of aggregation and separation of polymer from solution, which are present solution polymerization, are also absent. These allow for significant reduction in both construction-operation cost and environmental pollution.

The solvent-free polymerization of dienes in the gas-phase has not yet been established on the industrial scale, because there are not yet technically accomplished substantial ecological, economical, energetic and safe advantages. Studies on gas-phase polymerization of butadiene with neodymium containing Ziegler-Natta catalysts in laboratory scale were firstly carried out by the Berlin Technical University in 1993 [1–6]. The productivity of these heterogenized neodymium catalyst systems was ~500.0 kg PBD/g Nd•h. Many firms have also reported about gas-phase processes for diene monomers and several have established significant patent portfolios including Bayer, Bridgestone, and Union Carbide [7–15]. To date, Bayer appears to be the closest to commercializing a gas-phase process in the production of polybutadiene rubber [7–12]. However, catalytic systems (containing lanthanides, cobalt, or nickel) used in gas-phase polymerization of butadiene are characterized by shortcomings related to their low catalytic activity, ability to accelerate oxidative aging and degradation of PBD (PBD = Polybutadiene) [1–13].

Earlier we have developed highly active and efficient homogeneous bifunctional catalyst-stabilizers on the bases of Ni- and Co-dithioderivatives (such as O,O-disubstituted dithiophosphates, N,N-dithiocarbamates, xhantogenates) in combination with aluminum organic compounds (such as, dialkyl aluminum chlorides, aluminoxanes, and alkyl aluminum dichlorides) for butadiene polymerization and simultaneous stabilization of the end polymer [14–18]. These catalysts show high activity and stereo-selectivity in homogeneous polymerization of butadiene. Their productivities reached 5.0–109.0 kg PBD/g Me•h, intrinsic viscosities were in the range of 0.08–3.5 dL/g, and their 1.4-cis contents were between 80.0–96.0%. After polymerization, such catalysts were not washed out and remaining in the polymer effectively stabilizes them against thermo- and photo-oxidative ageing in storage without additional involvement of antioxidants.

Heterogenization of these new bifunctional cobalt-containing catalytic dithiosystems allow to reach very high process productivities of – 650.0–2200.0 kg PBD/g Co•h., which is much higher than those of solution processes using the same homogenous catalysts (5.0–109.0 kg PBD/g Co•h.) and the known gas-phase process using the heterogenized neodymium-catalysts (500.0 kg PBD/g Nd•h.).

In this article the results of gas-phase polymerization of butadiene are presented in the presence of new heterogenized bifunctional cobalt-containing catalytic dithiosystems, firstly developed by our group.

6.2 EXPERIMENTAL PART

Butadiene (99.8%, wt.), aluminum organic compounds (85.0–90.0%, wt., in benzene) were obtained from Aldrich.

Organic dithioderivatives (dithiophosphates and dithiocarbamates) of cobalt have been synthesized according to [14].

Where necessary, manipulations were carried out under dry, oxygen-free argon or nitrogen in a Schlenk-type apparatus with appropriate techniques and gas tight syringes. For the preparation of homogeneous metallocomplex catalysts, the desired volume of toluene, monomer, DEAC (Diethyl alu minum chloride) (or TEA-triethyl amine), and cobalt solutions were added to the 100 mL-reactor under magnetic stirring at oxygen-free atmosphere and temperature control by the usual order of addition of catalyst components: solvent, cobalt component, aluminum organic compound (at –78°C), and finally monomer.

Many types of solid materials have been investigated as support materials for the immobilization of metallocomplex catalysts: inorganic materials (SiO_2, $MgCl_2$, Al_2O_3, zeolite, etc.) and polymeric materials (polystyrene, polysiloxane, etc.). The most common support used is silica gel, due to its low cost and ease of functionalization. The silica gel surface contains hydroxyl and siloxane functional groups, which are useful in surface modification and catalyst immobilization [19, 20].

There are many ways that metallocomplex catalysts have been immobilized on supports [19, 20] and we have used the three main techniques for metallocomplex catalysts immobilization:

1) "*Direct deposition*" (also called grafting or impregnation) method. This is a physi-sorbtion process of a complex such as a coordination metal compound onto the support's surface. This is the most convenient method of metallocomplex catalysts immobilization. In a typical process, some pretreated (calcined and partially thermally dehydroxylated at 120–200°C under vacuum) silica gel is stirred with a solution of a cobalt compound of metallocomplex catalyst in inert atmosphere at room temperature or elevated temperatures for a period of time. Then, the slurry is filtered, and the remaining solid product is washed with solvent several times to remove weakly adsorbed metallocomplex compound molecules.

The washed product is dried under vacuum to remove the solvent. The metallocomplex compound is believed to react with hydroxyl groups on the silica gel surface and bond to the surface via an M-O-Si bond.

2) *"Pre-alumination"* immobilization method. This is the process in which the support material contacts with a co-catalyst (either methyl aluminoxane or an alkyl aluminum halides) before impregnating with a metallocomplex compound. In one process, the silica gel is stirred with a solution of co-catalyst (MAO- methyl aluminoxane) and then filtered (Scheme 1). The solid portion is washed and dried in vacuum to obtain the MAO-modified silica gel. A minor modification of this process is to add n-decane to the slurry of silica gel and MAO in toluene to precipitate the MAO onto the silica gel. In another process, supported MAO is generated in situ by reacting TMA with water in the presence of silica gel, or by reacting TMA with water adsorbed on the surface of silica gel without dehydration. The subsequent metallocomplex catalyst cobalt compound impregnation process is similar to that of the direct deposition method.

   When silica is treated with MAO, the surface hydroxyl groups react with MAO and release CH_4, so MAO is bonded to silica gel through Si-O-Al bonds. The metallocomplex compound molecules are thus immobilized on the silica gel by ionic interaction with the bonded MAO and become active catalytic species at the same time. The ionic interaction is weak, so the active species may be able to migrate over the MAO covered silica gel surface, resulting in a similar environment of the active species as that in homogeneous solution. Therefore, the polymers produced by these supported catalysts have similar properties as those produced by corresponding homogeneous metallocomplex catalysts.

3) The covalent tethering method to immobilize a homogeneous metallocomplex catalyst on silica gel that is similar to the pre-alumination method. In this method the homogeneous metallocomplex catalyst is prepared by the reaction of a solution of cobalt compound of metallocomplex and MAO (or alkyl aluminum halides) in the presence of monomer molecules, and then pre-aluminated silica gel is added to the solution.

The slurry is stirred and dried to form the supported catalyst. Since the cobalt compound of metallocomplex catalyst has been activated in solution by MAO in the presence of monomer molecules, the immobilization should occur between excess MAO and the silica gel surface. The structure of the supported catalyst prepared using this method should resemble that of the supported catalyst using the pre-alumination method.

The methods of heterogenization of homogeneous bifunctional catalyst-stabilizers and the calculated concentration of cobalt on support of heterogenized catalytic dithiosystems are shown in Table 6.1.

The molecular masses of high molecular 1,4-cis polybutadiene and 1,4-cis+1.2- polybutadiene were determined by viscosimetric method [21] with the relationships:

$$[\eta]_{30(\text{toluene})} = 3.05\times10^{-4} \times M^{0.725}; \ [\eta]_{30(\text{toluene})} = 15.6\times10^{-5} \times M^{0.75}$$

The molecular masses (Mw and Mn) and molecular mass distribution (Mw/Mn) of polybutadienes were measured by a Gel Permeation Chromatograph (GPC), constructed in Czech Republic with a 6,000 A pump, original injector, R-400 differential refractive index detector, styragel columns with nominal exclusions of 500, 10^3, 10^4, 10^5, and 10^6. The GPC Instrument was calibrated according to the universal calibration method by using narrow molecular weight polystyrene standards [22].

The microstructure of the polybutadiene was determined by the means of an FT-IR spectrometer (Nicholet NEXUS 670 with spectral diapason from 400 cm^{-1} to 4000 cm^{-1}, as a film on KBr, received from toluene solution) [23, 24].

6.3 RESULTS AND DISCUSSION

The catalytic activity of some samples of heterogenized bifunctional cobalt-containing catalytic dithiosystems in gas-phase polymerization of butadiene were investigated and the results are summarized in Table 6.2 in comparison with the known homogeneous bifunctional cobalt-containing catalytic dithiosystems and the gas-phase process of Berlin Technical University using heterogenized neodymium-catalysts.

TABLE 6.1 Heterogenization Methods of Cobalt-Containing Catalytic Dithiosystems

Catalyst	Method of heterogenization	Support preparation	Heterogenization conditions	Concentration of cobalt on support (calculated)
1	2	3	4	5
CAT.1	Direct deposition	Silica gel dehydrated at 150°C for 3 h.	The toluene solution of X-Co is physisorb onto the support for 60 min. The slurry is filtered; the remaining solid product is washed for a while with toluene and dried under vacuum and inert atmosphere.	$[Co] = 1.0\times10^{-7}$ mol/g; Al:Co = 100:1 $[X\text{-}Co] = 6.77\times10^{-5}$ g/g support or $[Co] = 5.9\times10^{-6}$ g/g support; Co = 0.00059%
			Toluene solution of DEAC was added and treated for 60 min. The slurry is filtered; the remaining solid product is washed for a while with toluene and dried under vacuum and inert atmosphere.	
CAT.2	Direct deposition	Silica gel dehydrated at 150°C for 3 h and then at 450°C for 2 h.	The toluene solution of X-Co+DEAC is physisorb onto the support for 60 min. The slurry is filtered; the remaining solid product is washed for a while with toluene and dried under vacuum and inert atmosphere.	$[Co] = 1.0\times10^{-7}$ mol/g; Al:Co = 100:1 $[X\text{-}Co] = 6.77\times10^{-5}$ g/g support or $[Co] = 5.9\times10^{-6}$ g/g support; Co = 0.00059%

TABLE 6.1 (Continued)

Catalyst	Method of heterogenization	Support preparation	Heterogenization conditions	Concentration of cobalt on support (calculated)
CAT.3	Direct deposition	Silica gel dehydrated at 650°C for 5 h.	The toluene solution of X-Co+MAO is physisorb onto the support for 60 min. The slurry is filtered; the remaining solid product is washed for a while with toluene and dried under vacuum and inert atmosphere.	[Co] = 1.0×10^{-8} mol/g; Al:Co = 1000:1 [X-Co]=6.77×10^{-6} g/g support or [Co] = 5.9×10^{-7} g/g support; Co = 0.000059%
CAT.4	Pre-alumination	Silica gel dehydrated at 200°C for 6 h and at 25°C the toluene solution of DEAC ([Al]= 4 mmol/g support) was added, the slurry is filtered, the remaining solid product is washed for a while with toluene and dried under vacuum and inert atmosphere.	The toluene solution of X-Co+DEAC+ butadiene complex is physisorb onto the support for 60 min. The slurry is filtered; the remaining solid product is washed for a while with toluene and dried under vacuum and inert atmosphere.	[Co] = 1.0×10^{-7} mol/g; Al:Co = 100:1 [X-Co]=6.77×10^{-5} g/g support or [Co] = 5.9×10^{-6} g/g support; Co = 0.00059%

TABLE 6.1 (Continued)

Catalyst	Method of heterogenization	Support preparation	Heterogenization conditions	Concentration of cobalt on support (calculated)
CAT.5	Pre-alumination	Silica gel dehydrated at 200°C for 2 h and at 25°C was added the toluene solution of MAO ([Al]= 4 mmol/g support), the slurry is filtered, the remaining solid product is washed for a while with toluene and dried under vacuum and inert atmosphere.	The toluene solution of X-Co+DEAC+butadiene complex is physisorb onto the support for 60 min. The slurry is filtered; the remaining solid product is washed for a while with toluene and dried under vacuum and inert atmosphere.	[Co]=1.0×10^{-7} mol/g support; Al:Co = 100:1 [X-Co]= 6.77×10^{-5} g/g support or [Co]= 5.9×10^{-6} g/g support; Co = 0.00059%
CAT.6	Pre-alumination	Silica gel dehydrated at 120°C for 2 h and at 25°C was added the toluene solution of TEA ([Al]= 25 mmol/g support), after 2 h the slurry is filtered, the remaining solid product is washed for a while with toluene and dried under vacuum and inert atmosphere.	The toluene solution of X-Co+MAO+butadiene complex is physisorbed onto the support for 60 min. The slurry is filtered; the remaining solid product is washed for a while with toluene and dried under vacuum and inert atmosphere.	[Co]= 1.0×10^{-8} mol/g support; Al:Co = 1000:1; [X-Co]= 6.77×10^{-6} g/g support or [Co] = 5.9×10^{-7} g/g support; Co = 0.000059%

TABLE 6.1 (Continued)

Catalyst	Method of heterogenization	Support preparation	Heterogenization conditions	Concentration of cobalt on support (calculated)
CAT.7	Pre-alumination	Al_2O_3 dehydrated at 120°C for 6 h and at 25°C the toluene solution of DEAC ([Al]= 4 mmol/g support) was added, the slurry is filtered, the remaining solid product is washed for a while with toluene and dried under vacuum and inert atmosphere.	The toluene solution of X-Co + DEAC+butadiene complex is physisorb onto the support for 60 min. The slurry is filtered; the remaining solid product is washed for a while with toluene and dried under vacuum and inert atmosphere.	[Co] = 1.0×10^{-7} mol/g; Al:Co = 100:1 [X-Co]=6.77×10^{-5} g/g support or [Co] = 5.9×10^{-6} g/g support; Co = 0.00059%
CAT.8	Pre-alumination	Zeolite NaY dehydrated at 160°C for 6 h and at 25°C the toluene solution of DEAC ([Al]= 4 mmol/g support) was added, the slurry is filtered, the remaining solid product is washed for a while with toluene and dried under vacuum and inert atmosphere.	The toluene solution of X-Co + DEAC+butadiene complex is physisorb onto the support for 60 min. The slurry is filtered; the remaining solid product is washed for a while with toluene and dried under vacuum and inert atmosphere.	[Co] = 1.0×10^{-7} mol/g; Al:Co = 100:1 [X-Co]=6.77×10^{-5} g/g support or [Co]=5.9×10^{-6} g/g support; Co = 0.00059%

TABLE 6.1 (Continued)

Catalyst	Method of heterogenization	Support preparation	Heterogenization conditions	Concentration of cobalt on support (calculated)
CAT.9	Pre-alumination	$MgCl_2$ dehydrated at 200°C for 6 h and at 25°C the toluene solution of DEAC ([Al]= 4 mmol/g support) was added, the slurry is filtered, the remaining solid product is washed for a while with toluene and dried under vacuum and inert atmosphere.	The toluene solution of X-Co + DEAC+butadiene complex is physisorb onto the support for 60 min. The slurry is filtered; the remaining solid product is washed for a while with toluene and dried under vacuum and inert atmosphere.	[Co] = 1.0×10^{-7} mol/g; Al:Co = 100:1 [X-Co]=6.77×10^{-5} g/g support or [Co]=5.9×10^{-6} g/g support; Co = 0.00059%
CAT.10* (BTU)	Pre-alumination	Silica gel (15 g) dehydrated at 200°C for 2 h and at 25°C was added the toluene solution of MAO (1.5 g), the slurry is filtered, the remaining solid product is washed for a while with toluene and dried under vacuum and inert atmosphere.	Cyclohexene solution of homogeneous catalyst Nd $(Vers.)_2$+ TEA+EASC (0.5 g) is physisorbed onto the support for 60 min. The slurry is filtered; the remaining solid product is washed for a while with toluene and dried under vacuum and inert atmosphere.	[Nd]=3.5×10^{-2} mol/L cat support; BD:Nd =12400; Al:Nd = 170; Cl:Nd = 1.7
CAT.11**	Homogeneous	–	Homogeneous catalyst X-Co+DEAC	[Co] = 1.0×10^{-4} mol/L; Al:Co = 100:1

(*) Heterogenized neodymium-catalyst of Berlin Technical University (BTU) [8, 9]: [Nd] = 3.2×10^{-2} mol/L_{cat}; catalyst productivity: 500.0 kg PBD/g Nd.h;
(**) Homogeneous bifunctional catalyst-stabilizers of IPCP of ANAS [13–16]: X-Co+DEAC; [Co] = 1.0×10^{-4} mol/L; Al:Co = 100:1; T = 25°C; solvent: toluene.

TABLE 6.2 Catalytic Activities of Heterogenized X-Co+AOC (Aluminum Organic Compound) Catalyst Systems in the Gas-Phase Polymerization of Butadiene (Reaction Conditions: [Co]= 1.0×10^{-7} mol/g Support; T = 25°C)

Catalyst	Reaction time (min)	Productivity (kg PBD/g Co.h)	Intrinsic viscosity [η] (dL/g)	Molecular mass		Microstructure (%)		
				$M_w\times10^{-3}$	M_w/M_n	1.4-cis	1.4-trans	1.2
CAT.1	60	340	3.4	330	2.75	90	6	4
CAT.2	45	510	3.8	466	2.6	93	5	2
CAT.3	90	1250	4.9	950	2.2	94	4	2
CAT.4	60	1450	3.3	310	2.1	91	6	3
CAT.5	45	1700	3.8	466	2.6	93	5	2
CAT.6	90	950	4.1	500	2.5	92	5	3
CAT.7	90	1150	2.8	300	2.4	90	7	3
CAT.8	90	850	2.5	250	2.5	91	8	1
CAT.9	90	900	2.6	270	2.3	92	6	2
CAT.10* (BTU)	60	500	5.2	1500	2.3	97	2	1
CAT.11**	15	57	3.5	450	2.5	91	5	4

(*) (**) as in Table 1 footnotes.

The data demonstrates that the pre-alumination method of immobilizing the bifunctional cobalt containing catalytic dithiosystem with X-Co shows a very high activity in the gas-phase polymerization of butadiene. Results of Table 6.2 show that at reaction conditions: [Co] = 1.0×10^{-7} mol/g support; T = 25°C catalysts productivity reaches ~1700.0 kg PBD/g Co.h., which is much higher than of the solution process using the same homogenous catalyst X-Co+DEAC (57.0 kg PBD/g Co.h.) and gas-phase process of Berlin Technical University using the heterogenized neodymium-catalysts (500.0 kg PBD/g Nd.h.).

High catalytic activities are showed by methods of heterogenization in CAT.4 and CAT.5. Therefore the investigations of activity and stereo selectivity of cobalt catalysts with various organic dithioderivatives were investigated using these heterogenization methods. Results are shown in Table 6.3.

Table 6.3 shows that by changing ligands in cobalt compounds it is possible to change the catalyst productivity and stereo-regularity in large diapason. The highest catalytic activity and stereo selectivity was obtained using X-Co, DCDTPh-Co (Cobalt O,O'-di-4-methylphenyl dithiophosphate) and NGDTPh-Co (Cobalt *O, O'*-di- (2,2′-methylene-bis-4-methyl-6-*ert.*butylphenyl) dithiophosphate) as a component of heterogenized catalyst. Their productivities reached 1340.0–1700.0 kg PBD/g Co.h, intrinsic viscosities were in the range of 2.5–3.8 dL/g, and 1.4-cis contents were between 91.0–96.0%.

Heterogenized DEDTC-Co+DEAC (Cobalt diethyl dithiocarbamate) catalytic system allows for the production of high molecular 1,4-cis+1,2-PBD with 1,4-cis content of 62.0%, 1,2-content of 35.0%, 1,4-trans content of 3.0%, intrinsic viscosity pf 1.94 dL/g and productivity of 850.0 kg PBD/g Co.h.

Changing of ligands in cobalt compound is similar to those of homogeneous catalyst for polymerization of butadiene [16, 25, 26]. High activity of X-Co (Cobalt O,O'-di- (3,5-di-*tert.* butyl-4-hydroxyphenyl) dithiophosphate), DCDTPh-Co and NGDTPh-Co in butadiene polymerization can be explained by the high solubility of these compounds in toluene and the formation of stable metallocomplex nanosized active centers, covalently tethered on support, which allows obtaining polymers with very high productivities than those in solvent polymerization.

TABLE 6.3 Catalytic Activities of Heterogenized DTC-Co + AOC Catalyst Systems in the Gas-Phase Polymerization of Butadiene (Reaction Conditions: [Co] = 1.0×10–^{7} mol/g Support; T = 25°C)

Cobalt Dithio compound (DTC-Co)	Method of heterogenization	Reaction time (min)	Productivity (kg PBD/g Co.h)	Intrinsic viscosity [η] (dL/g)	Molecular mass		Microstructure (%)		
					$M_w \times 10^{-3}$	M_w/M_n	1.4-cis	1.4-trans	1.2-vinyl
DCDTPh-Co	CAT.4	60	1340	2.5	250	1.56	93	5	2
DCDTPh-Co	CAT.5	60	1650	2.6	270	1.54	95	4	1
DTBPhDTPh-Co	CAT.4	60	750	3.1	320	1.72	93	5	2
DPhDTPh-Co	CAT.4	60	810	2.9	300	1.94	91	6	3
TBPh-Co	CAT.4	60	920	2.4	260	1.93	92	6	2
4-m,6-TBPh-Co	CAT.4	60	650	2.8	290	1.93	93	5	2
X-Co	CAT.4	60	1450	3.3	310	2.10	91	6	3
X-Co	CAT.5	45	1700	3.8	466	2.60	93	5	2
NGDTPh-Co	CAT.4	60	1300	3.2	330	1.74	95	4	1
NGDTPh-Co	CAT.5	60	1600	2.5	290	1.65	96	3	1
DEDTC-Co	CAT.4	90	850	3.3	350	1.94	62	35	3

As was shown earlier the direct deposition method is simple and convenient, but the supported catalysts prepared using this method usually have low polymerization activities due to the influences of the silica surface, that is, steric hindrance and poisoning effect.

In the supported homogeneous catalysts prepared by the using of pre-alumination method, active sites have appreciable freedom of movement on the support, so the catalytic properties of the supported active sites resemble those of the homogeneous analogs and the polymers produced are similar to those produced by the homogeneous analogs. However, the nature of the weak ionic interactions that bind the active sites to the MAO-covered support is not clear and is difficult to study.

An important aspect of the immobilized homogeneous catalysts is the stability of the active species on the supports. In the gas phase polymerization process, the migration of unstable active species from the interior of the catalyst particles to the surface will result in polymer products of poor molecular weight distribution and micro-structural dispersion. Both poor morphology and reactor fouling are manifestations of "catalyst leaching." Catalyst leaching has been found in the first and second methods of catalyst immobilization. No leaching of active species from the support was observed in the covalent tethering method [26].

6.4 CONCLUSION

A new high activity and stereoselective heterogenized bifunctional cobalt-containing catalytic dithiosystem for butadiene gas-phase polymerization has been developed. Based on organic dithiocompounds of cobalt (O,O-dithiophosphates, N, N-dithiocarbamates) in combination with aluminum organic compounds (DEAC, TEA, MAO), heterogenized by the methods of pre-alumination and covalent tethering on silica gel support, these catalysts provide the preparation of high molecular mass 1,4-cis (91.0–96.0 %) or 1,4-cis+1.2 (1.4-cis – 62.5 and 1.2 – 35.0 %) polybutadienes with productivities of 650.0–2200.0 kg PBD/g Co.h and intrinsic viscosities of 2.5–3.8 dL/g.

Synthesized polybutadienes can be used in the production of tires, technological rubbers and impact resistant polystyrenes.

KEYWORDS

- **butadiene**
- **catalytic dithiosystems**
- **cobalt**
- **gas-phase polymerization**
- **heterogenization**

REFERENCES

1. Silvester G. Gas phasen polymerisation von butadiene, *Gummi Asbest Kunststoffe,* 49, 60, 1996 (in Germany).
2. Eberstein C. Gasphasen polymerisation von butadiene – kinetik, partikelbildung und modellierung, Ph. D. thesis, TU Berlin, 1997 (in German).
3. Spiller C. Untersuhungen zum stoff- und warmetransport bei der gasphasenpolymerisation von 1,3-Butadien, Ph. D. thesis, TU Berlin, 1998 (in German).
4. Zoellner K., Reichert K. H. Experimentelle untersuchungen der gasphasen polymerisation von butadiene im labor reactor, *Chemie Ingenieur Technik,* 4, 396–400, 2000 (in German).
5. Bartke M., Wartmann A., Reichert K.-H. Gas phase polymerization of butadiene. Data acquisition using minireactor technology and particle modeling, Journal of Applied Polymer Science, 87, 270–279, 2003.
6. Zoellner K., Reichert K. H. Gas phase polymerization of butadiene – kinetics, particle size distribution, modeling, Chemical Engineering Science, 56, 4099–4106, 2001.
7. Catalyst, its production and its use for the gas-phase polymerization of conjugated dienes /Bayer AG, Leverkusen, DE, Pat. USA 5908904 (1999).
8. Method for the production of diene rubbers in the gas phase /Bayer AG, Leverkusen, DE, Pat. USA 5914377, 1999.
9. Gas-phase polymerisation of conjugated dienes in the presence of rare earth allyl compounds /Bayer AG, Leverkusen, DE, Pat. USA 5958820, 1999.
10. Supported cobalt catalyst, production thereof and use thereof for the polymerization of unsaturated compounds / Bayer AG, Leverkusen, DE, Pat. USA 6093,674, 2000.
11. Gas phase production of polydienes with pre-activated nickel catalysts / Union Carbide Chemicals & Plastics Technology Corporation, Danbury, CT., Pat. USA 6004677, 1999.
12. Start-up process for gas phase production of polybutadiene /Union Carbide Chemicals & Plastics Technology Corporation, Danbury, CT, Pat. USA 6255,420, 2001.

13. Gas phase anionic polymerization of diene elastomers /Bridgestone Corporation, Tokyo, JP, Pat. USA 6258,886, 2001.
14. Nasirov F. A. Development of scientific foundation for creation of bifunctional catalytic dithiosystems for preparation and stabilization of polybutadienes, Doctoral Dissertation (Prof. Doctor, Chemistry). Baku: IPCP of Azerbaijan National Academy of Sciences, 376 p., 2003 (in Russian).
15. Azizov A. G., Djanibekov N. F., Nasirov F. A., Aliev V. S., Gadjiev R. K., Novruzova F. M. Bifunctional catalyst for obtaining and stabilization of polybutadienes, Pat. 2030210, RF, 1987 (in Russian).
16. Nasirov F. A. Bifunctional Nickel- or Cobalt containing catalyst-stabilizers for polybutadiene production and stabilization (Part I): Kinetic study and molecular mass stereoregularity correlation, *Iranian Polymer Journal,* 12, 4, 217–235, 2003.
17. Nasirov F. A. Bifunctional Nickel- or Cobalt-containing Catalyst-stabilizers for Polybutadiene Production and Stabilization. (Part II) Antioxidative Properties and Mechanism of Catalyst-stabilizer's Stabilizing Action in the Processes of Thermo- and Photooxidative Ageing of Polybutadiene, Iranian Polymer Journal, 12, 5, 281–289, 2003.
18. Nasirov F. A. New catalysts and technologies of producing stereoregular polybutadienes. In.: Synthesis and Applications of Reactive Oligomers and Polymers. In: Synthesis and Application of Reactive Oligomers and Polymers. Azizov A.G., Abdullayev Y. G., Asadov Z. G., Vezirov S. S., Gadjiyev T. A., Nasirov F. A. /Under the scientific editorship of Azizov A.G. Baku, IPCP, Azerbaijan Academy of Sciences, 274, 107–144, 1998 (in Russian).
19. Xu C. The use of functionalized zirconocenes as precursors to silica-supported zirconocene olefin polymerization catalysts, PhD in Chemistry, Virginia Tech University, 186 p., 2001.
20. Mckittrick M. W. Single-site olefin polymerization catalysts via the molecular design of porous silica, PhD in Chemical and Biomolecular Engineering, Georgia Institute of Technology, 195 p., 2005.
21. Rafikov S. P., Pavlova S. A., Tvyordokhlebova I. I. Methods of determination molecular weight and polydispersity of high molecular materials. Moscow: Academy of Sciences of USSR, 336 p., 1963.
22. Deyl Z., Macek K., Janak J. (Eds.). Liquid Column Chromatography, Elsevier, Amsterdam Scientific v.1, 2, 1975.
23. Haslam J., Willis H. A. Identification and analysis of Plastics, London, Iliffe Books, Princeton, New Jersey: D.Van Nostrand Co., 172–174, 1965.
24. Bellami L. J. The Infra-red Spectra of Complex Molecules, London, Methuen and Co, New York: J. Wiley, 592 p., 1957.
25. Nasirov F. A. Novruzova F. M., Salmanov S. S., Qasimzade E. A., Djanibekov N. F. Method of gas phase polymerization of butadiene. Pat. 20060065, Azerbaijan, 2006 (in Azerbaijanian).
26. Nasirov F. A., Novruzova F. M., Salmanov S. S., Azizov A. H., Janibayov N. F. Gas Phase Polymerization of Butadiene on Heterogenized Cobalt-containing Catalytic Dithiosystems, Iranian Polymer Journal 18 (7), 521–533, 2009.

CHAPTER 7

ELECTRIC CONDUCTING PROPERTIES OF ELECTROLYTES BASED ON SOME OLYORGANOSILOXANES WITH DIFFERENT FUNCTIONAL PENDANT GROUPS

J. ANELI,[1] O. MUKBANIANI,[1,2] T. TATRISHVILI,[1,2] and E. MARKARASHVILI[1,2]

[1]*Institute of Macromolecular Chemistry and Polymeric Materials, I. Javakhishvili Tbilisi State University, I. Chavchavadze Ave. 13, 0179 Tbilisi, Georgia, E-mail: jimaneli@yahoo.com*

[2]*Iv. Javakhishvili Tbilisi State University, Department of Chemistry, I. Chavchavadze Ave. 1, 0179 Tbilisi, Georgia*

CONTENTS

ABSTRACT

The electrical conductivity of solid polymer electrolyte membranes based on three type of polysiloxanes containing propyl butyrate, acetyl acetonate and allyl cyanide as pendant groups and lithium trifluoromethylsulfonate (triflate) or lithium bis(trifluoromethylsulfonyl)-imide were investigated by impedance spectroscopy. It is experimentally shown that the dependence of conductivity of polyelectrolytes has extreme character – corresponding curves are characterized with maximums, the intensity and location of which are determined by types of polymer matrix and Li-salt. This result is explained in terms of formation of ion-pairs at relatively high concentrations of anions, the mobility of which is lower in comparison with monoions. Temperature dependence of the electrolytes conductivity describes by Vogel-Tamman-Fulcher (VTF) or Arrhenius formula. These dependences are defined mainly with the value of ion mobility in the polymer matrix. The electrical conductivity of these materials at room temperature belongs to the range of 7×10^{-9} to 4×10^{-4} S/cm. The voltamogrames of investigated polyelectrolytes for small values of voltages have been obtained.

7.1 INTRODUCTION

Hybrid organic-inorganic materials, where molecular organic and inorganic fragments are combined, have been considered potentially attractive for the purpose of developing of new materials with a broad spectrum of interesting properties. In comparison with organic and inorganic constituents and polymers separately, hybrid organic-inorganic materials have a lot of advantages [1–4].

Polymer electrolytes (PE) play an important part in electrochemical devices such as batteries and fuel cells. To achieve optimal performance, the PE must maintain a high ionic conductivity and mechanical stability at both high and low relative humidity. The polymer electrolyte also needs to have excellent chemical stability for long product life and robustness.

According to the prevailing theory, ionic conduction in polymer electrolytes is facilitated by the large-scale segmental motion of the polymer backbone and primarily occurs in the amorphous regions of the polymer electrolyte. Crystallinity restricts polymer backbone segmental motion and significantly reduces conductivity. Consequently, polymer electrolytes with high conductivity at room temperature have been sought through polymers, which have highly flexible backbones and have largely amorphous morphology.

The interest to polymer electrolyte was increased also by potential applications of solid polymer electrolytes in high energy density solid-state batteries, gas sensors and electrochromic windows.

Conductivity of 10^{-3} S/cm is commonly regarded as a necessary minimum value for practical applications in batteries [4, 5]. At present, polyethylene oxide (PEO)-based systems are most thoroughly investigated reaching room temperature conductivities of 10^{-7} S/cm in some cross-linked salt in polymer systems based on amorphous PEO–polypropylene oxide copolymers. However, conductivity with such value unfortunately is low resulting from the semi crystalline character of the polymer as well as from the increase on the glass transition temperature of the system. It is widely accepted that amorphous polymers with low glass transition temperatures T_g and a high segmental mobility are important prerequisites for high ionic conductivities. Another necessary condition for high ionic conductivity is a high salt solubility in the polymer, which is most often achieved by donors such as etheric oxygen or imide groups on the main chain or on the side groups of the PE. It is well established also that lithium ion coordination takes place predominantly in the amorphous domain and that the segmental mobility of the polymer is an important factor in determining the ionic mobility. Great attention was pointed to PEO-based amorphous electrolyte obtained by synthesis of comb-like polymers, by attaching short ethylene oxide unit sequences to an existing amorphous polymer backbone.

Comb-like polysiloxanes solid PE systems incorporating different lithium salts nowadays attract much more attention because of it relatively high conductivity of about 10^{-4} S/cm [5–7] and 5×10^{-4} S/cm for double comb polysiloxanes PE having two oligoether side groups per silicon and dissolved lithium bis(trifluoromethylsulphonyl)imide.

Synthesis and conductivity studies were published for another series of oligoether-substituted mono-comb polisiloxane PE, which was additionally cross-linked by α,w-diallylpolyethyleneglicol [8, 9]. Careful analysis of these and other results conducted in Ref. [10] suggests that there is no significant conductivity loss at transfer from double-comb to mono-comb polydimethylsiloxanes.

It was observed that the dependence of materials conductivity σ on the inverse temperature is described by one of the following types of regularities: (a) Vogel-Tammann-Fulcher (VTF) and (b) Arrhenius formula with two activation energies [11, 12].

The aim of presented work is an obtaining of new solid polymer electrolyte membranes on the base of comb-like methylsiloxane matrix with regular arrangement of propyl butyrate, propyl acetoacetate, triethoxysilane and allyl cyanide pendant groups separately and investigation of their electric conducting properties.

7.2 EXPERIMENTAL PART

7.2.1 OBJECTS OF STUDY

Polyelectrolytes on the basis of some siliconorganic polymers and two type of Li-salt were the objects for the investigations of some electrical- physical properties in dependence of type of polymers and ions and concentration of lasts. Following types of polymers have been synthesized: (a) polysiloxane containing propyl butyrate side groups; (b) polysiloxane containing propyl acetoacetate and triethoxysilane pendant groups at silicon; (c) polysiloxane containing allyl cyanide side groups. The synthesis of these polymers held earlier and now these materials are in the press for publication. Li-salts of types lithium trifluoromethylsulfonate (triflate) – salt S_1 and lithium bis(tr ifluoromethylsulfonylimide) –salt S_2 were used in the investigated electrolytes. On the basis of these polymers and Li-salts there were obtained the membranes – films in the form of thin (0.2–0.3 mm) discs with diameter 10 mm at using following method: 0.75 g of the base compound was dissolved in 4 mL of dry THF and thoroughly mixed for half an hour before the addition catalytic amount of acid (one drop of 0.1 N HCl solution in ethyl alcohol) to initiate the cross-linking process. After stirring for another 3 h required amount of lithium triflate from the previously prepared stock solution in THF was added to the mixture and stirring continued for further 1 h. The mixture was then poured onto a teflon mould with a diameter of 4 cm and solvent was allowed to evaporate slowly overnight. Finally, the membrane was dried in an oven at 70°C for 3 d and at 100°C for 1 h. Homogeneous and transparent films with average thickness of 200 μm were obtained in this way. These films were insoluble in all solvents, only swollen in THF. The salts S1 and S2 were introduced to the polymers with concentrations 5, 10, 15, 20 and 25 wt % of each salt. The polyelectrolytes presented in this paper will be noted as:

P_1S_1 and P_1S_2 _ polysiloxane containing propyl butyrate pendant groups with salt S_1 and S_2;

P_2S_1 and P_2S_2 _ polysiloxane containing acetil-acetate pendant groups with salt S_1 and S_2;

P_3S_1 and P_3S_2 _ polysiloxane containing allyl cyanide pendant groups with salt S_1 and S_2.

7.2.2 METHODS OF MEASURES OF ELECTRIC CONDUCTING CHARACTERISTICS OF MATERIALS

The total ionic conductivity of samples was determined by locating a membrane between two disks like brass electrodes with 10 mm diameter. The electrode/electrolyte assembly was secured in a suitable constant volume support, which allowed extremely reproducible measurements of conductivity to be obtained between repeated heating-cooling cycles. The cell support was placed in an oven and the change of sample temperature was measured by a thermocouple close to the electrolyte disk. The bulk conductivities of the electrolytes were obtained during a heating cycle using the impedance technique (Impedance meter BM 507 –TESLA for frequencies 50 Hz-500 kHz) over a temperature range between 20 and 110°C. Voltamogrames measured at room temperatures were obtained by use of simple arrangement allowing the registration of the value of current passing true the electrolyte sample at gradual increase of the voltage.

7.3 RESULTS AND DISCUSSION

7.3.1 POLYSILOXANE CONTAINING PROPYL BUTYRATE PENDANT GROUPS

First of all it was defined a character of dependence of compounds electric conductivity on the concentration of Li salts. The curves on Figure 7.1

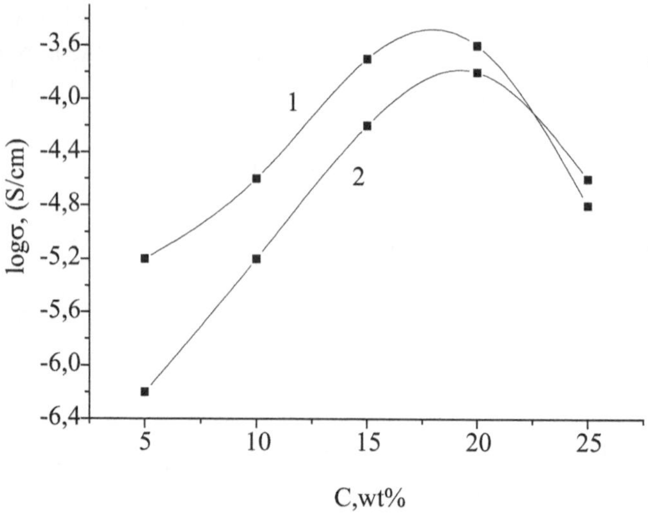

FIGURE 7.1 Ionic conductivity of the systems based on polymer P_1 as a function of salt S_1 (1) and S_2 (2) concentrations at 25°C temperature number. The maximum of ionic conductivity can be described by two opposing process, which are in accordance with conceptions noted in Ref. [9]: (a) increasing of the number of charge carriers (ions) in result of increasing of the salt concentration; (b) increasing of possibility of the formation of ion pairs, thanks to which the ions migration will be prevented in the electrolyte network.

show that these dependences have extreme character, the conductivity rises with an increase of the salt concentration, reaches a maximum value and after declines. Obviously this fact may be described with increasing of charge carrier's.

In accordance with Figure 7.1 noted above maximums on the curves for systems P_1S_1 and P_1S_2 appear at concentrations near 17–18 wt.% of salt, respectively.

Quantitative difference between curves in the Figure 7.1 leads to conclusion that in the electrolyte P_1S_1 ions of the salt S_1 having relatively small anions is characterized with more high mobility, than that for the salt S_2. Therefore the maximum of conductivity for P_1S_1 is higher to some extent than for P_1S_2 having anions with less mobility. Decreasing of conductivity of both electrolytes at relatively high concentrations of both types of salts is due to mentioned above phenomenon – increasing of probability of formation of ion pairs at high ion concentrations.

On the basis of well known experimental results about effect of the length of side groups on the value of conductivity of polyelectrolytes it would be expected that the conducting complexes containing the molecules with long side groups would be more conductive than ones with relatively short side chains. However the literature data obtained by other authors and by us shown that, as a rule, it is no direct correlation between side chains lengths and conductivity of polyelectrolytes [12,13]. Main reason of these deviations we can find not only in the side chain lengths. There are also other factors (e.g., molecular morphology, content of salts, distribution of salt molecules between polymer chains and the character of interactions between them, etc.), which influence simultaneously on the electrical conductivity of polyelectrolytes and creates the difficulties in unique establishment of real nature of the polyelectrolytes conductivity. At establishment of the effect of microstructure of the polymer system on the mobility of Li ion it must be took into account the effect of the free volume between macromolecules and salt molecules. Therefore it is very difficult to define effect of influence of the length of side chains on the free volume, because at increasing of side chain length on last apparently has non-linear character. The exact estimation of the free volume with different methods would be introduced some definition to this problem.

Usually coming from practical interests the dependence of the conductivity on temperature of polyelectrolytes one defined mainly in the range of about 30–90°C [6, 7]. The conductivities of the investigated compounds defined on these limits are given in the Table 7.1.

The dependence log σ–1/T has nearly linear character and obeys to Arrhenius low (Figures 7.2 and 7.3). These curves were designed after several measuring of this dependence and the data of conductivities were obtained after averaging of them.

As it is seen from Figures 7.2 and 7.3 conductivity of the investigated compounds rises with an increase of temperature. Main factor which defines a such character of these dependences it must be fond in increasing of charge carriers (ions) mobility at increasing of temperature. This conclusion on the temperature dependence of conductivity of PE is in good agreement with experimental data presented on Figure 7.2 and the Table 7.1.

It was interesting to define the dependence of current – voltage for investigated electrolytes (PE). On the Figure 7.4 the curves of I – V functional dependences (so called the voltamograms) for some obtained membranes are presented.

TABLE 7.1 The Conductivities of Investigated Compounds PS1 and PS2, Defined at 25 and 90°C

Salt			S_1					S_2		
Concentr. Wt.%	5	10	15	20	25	5	10	15	20	25
S(30°C), S/cm	6.3×10^{-6}	1.2×10^{-5}	3.2×10^{-4}	3.5×10^{-4}	1.6×10^{-5}	6.4×10^{-7}	6.6×10^{-6}	1.5×10^{-5}	1.3×10^{-4}	7.2×10^{-6}
S(90°C), S/cm	2.5×10^{-4}	5.2×10^{-5}	7.9×10^{-4}	1.1×10^{-3}	2.4×10^{-4}	3.3×10^{-5}	2.3×10^{-5}	6.2×10^{-5}	3.4×10^{-4}	2.6×10^{-5}

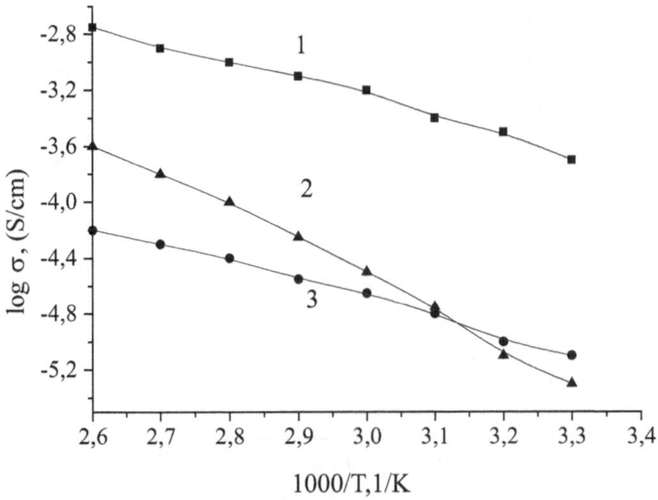

FIGURE 7.2 Arrhenius plots of ionic conductivity for polymer electrolytes P_1S_1 containing 20 (1), 5 (2) and 10 wt% (3) of salt S_1.

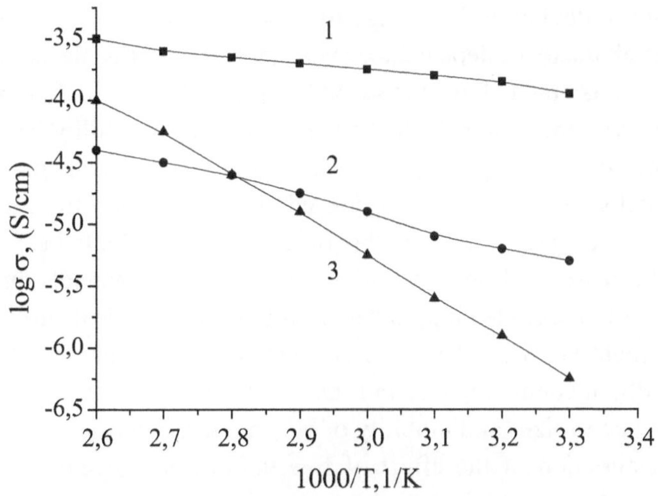

FIGURE 7.3 Arrhenius plots of ionic conductivity for polymer electrolytes P_1S_2 containing 20 (1), 10 (2) and 5 wt% (3) salt S_2.

The dependences I – U, that is, voltamograms (Figure 7.4) show that in the used interval of change of voltage the value of the current increases

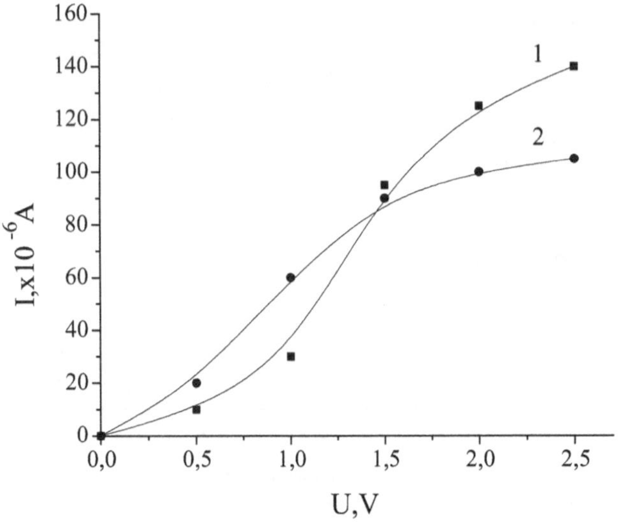

FIGURE 7.4 Voltamograms for membranes containing 20% S_1 (1) and same amount of S_2 (2).

with definite deviation from linearity. The behavior of the curves corresponds to character of dependence of carriers transport on the nature of salt ions. As it was noted above the second salt ions differ from first ones with more high volume to some extent and relatively low mobility. Therefore at increasing of the voltage accelerated ions after charge-phonons scattering are gradually slowed, which is reflected on the deviation from the linearity of A – V characteristics. At this time appears the Joule heat, which is one of the reasons of destruction of the conducting channels due to heat scattering of charges leading to decreasing of the electrical current passed through material. This deviation appears for the membrane PS_2 earlier than for PS_1 at containing one and same concentration of the salts because of difference in sizes and mobility of the compared ions.

For estimation of the effect of free volume in polyelectrolyte compounds on the ion conductivity we provided the experiments on investigation of the influence of external pressure on the conductivity for the investigated polyelectrolytes (Figure 7.5).

The curves on the Figure 7.5 show that this dependence of resistance on the pressure has exponential character in relatively narrow interval of pressures. This result is non-expected to some extent, because, as it is well

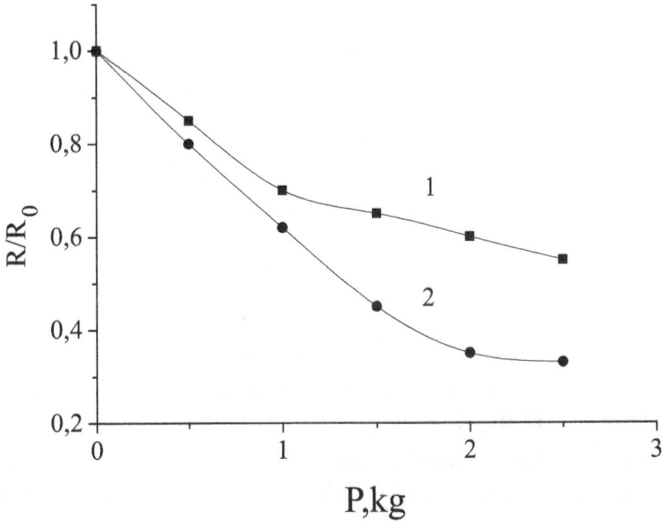

FIGURE 7.5 Dependence of the resistance of electrolytes based on P_1S_1 with 10 (1) and 20 (2) wt % S_1. R_0 – resistance of the samples at normal pressure.

known, the free volume in the polymer matrix under increasing pressure decreases and, consequently, the transport of Li ions must be decreased. However it may be suggested that at same time the segmental mobility of side chains of oligoesters significantly decreases, which in general presents the definite barrier on the way of Li ions thanks to interactions between charge carriers and phonons corresponding to side group vibrations. Therefore the noted barrier will be decreased at increasing of pressure. Probably from two effects (decreasing of free volume and simultaneously the phonon-charge carrier's interaction) the second is more effective than first one in the considered case. Generally the conductivity of polyelectrolytes will be depended significantly on the ratio of these factors.

7.3.2 POLYSILOXANE CONTAINING PROPYL ACETOACETATE AND TRIETHOXYSILANE PENDANT GROUPS AT SILICON

It was interesting also the investigation of the electric physical properties of analogical polyelectrolytes with another type of pendant groups – allyl acetoacetate and vinyltriethoxysilane.

On the Figure 7.6 the temperature dependences of the polyelectrolytes based on polysiloxane containing propyl acetoacetate and triethoxysilane pendant groups at silicon atom P_2S_1 and P_2S_2 are presented. The curves of these dependences show that the values of the electrical conductivity of membranes and the range of their change, significantly depends on both polymer type and salt concentration. In case of PE – P_2S_1 containing 10 wt. % of S_1 the conductivity of the PE is rather low in all the temperature range. However, the conductivity of the same PE with relatively high concentration (20 wt. %) gets the significances on two order higher than for P_2S_1 containing 10 wt. % of this salt. Difference between conductivities of P_2S_1 and P_2S_2 may be described by difference in the electrical states of the PEs. Probably in the first PE possibility of creation of donor-acceptor complex is lower than in second one and consequently the corresponding conductivity for second PE is higher (but at low concentrations of the salts) besides of relatively high mobility of the anions of the first salt. However the picture is changed at more high content of the salts. In accordance with the conception expressed in Ref. [13] at increasing of the concentration of salts with relatively big anions the probability of creation of so called pairs of the ions increases and consequently the current density decreases more intensively than in case of little anions.

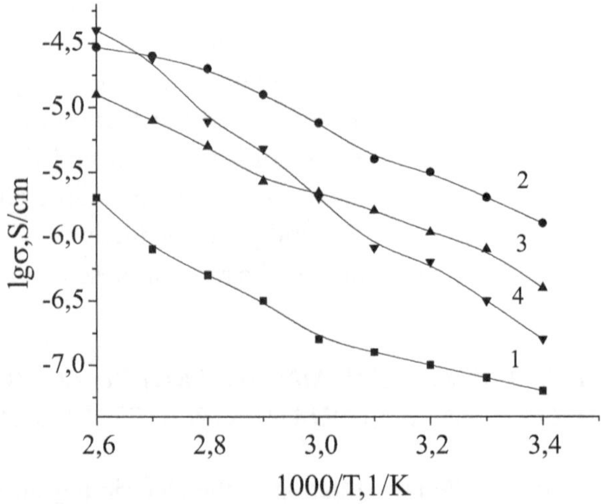

FIGURE 7.6 Dependence of electrical conductivity on the temperature for membranes P_2S_1, salt concentration in wt. % – 10 (1) and 20 (2); P_2S_2: 10 (3) and 20 (4).

It was interesting to define the behavior of the conductivity of PE – P_2S_2 with 20 wt. % of the salt. In this case, low conductivity of this PE at low temperatures increases with the rise of temperature, because of damage of charge pairs and increasing of mobility of corresponding ions. The wide interval of changes of conductivity of PE – P_2S_2 must be due to gradual structural changes of this material (gradual increasing of the segmental mobility of the polymer side groups).

The dependence of conductivity of PE based on polymer P_2 on salt concentration at change of temperature is presented on the Figure 7.7. The curves shown on this figure differ one from another with relatively narrow interval of values. The low electrical conductivity of these compounds shows, that the microstructure is less suitable for the charge transfer in these polymer matrixes. In general the conductivity values at the edge temperatures are shown in the Table 7.2.

The voltamograms of compounds P_1S_1, P_1S_2, P_2S_1 and P_2S_2 presented on the Figures 7.8 and 7.9 show that these dependences in the used interval of the changes of voltage practically have linear character, which is due to no significant formation of the Joul heat effect after charge transfer in the polymer matrix.

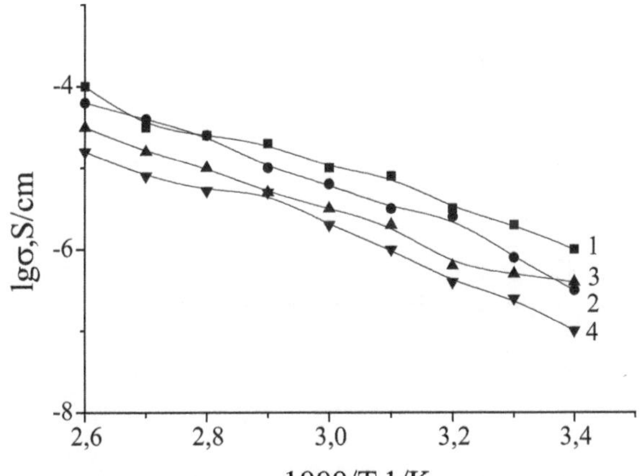

FIGURE 7.7 Dependence of electrical conductivity on the temperature of the electrolyte membranes P_2 S_1, salt concentration in wt. % – 20 (1) and 10 (2); P_2S_2: 10 (3) and 20 (4).

TABLE 7.2 Dependence of Conductivities For Compounds Based on Polymers P_1 and P_2 at Initial and Last Measuring Temperatures

Membrane	Polymer	Salt	Salt Concentr., wt.%	σ, (20°C), S/cm	σ, (115°C), S/cm
P_1S_1	P_1	S_1	10	9.0×10^{-8}	2.5×10^{-6}
P_1S_1	P_1	S_1	20	2.2×10^{-6}	5.2×10^{-5}
P_1S_2	P_1	S_2	10	$6,1\times10^{-7}$	2.3×10^{-5}
P_1S_2	P_1	S_2	20	2.5×10^{-7}	5.9×10^{-5}
P_2S_1	P_2	S_1	10	5.4×10^{-7}	7.2×10^{-5}
P_2S_1	P_2	S_1	20	6.5×10^{-7}	$9,1\times10^{-5}$
P_2S_2	P_2	S_2	10	6.2×10^{-7}	$4,1\times10^{-5}$
P_2S_2	P_2	S_2	20	8.2×10^{-8}	2.5×10^{-5}

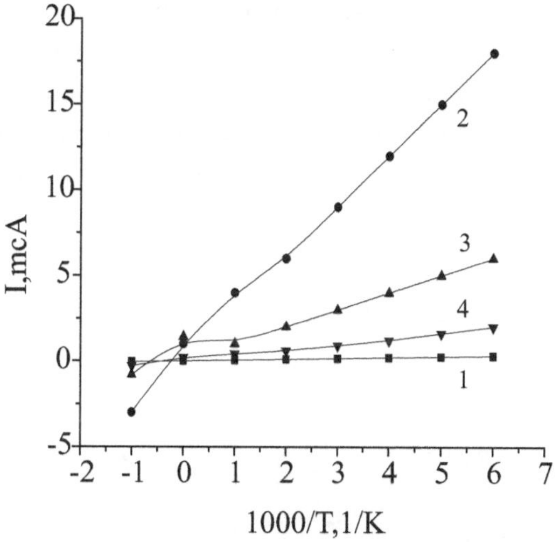

FIGURE 7.8 Voltamograms of compounds on the basis of polymer P_2, B-10 wt.% S_1, C-20 wt. % S_1, D-20 wt.% S_2 and E-10 wt.% S_2.

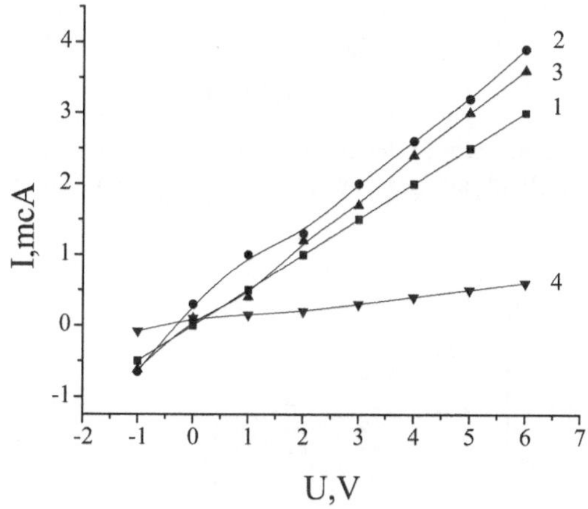

FIGURE 7.9 Voltamograms of compounds on the basis of polymer P_2 – B-10 wt.% S_1, C–20 wt.% S_1, D–20 wt. % S_2 and E –10 wt.% S_2.

7.3.3 POLYSILOXANE CONTAINING ALLYL CYANIDE PENDANT GROUPS

The next study on definition of behavior of dependence of electric conductivity on the temperature of comb like polyelectrolytes was conducted on the silicon-organic polyelectrolytes containing allyl cyanide pendant groups. Obtained on the base of synthesized polymers the membranes containing Li-salts of two types were investigated in the temperature range 25–90°C.

Below the dependences of the membranes conductivity on temperature are presented (Figures 7.10 and 7.11). The values of conductivity of membranes at initial and final temperatures were measured separately (Table 7.2).

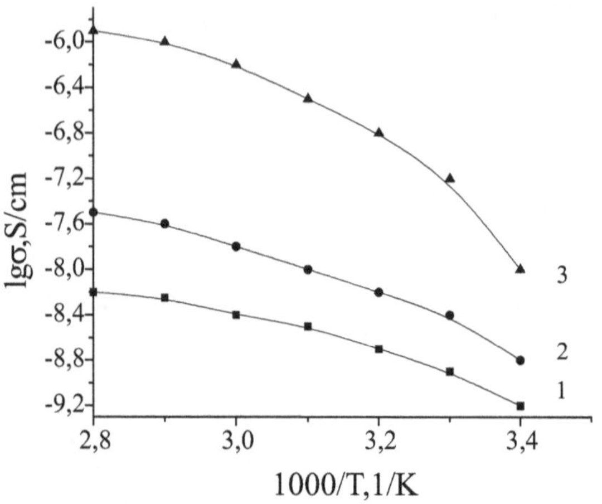

FIGURE 7.10 Arrenhius plot of the dependence of electric conductivity of electrolytes based on the polymer Pand salt S_1 at concentrations 5 (B), 20 (C) and 10 wt % (D).

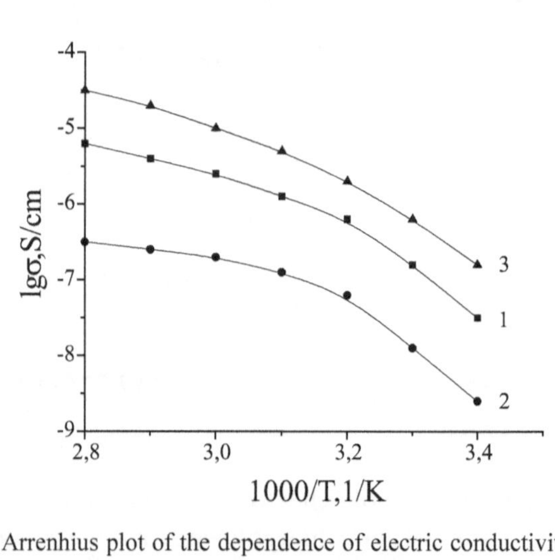

FIGURE 7.11 Arrenhius plot of the dependence of electric conductivity of electrolytes based on the polymer P and salt S_2 at concentrations 20 (B), 5(C) and 10 wt % (D).

Analysis of the curves of Figures 7.10 and 7.11 leads to analogical opinion expressed above at consideration of temperature dependences obtained for other polyelectrolytes (e.g., extreme dependence of conductivity of PE on the concentration of salts).

7.4 CONCLUSIONS

Electrical conductivity of membranes and the range of their change significantly depend on both polymer type and salt concentration. It is unusual fact that for all electrolytes dominated the decreasing of the conductivity at increasing of salt concentration. As main reason of such behavior of the salt concentration dependence of conductivity was described in terms of formation of so called ion pairs in polymer matrix as separate charges. The mobility of these pairs is rather small, than the separated ion and correspondingly the polymer electrolytes with such pairs are characterized by lower conductivities in comparison with electrolytes without such pairs. It may be supposed that in our polyelectrolytes these pairs are formed already at 5 wt % concentrations of the salts. Difference between conductivities of electrolytes based on one and same polymers with two types of salts (S_1 and S_2) may be described by distinction in structure of salts and their anions. The character of the electrolytes conductivity at high concentrations of the salts changes essentially: the conductivity of electrolytes with S2 exhibit lower increase of temperature rise than for electrolytes with S1 because of increasing of probability of formation of the anion pairs in the electrolytes with S2 (big anions) leading to decreasing of conductivity of the electrolyte in comparison with one containing S1 (small anions). At cyclic heat loading of the investigated polyelectrolytes the behavior of the curves conductivity vs. temperature is repeated with small hysteresis, which witnesses about stable microstructure of obtained polyelectrolytes in the used range of the temperatures.

The electric physical properties of the polyelectrolytes based on synthesized polymers and two types of the Li-salts allow consider these materials as the perspective basis for production of the Li-battery.

ACKNOWLEDGEMENTS

The financial support of the Georgian National Science Foundation Grant #STCU 5055 is gratefully acknowledged.

KEYWORDS

- **polysiloxanes**
- **polymer electrolyte**
- **membranes**
- **ionic conductivity**

REFERENCES

1. Chauhan, B. P. S., Rathore, J. S., Glloxhani, N. First example of 'palladium-nanoparticle'-catalyzed selective alcoholysis of polyhydrosiloxane: a new approach to macromolecular grafting. Appl. Organomet Chem., 19, 542–550, 2005.
2. Chauhan, B. P. S., Boudjouk, P. Dehydrogenative condensation of SiH and SH bonds. A metal-catalyzed protocol to stable thiopolysiloxanes. Tetrahedron Lett., , 41(8), 1127–1130, 2000.
3. Mukbaniani, O., Tatrishvili, T., Titvinidze, G., Mukbaniani, N. Hydrosilylation reactions of methylhydridesiloxane to styrene and α-methylstyrene. J. Appl. Polym. Sci., 101(1), 388–394, 2006.
4. Mukbaniani, O., Tatrishvili, T., Titvinidze, G., Mukbaniani, N., Lezhava, L., Gogesashvili, N. Hydrosilylation Reaction of Methylhydridesiloxane to Phenylacetylene. J. Appl. Polym. Sci., 100, 2511–2515, 2006.
5. Iwahara, T., Kusakabe, M., Chiba, M., Yonezawa, K. Synthesis of novel organic oligomers containing Si—H bonds. J. Polym. Sci., Part A: Polym. Chem., 31(10), 2617–2631, 1993.
6. Jonas, G., Stadler, R. Carbohydrate modified polysiloxanes II. Synthesis via hydrosilation of mono-, di-and oligosaccharide allylglycosides Acta Polym., 45(1), 14–20, 1994.
7. Marciniec, B., Gulinski, J., Kopylova, L., Maciejewski, H., Grundwald-Wyspianska, M., Lewandowski, M. Appl. Catalysis of hydrosilylation: Part XXXI. Functionalization of poly(methylhydro)siloxanes via hydrosilylation of allyl derivatives. Organomet. Chem., 11, 843–849, 1997.
8. Toshiaki Murai, Takehiko Sakane and Shinzi Kato. Cobalt Carbonyl Catalyzed Hydrosilylation of Nitriles: A New Preparation of N,N-Disilylamines. J. Org. Chem., 55, 449–453, 1990.
9. Ganicz Tomasz, Mizerska Urszula, Monszner Monika, O'Brien Michail, Perry Robert, Stanczyk Wlodzimierz. The effectiveness of rhodium (I), (II) and (III) complexes as catalysts in hydrosilylation of model olefin and polyether with triethoxysilane and poly(dimethylsiloxane-*co*-methylsiloxane). Applied Catalysts A; General, 259, 49–55, 2004.
10. Nozaruka, S., Konotsune, S. Bull. Chem, Soc. Japan, 29, 322–326, 1956.

11. Ki Hong Min, Dae Beom Kim, Yong Ku Kang, Dong Hack Suh. "Ionic conductivity and morphology of semi-interpenetrating-type polymer electrolyte entrapping poly(siloxane-g-allyl cyanide)." Journal of Applied Polymer Science, 107, 1609–1615, 2008.
12. Hooper, R., Lyons, L. J., Moline, D. A., West, R. Novel siloxane polymers as polymer electrolytes for high energy density lithium batteries. Silicon Chem., 1(2), 121–128, 2002.
13. Hooper, R., Lyons, L. J., Mapes, M. K., Schumacher, D.; Moline. D. A., West, R. Highly Conductive Siloxane Polymers. Macromolecules, 34(4), 931–936, 2001.

CHAPTER 8

THE IMPROVEMENT OF THE QUALITY OF LUBRICATING OILS BY POLYMERIC COMPOUNDS

VAGIF MEDJID FARZALIYEV and ALADDIN ISLAM AKHMEDOV

Institute of Chemistry of Additives Named After ACAD. A.M. Quliyev of National Academy of Sciences of Azerbaijan Republic, Baku, Azerbaijan
E-mail: aki05@mail.ru

CONTENTS

ABSTRACT

In this article there are given the results of the investigations conducted at the Institute of Chemistry of Additives of the National Academy of Sciences of Azerbaijan Republic in the direction of preparation of polymeric compounds, which have different functions in the composition of lubricating oils depending on the meaning of their molecular mass and chemical composition. The results of the conducted investigations have shown that for the purposes of the improvement of thermal stability of viscous additives, initial monomers used for their analysis are underwent to copolymerization with vinyl aromatic and carboxylic compounds. With the purpose of the development of polyfunctional polymeric compounds is conducted functionalization of oligomers of the high α-olefins. This way gives the possibility to develop lubricating compositions of the more simple structure, which leads to economic effect. On the base of oligomers of high α-olefins sulfonate, succinimide and thiophosphate additives were synthesized. Introduction of synthesized additives into the structure of lubricating compositions will give the possibility to develop lubricating compositions of various purposes.

8.1 RESULT AND DISCUSSION

Polymeric compounds are widely used in different fields of national economy and in Petro chemistry, concretely in the improvement of the quality of lubricating oils.

Lubricating oils have compositional structure and contain additives of different functional condition. Between these additives polymeric additives take special place because by chemical structure they close to petroleum oils (hydrocarbons) and that is why they have no problem in solubility, not volatile and easily under go to chemical modification.

Polymeric additives by functional action are divided into two groups: viscous additives and polyfunctional polymeric additives. Viscous additives by their function are divided on thickener additives and modificators of viscous index. Thickener additives are used for increase of initial viscosity and for increase only of viscosity index of petroleum oils in concentration 0.5–1%.

Polyfunctional polymeric additives are obtained by functionalization of oligomers of α-olefins and with increase of viscosity index of oil improve and other exploitation characteristic including detergent-dispersant, anti oxidative, anti corrosive and other properties.

As viscous additives, mainly, polyisobutylenes and polyalkylmethacrylates were used. However, with the tightening conditions of exploitation of lubricating oils in the structure of the new models of auto tractor equipment, the above-mentioned additives had stopped to satisfy the increasing requirements. That is why, the investigations for the development of more contemporary viscous additives, which have high stability to thermal and mechanical influences had begun. In this direction the investigations at the Institute of Chemistry of Additives were conducted. The main point of these investigations is consisted in copolymerization of the initial high methacrylates with vinyl aromatic monomers [1–3].

It was shown that by copolymerization of decyl methacrylate with dicyclopentadiene or with indene are obtained viscous additives which are characterized not only by high stability to thermal influences, but they have anticorrosive and depressor properties in the composition of petroleum oil, that is only by copolymerization can be obtained additives, which improve a number of characteristics of lubricating oils. By the right selection of copolymerized pairs of monomers can be obtained polyfunctional additives without functional group.

Viscous additives of ester type on the base of compound allyl esters of the high carbonic acids were obtained [4].

Experiment of the usage of viscous additives had shown that polymeric compounds of ester type by the improvement of viscous-temperature, mainly, low – temperature properties of lubricating oils have advantages before hydrocarbon polymers. From another side, compound ester polymers are obtained by ecological technology – the methods of radical polymerization. Hydro carbonaceous polymers are obtained by catalytic method and arises problem to release product from the remains of catalyst.

By copolymerization of high methacrylates with α-allylphenol viscous additives, improving viscous-temperature and anti oxidative properties of oils were synthesized [5].

More interesting work is the development of polyfunctional polymeric additives by functionalization of oligomers of α-olefins. It gives the possibility to develop lubricating compositions of the more simple composition and it leads to economic effect.

Between the additives to motor oils detergent – dispersant are wide – spread additives – about 50–60% of all additives comes on their portion. Oligoalkylsulfonates are wide – spread additives.

For their synthesis oligomers of different α-olefins are underwent to sulfonation.

For increase of efficiency of sulfonation in the structure of initial oligomers of α-olefins are created additional chemical centers by cooligomerization of high α-olefins with styrene or indene, into molecule include aromatic fragments, which make easier for passing of the reaction of sulfonation [6]. By the further reaction of sulfoacid with corresponding reagents sulfonated or sulfamide additives are obtained.

Synthesis of polyalkenylsuccinimide additives on the base of cooligomer of hexene-1 with dicyclopentadiene gives the possibility to obtain compound, which in the composition of oils improves their detergent-dispersant, viscous-temperature and anticorrosive properties [7]. The well-known succinimides do not have anticorrosive properties, but viscous-temperature properties improve slightly.

The research in the field of development of additives containing such active elements as phosphor, sulfur, nitrogen, metal and other is more interesting. This variant of synthesis of polyfunctional additives is carried out by reaction of α-olefinic oligomers with sulfides of phosphor [8].

The obtained dioligoalkylphosphiric acid by neutralization with various reagents is turned into ash (metal containing) and ash less additives. By including of the obtained additives into the structure of lubricating compositions may be developed lubricating structures of various purposes.

KEYWORDS

- **additives**
- **functionalization**
- **lubricating compositions**
- **polymeric compounds**

REFERENCES

1. Akhmedov, A. I., Farzaliyev, V. M., Aliguliyev, R. M. Polymeric additives and oils. Baku: Elm, 175, 2000.
2. Farzaliyev, V. M., Akhmedov, A. I., Hamidova, C.Sh., Copolymer allylnaphtenate with butylmethacrylate as viscous additives. Az. Patent I 20040053, SM, № 4, 73, 2004.

3. Farzaliyev, V. M., Askerova Kh.A., Akhmedov, A. I. and others. Copolymers of alkylakrylates with styrene as viscous additives. Azerbaijan Chemical Journal. №4, 16–18, 2009.
4. Akhmedov, A. I., Hamidova, C.Sh., Isakov, E. U. Copolymers allylnaphtenates with styrene as viscous additives. Az. Patent I 20080059, SM, №10, 74, 2008.
5. Farzaliyev, V. M., Akhmedov, A. I., Hasanova, E. I. Synthesis of copolymers of decylmethacrylate with α-allylphenol and investigation of them as viscous additives to petroleum oils. Russian Journal of Applied Chemistry. 85 (10), 1717–1719, 2012.
6. Akhmedov, A. I., Hamidova, C.Sh. Polyfunctional additives to oils on the base of sulfonated oligomers. Russian Journal of Applied Chemistry, 80 (2), 347–348, 2012.
7. Akhmedov, A. I. Oligomers on the base of α-olefins C_6 – raw material for additives and oils. Russian Journal of Chemistry and Technology of Fuel and Oils, 3, 35–37, 2002.
8. Mekhtiyeva, S. T., Akhmedov, A. I., Mamedov, E. I. Phosphor and sulfur containing polymeric additives. Journal of Azerbaijan Oil Industry, № 2, 39–42, 2013.

CHAPTER 9

THERMOOXIDATIVE DEGRADATION OF THE LOW-DENSITY POLYETHYLENE IN THE PRESENCE OF FULLERENES C_{60}/C_{70}

ELDAR B. ZEYNALOV

Institute of Petrochemical Processes Named After ACAD. Y.G. Mamedaliyev, Azerbaijan National Academy of Sciences, Khojaly Aven 30, AZ1025 Baku, Azerbaijan, E-mail: zeynalov_2000@yahoo.com

CONTENTS

ABSTRACT

The chapter describes results of thermooxidative degradation of low-density polyethylene (LDPE) composites containing additives of fullerene C_{60}/C_{70}. It has been established that fullerenes C_{60}/C_{70} exhibit high stabilizing activity comparable with that of strong commercial basic antioxidants. Mechanism of the stabilizing action relates mostly to scavenging of alkyl macroradicals. Fullerenes C_{60}/C_{70} and fullerene soot are recommended for use in the polymer industry as active basic components of stabilizers package to block the processes of thermal and thermo-oxidative degradation of polyolefines.

9.1 INTRODUCTION

Intrinsic durability of most polymers especially of polyolefines to thermooxidative ageing is extremely low, and therefore usage of stabilizers package is necessary to provide the long-term retention and high performance of polymeric materials [1, 2]. Up to date the package containing basic (primary), secondary and light stabilizers is used to ensure the effective stabilization of polyolefines [2–4]. Scavengers of peroxy- and alkyl radicals such as sterically-hindered phenols, secondary aromatic amines, hydroxylamines, benzophuranones are commonly related to the basic stabilizers. They exhibit high stabilization effects and as of today are most important stabilizers for polyolefines.

Secondary stabilizers are compounds acting on mechanism of non-radical decomposition of hydroperoxides to afford molecular inert products. Phosphites, sulphides and tioesters and oths. are usually attributed to secondary stabilizers.

Besides these two main types of stabilizers sterically-hindered amine stabilizers on the basis of 2,2',6,6'-tetramethyl-piperidine has been recently offered as additives with dual (light and thermal) effects of stabilization.

Usually the stabilization occurrence is governed by two main factors – by chemical structure and behavior of a stabilizer in condensed substrate environment. Hence, an approach used in this work is preliminary assessment of efficiency and mechanism of stabilizer's action on model oxidative reaction followed by speeded-up tests of stabilizers directly in polymeric materials.

As it has been established in model reactions of cumene and styrene oxidation the fullerenes C_{60} and C_{70} demonstrate high antioxidative activity and they therefore can be ranked as promising class of polymer stabilizers acting as radical quenchers [5–7]. According to the mechanism, the fullerenes, along with quinones, nitroxyl radicals and condensed aromatics are related to the basic stabilizers [4, 8].

Determination of activity and mechanism of fullerenes antioxidant action in polymer materials may be also carried out by thermal analysis of the composites.

9.2 THEORETICAL PRESUMPTIONS

The preliminary theoretical analysis of electronic structure of fullerenes indicates their potential antioxidative capacity due to:

- high value of fullerenes electronic affinity 2.7–3.4 eV that significantly exceeds that (0.5eV) for alkenes [9];
- multiple additions of free radicals to fullerenes [10];
- higher mobility of the molecular clusters in condensed mediums (e.g., in polymers)

9.3 OBJECTS, PREPARATION PROCEDURES

Commercial low-density (0.9185 g/cm^3) polyethylene (LDPE) was employed in experiments. This polymer is large-tonnage polymer of a great practical importance. The problem of LDPE stabilization is interesting since the LDPE is partly crystal polyolefine usually characterized by different solubility of oxygen and antioxidant distribution in matrixes.

Fullerene C_{60}/C_{70} (85/15 %) were provided by Xzillion GmbH & Co. KG (recently renamed to Proteome Sciences R&D GmbH & Co. KG).

The LDPE and fullerene C_{60}/C_{70} compounding was made on laboratory rollers at temperatures150–160°C for 5–7 minutes.

The thermic analysis of composites was conducted in a dynamic mode on derivatograph Q-1500D of "Paulik-Paulik-Erdei" system, amount of a sample = 100mg, rate of heating was 5° C/min., the reference sample was α-Al_2O_3.

9.4 RESULTS OF THERMAL ANALYSIS OF THE LDPE COMPOSITES

According to the results of differential thermal analysis (DTA), the substantial changes upon introducing fullerenes in LDPE are just observed in the plot of high temperatures. Endo-effects of melting of the LDPE crystalline phase observed at 378°K are practically not changed. This is evidence that fullerenes are preferably located in amorphous phase of the polyethylene. Little exothermic peak observed at 508K is seemingly related to recrystallization processes. The heat effects of exothermic thermo-oxidative processes at 600K and higher are notably sensitive to the fullerene presence. In the presence of C_{60}/C_{70} the wide and intensive peak of the polyethylene decomposition is transformed into narrow exothermic peak having square, which by an order less than that for initial neat polyethylene samples.

These results obtained are verified by thermogravimetric analysis data. The weight loss degree at different temperatures as well as half-decay period $\tau/2$ of the composites are accumulated in Tables 9.1 and 9.2:

The data given in Tables 9.1 and 9.2 show the influence of additives on thermo-stability of LDPE in the air atmosphere. It should be noted that there was observed an excellent compatibility of additives with LDPE polymer matrix.

The LDPE resistance to oxidation is low and the polymer decomposition starting point is near 200°C. It indicates that the polymer has labile bonds and domains that lead to the substantial weight loss of the samples even at moderate temperatures.

Anyway, the stabilizing effect of fullerene C_{60}/C_{70} is evidently equal to that of strong commercial antioxidants. Thus, introduction of fullerene C_{60}/C_{70} to the polymer matrix significantly increases the thermostability of LDPE.

It is assumed that fullerenes prevent the oxidation chain processes just at the initial stages according to the reactions:

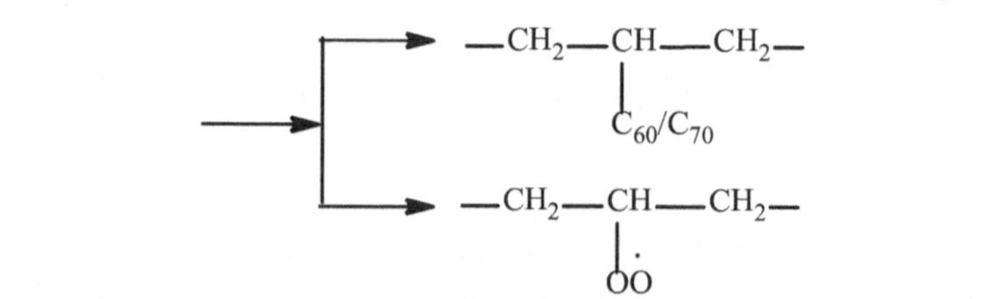

TABLE 9.1 Thermal Analysis of LDPE Composites, Comparative Efficiency of Commercial Antioxidants and Fullerene C_{60}/C_{70}

№	Polymer composite LDPE	Tнр°C	T_{10}°C	T_{20}°C	T_{50}°C	Half-decay period τ/2 of the composites
0	Virgin sample	225	330	370	380	68
1	+0.5 wt.% Phenantrene 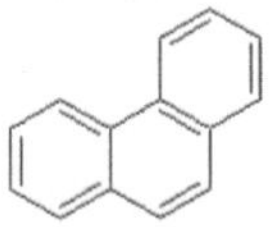	235	325	375	420	72
2	+0.5 wt.% Irganox 1010 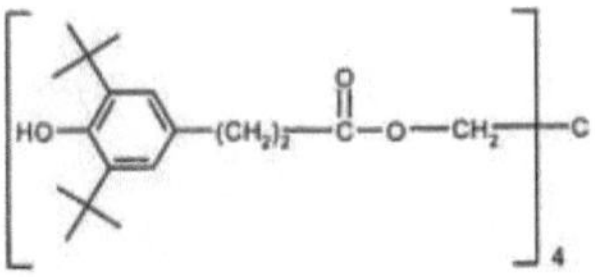	240	350	380	410	76
3	+0.5 wt.% Agerite White 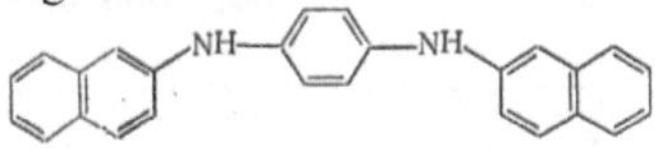	260	375	390	430	76
4	+0.5 wt.% Fullerene C_{60}/C_{70} 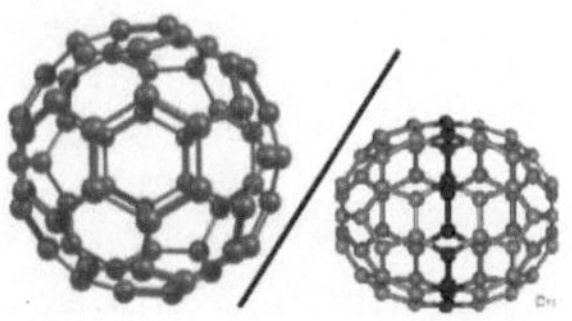	260	330	370	380	77
5	+0.25 wt.% Fullerene C_{60}/C_{70} + +0.25 wt.% Agerite White	295	370	395	455	80.9

and thereby they preserve polymer oxidative disruption. Total LDPE stabilization can be realized in the simultaneous presence of peroxy

(AH) – and alkyl radical acceptors. In this case the reaction scheme may be presented as:

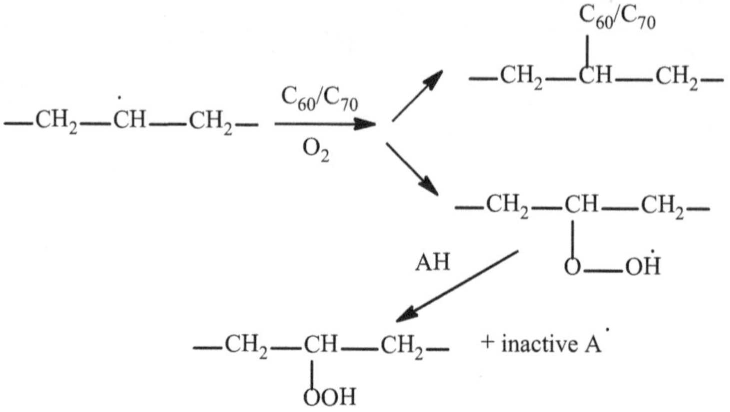

TABLE 9.2 Heat Resistance of LDPE Composites

№	LDPE composites	Weight loss at %:								
	LDPE	220	225	250	275	300	325	350	375	400
0	Without additives	0	0	2.8	3.8	4.5	7.0	14	22	57
1	+0.5 wt.% Phenantrene 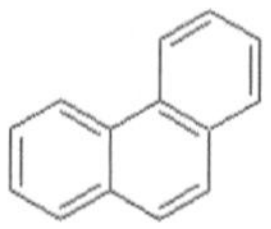	0	0	2	4	6	10	12	20	38
2	+0.5 wt.% Irganox 1010 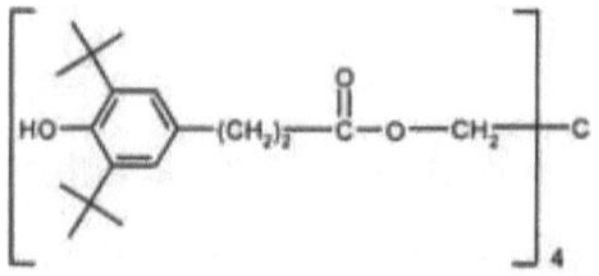	0	0	0	1	1.5	3	6	10	37

TABLE 9.2 Continued

№	LDPE composites	Weight loss at %:								
	LDPE	**220**	**225**	**250**	**275**	**300**	**325**	**350**	**375**	**400**
3	+0.5 wt.% Agerite White 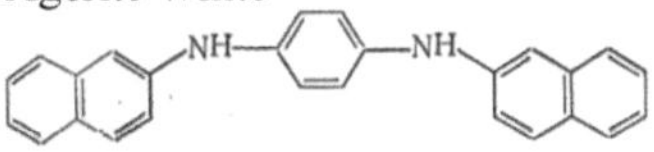 	0	0	1	1.5	3.5	6	10	15	42
4	+0.5 wt.% Fullerene C_{60}/C_{70} 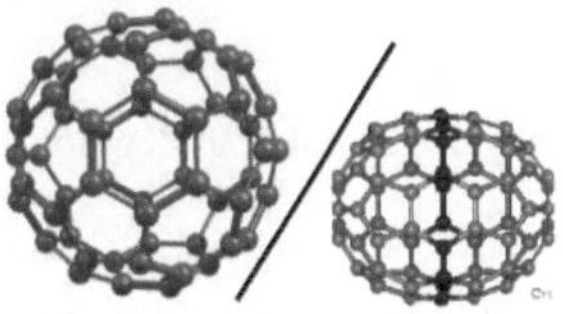	0	0	0	2	3	6	15	35	53
5	+0.25 wt.% Fullerene C_{60}/C_{70} + 0.25 wt.% Agerite White	0	0	0	0	1.0	1.2	4	12	48

9.5 CONCLUSIONS AND OUTLOOK

1. Fullerene C_{60}/C_{70} has been considered as active component of stabilizers package for low-density polyethylene (LDPE) composites
2. Fullerenes C_{60}/C_{70} exhibit high stabilizing activity purely comparable with activity of such strong commercial basic antioxidants as Irganox 1010 or Agerite White.
3. Mechanism of stabilizing action of fullerenes has been furnished. Fullerene introduced in a polymer matrix as physical additive is chemically bonded with macromolecule, keeping the remaining spin and forming stable radical capable to multiple breakage of oxidation chains.
4. Fullerenes C_{60}/C_{70} are recommended for polymer industry as active basic components of stabilizers package to block the processes of thermal and thermo-oxidative degradation of polyolefines.

KEYWORDS

- **fullerene additives**
- **low-density polyethylene**
- **scavenger of alkyl macroradicals**
- **thermal analysis**
- **thermooxidative degradation**

REFERENCES

1. Emanuel, N. M., Buchachenko, A. L. Chemical physics of polymer degradation and stabilization. Utrecht: VNU Science Press, 1987, 337.
2. Zweifel, H. Stabilization of polymeric materials. Berlin: Springer-Verlag, 1998, 219.
3. Allen, N. S. Degradation and stabilization of polyolefins. London: Elsevier Applied Sciences, 1983, 247p.
4. Zeynalov, E. B. Antioxidants for polymeric materials. Baku: Elm, 2009, 342.
5. Zeynalov, E. B., Koβmehl, G. Fullerene C_{60} as antioxidant for polymers. Polymer Degradation and Stability, 71(2), 197–202, 2001.
6. Zeynalov E. B., Allen, N. S., Salmanova, N. I. Radical scavenging efficiency of different fullerenes C60 – C70 and fullerene soot. Polymer Degradation and Stability, 94(8), 1183 – 1189, 2009.
7. Zeynalov E. B., Magerramova, M. Ya., Ishenko, N. Ya. Fullerenes C_{60}/C_{70} and C_{70} as antioxidants for polystyrene. Iranian Polymer Journal 13(2), 143–148, 2004.
8. Troitskii, B. B., Troitskii, L. S., Dmitriev, A. A., Yakhnov, A. S. Inhibition of thermo-oxidative degradation of polysterene by C_{60}. European Polym. Journ. 36(5), 1073–1084, 2000.
9. Cioslowski, J., In: Electronic structure calculations on fullerenes and their derivatives. New York: Oxford University Press, 1995, Chaps. 3–5.
10. Sokolov, V. I., Stankevich, I. V. The fullerenes – new allotropic forms of carbon: molecular and electronic structure, and chemical properties. Russ. Chem. Rev. 62 (5), 419–435, 1993.

CHAPTER 10

MUTUAL ACTIVATION AND HIGH SELECTIVITY OF POLYMERIC STRUCTURES IN INTERGEL SYSTEMS

TALKYBEK JUMADILOV, SALTANAT KALDAYEVA, RUSLAN KONDAUROV, BAKHYTZHAN ERZHAN, and BARNAGUL ERZHET

JSC "Institute of Chemical Sciences Named After A.B. Bekturov," 050010, Republic of Kazakhstan, Almaty, Sh. Valikhanov st. 106, Kazakhstan, E-mail: jumadilov@mail.ru

CONTENTS

ABSTRACT

The remote interaction between polyacrylic acid hydrogel (gPAA) and poly-2-methyl-5-vinylpyridine hydrogel (gP2M5VP) depending on their molar ratio in the aqueous medium is studied. It is established that at ratios of hydrogels (gPAA: gP2M5VP) 5:1 and 1:5 intergel system in aqueous medium has maximum specific conductivity. With increase of one hydrogel content in solution, there is an increase in hydrogels swelling proportionally to concentration of second component, which indicates to their mutual activation. For poly-2-methyl-5-vinylpyridine hydrogel ratio area at which the cationic hydrogel have low coefficients of swelling due to formation of intermolecular bonds between inter-node links of polibasis is detected.

Obtained results are explained as a result of remote interaction with the result that both hydrogels have high charge density formed without counterions along inter-node chains.

10.1 INTRODUCTION

Earlier in the systematic study of the interaction of hydrogels with different compounds it was found that in remote interactions of two hydrogels of different nature there is a significant change in their properties [1–4]. Subsequently, the growth of the sorption capacity of intergel system under certain molar ratios was shown for some salts [4]. It was assumed that the high sorption ability of intergel system at their remote interaction occurred due to their mutual activation ofhydrogels, in result uncompensated charges of inter-node links of individual hydrogels appear. However, there was no experimental confirmation of this hypothesis. In this regard, the goal of this work is to study the properties of intergel system solutions consisting of rare cross-linked polyacrylic acid and poly-2-methyl-5-vinylpyridine.

10.2 EXPERIMENTAL PART

10.2.1 MATERIALS

For measurement of electroconductivity conductometer "MARK 603" (Russia) was used, pH of solutions was measured on pH meter "Seven Easy" (METTLER TOLEDO, China). Swelling coefficient K_s was defined by weighting of hydrogel-swollen samples on electronic scales "SHIMADZU AY220" (Japan).

10.2.2 TECHNICS

Researches were carried out in distilled water medium. Hydrogels of polyacrylic acid were synthesised in the presence of crosslinking agent N, N-metilen-bis-acrylamid and redox system $K_2S_2O_8$, $Na_2S_2O_3$. The hydrogel of poly-2-methyl-5-vinylpyridine (gP2M5VP) was synthesized from a linear polymer in dimethylformamide in presence of epichlorohydrin at 60C.

Hydrogels synthesized in an aqueous medium is a intergel pair polyacrylic acid gel – poly-2-methyl-5-vinylpyridine gel (gPAA-gP2M5VP). Swelling coefficients of hydrogels are $K_{sw(gPAA)} = 10,1$ g/g, $K_{sw(gP2M5VP)} = 0.46$ g/g. Experiments were occurred at room temperature. Study of intergel systems was made by this way: each hydrogel was located in separated glass weighing filter pores of which are permeable for low-molecular ions and molecules, but it is not permeable for a dispersion hydrogels.

Then weighing filter with hydrogels were located in glasses with distilled water. Electroconductivity and pH of overgel liquid was measured by taking out of weighing filters with hydrogels from the glass. Swelling coefficient was calculated as the difference of weights of weighing bottle with hydrogel and empty weighing filter according to this equation:

$$K_{sw} = \frac{m_2 - m_1}{m_1}$$

where m_1 – weight of dry hydrogel, m_2 – weight of swollen hydrogel.

10.3 RESULTS AND DISCUSSIONS

Experimental data on electroconductivity of intergel systems' solutions depending on the starting time and molar ratios is shown in Figure 10.1. Figure 10.1 reflects the change in electrochemical properties of hydrogels constituting the intergel system.

High conductivity values at the maximum point point to the high concentration of charge carriers. In this case it may be H^+ ions in an aqueous medium, the concentration of which is determined by the degree of dissociation of the polyacrylic acid carboxyl groups and increases during swelling. However, there is presence of polybasic – poly-2-methyl-5- vinylpyridine

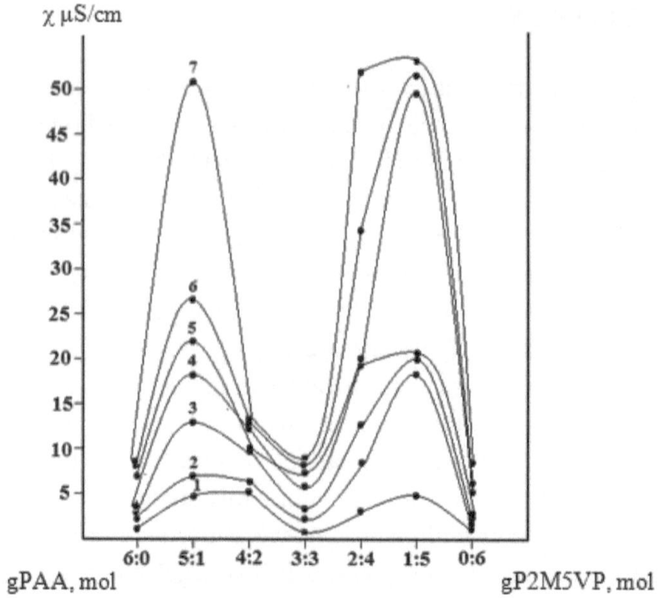

FIGURE 10.1 Curves of specific conductivity changing of polyacrylic acid hydrogel – poly-2-methyl-5-pyridine hydrogel system in dependence of time Curves' description: 1 – 0 min; 2 – 20 min; 3 – 40 min; 4 – 60 min; 5 – 120 min; 6 – 240 min; 7 – 360 min.

hydrogel, which can easily attach the H^+ ions and it moves into the charged state. The process should reduce the concentration of ionized particles in the solution.

Appearance of the electroconductivity maximum at a hydrogels ratio 5:1 can be explained by the low concentration of nitrogen atoms in the solution, since the concentration of vinylpyridine 5 times less than the concentration of carboxyl groups. Electroconductivity reducing with increasing content of vinylpyridine groups in solution can be explained by the increasing of the nitrogen atoms associated with a proton.

At equimolar ratio gPAA: gP2M5VP electroconductivity reaches minimum, due to the maximum protonization of vinylpyridine's links.

In the right part of the curve χ – gPAA: gP2M5VP electroconductivity reduction due to the binding of H^+ $C \geq NN$ groups could be expected. However, there is an increase in electroconductivity in the first part of the curve. On Figure 10.1 high electroconductivity values, which are obtained in maximum point at a ratio 1:5, point to the greatest number of ionized particles in the solution.

The interaction between the hydrogels in intergel system was studied by pH-metry. Figure 10.2 describes the change of H^+ and OH^- ratios in various ratios of hydrogels depending on the hydrogels remote interaction time. As it can be seen from the figure at the ratios gPAA – gP2M5VP = 4:2 and 2:4, there is a minimum concentration of H^+ ions and maximum concentration of OH^- ions in solution. When the composition is 3:3 and 1:5 of H^+ ion content is maximized.

To describe the obtained data it is necessary to analyze equilibrium in solution. In an aqueous medium the following processes take place:

1. –COOH groups of inter-node links dissociate by the scheme:

$$-COOH = COO^- + H^+$$

2. A nitrogen atom in the pyridine ring of ionizes:

$$\geq N + H_2O = \geq NH^+ + OH^-$$

3. Next, the nitrogen atom also interacts with a proton, which is split off from the carboxyl group:

$$\geq N + H^+ = \geq NH^+$$

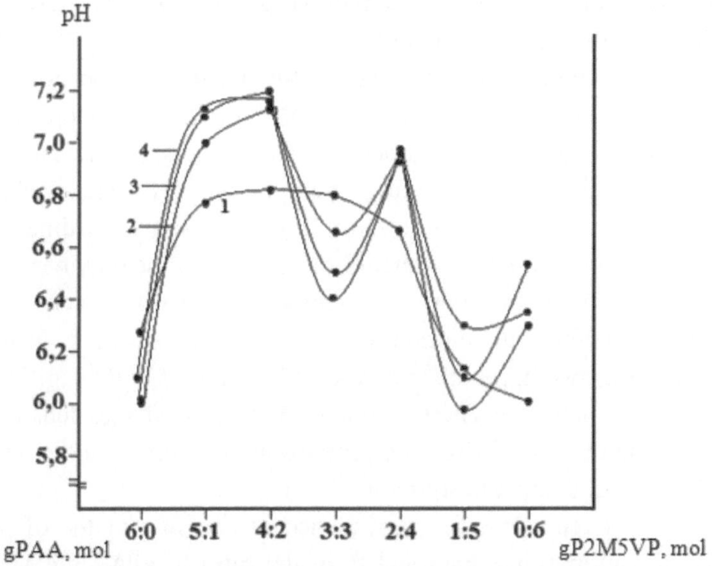

FIGURE 10.2 Curves of pH changing of polyacrylic acid hydrogel – poly-2-methyl-5-pyridine hydrogel system in dependence of time. Curves' description: 1 – 0 min; 2 – 20 min; 3 – 40 min; 4 – 60 min.

4. H^+ и OH^- groups formed by the interaction of the functional groups attach forming a water molecule:

$$H^+ + OH^- = H_2O$$

The result of these interactions is the lack of counter ions of charged functional groups of the hydrogels.

High pH values when the composition gPAA-gP2M5VP = 4:2 and 2:4 point to predomination of OH^- ions in the solution. This is possible if the occurrence of the second reaction in which hydroxyl anions are allocated in a solution, the third reaction also proceeds in parallel whereby the free proton bound to the pyridine ring and the concentration of positively charged ions in the solution is significantly reduced.

When the composition is 3:3 and 1:5, the high concentration of H^+ ions is observed. This is probably proceeds due to the fact that the hydroxyl ions and protons are neutralized by protons and the high concentration of positive ions in the solution remains due to a moderate degree of carboxyl groups' dissociation.

It is known that polyacrylic acid is a weak acid.

Figure 10.3 shows how the polyacryllic acid swelling coefficient depends from time in intergel system gPAK gP2M5VP – at different ratios of hydrogels gPAK:gP2M5VP mol:mol.

It is known that the swelling degree of hydrogel is proportional to the charge density along inter-node chains. Remote interaction should be shown in the volume – gravimetric properties of hydrogels. Figure 10.3 shows how swelling coefficient (K_{sw}) of polyacrylic acid hydrogel is changing in the presence of the second gel-poly-2-methyl-5-vinylpyridine. As can be seen from the figure, with increase of the time of the polymer network is in the solution the swelling coefficient also increases. Increase content of the main hydrogel in solution also increases the swelling. Moreover, with increasing swelling time at a ratio of 2:4 and 1:5 (gPAK:gP2M5VP) it increases dramatically. These results show that the charge density along the inter-node chains is increasing proportionally with the increase of the second hydrogel content in solution.

Figure 10.4 represents the dependence of swelling of the of poly-2-methyl -5- vinylpyridine hydrogel from the ratio of gPAA:gP2M5VP at different time. When swelling occurs for 20 minutes gP2M5VP swelling coefficient increases proportionally to the content of polyacrylic acid in

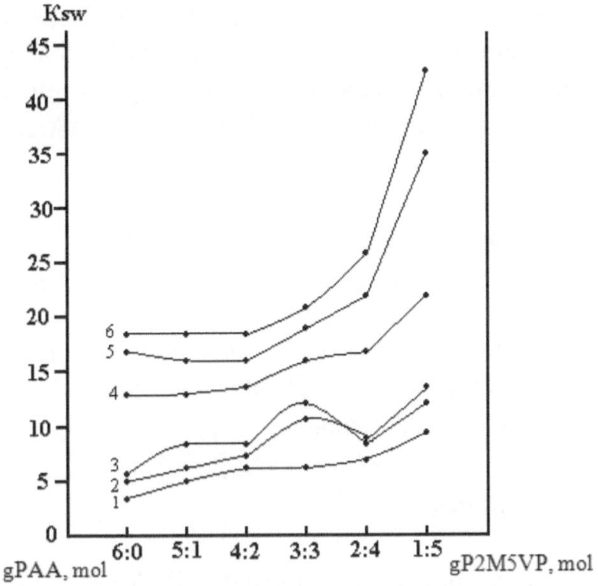

FIGURE 10.3 Polyacrylic acid hydrogel swelling coefficient dependence from polyacrylic acid hydrogel – poly-2-methyl-5-pyridine hydrogel system ratio in dependence of time (Curves' description: 1 – 20 min; 2 – 40 min; 3 – 60 min; 4 – 120 min; 5 – 240 min; 6 – 360 min).

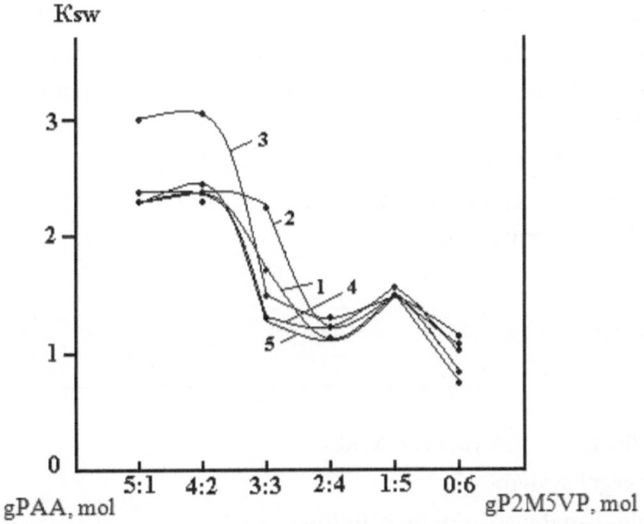

FIGURE 10.4 Poly-2-methyl-5-pyridine hydrogel swelling coefficient dependence from polyacrylic acid hydrogel – poly-2-methyl-5-pyridine hydrogel system ratio in dependence of time (Curves' description: 1 – 20 min; 2 – 60 min; 3 – 120 min; 4 – 240 min; 5 – 360 min).

an aqueous medium. Swelling increase of the main hydrogel in this case is caused by the formation of≥NH^+ groups by addition of H^+ ions by inter-node chains containing nitrogen atoms. A further increase of swelling time leads to the appearance of a minimum in the range of ratios gPAA:gP2M5VP and reaches maximum at a ratio of gPAA:gP2M5VP = 4:2.

Reduction of swelling in this area is probably due to the appearance of the intramolecular bonds of $\geq$ NH + … N $\leq$ between inter-node links.

With further increase of the gPAA concentration in the aqueous medium the swelling increases and K_{sw} increases significantly.

Unfolding of inter-node links and change of the local conformation is due to an increase of the charge density along the chain, which leads to the destruction of intramolecular bonds.

10.4 CONCLUSIONS

1. There is a remote interaction between the gPAA and gP2M5VP hydrogels in an aqueous medium, what was pointed by physicochemical methods during study of the intergel system.
2. As a result of the remote interaction there is additional activation of hydrogels, consisting in that the inter-node chains acquires additional charge without counterions.
3. Remote interaction provides conformational change in the inter-node chains of both hydrogels, which causes their considerable additional swelling.
4. There is a region of intramolecular bonds formation of gP2M5VP hydrogel at certain ratios of gPAA:gP2M5VP, which provides to folding of polymer networks.

KEYWORDS

- **hydrogels of polyacrylic acid**
- **intergel systems**
- **poly-2-methyl-5-vinylpyridine**
- **remote interaction**

REFERENCES

1. Bekturov, E. A., Jumadilov, T. K. New approaches to study of remote influence effect of hydrogels. News of national academy of sciences of Republic of Kazakhstan. 2009. №1. p. 86–87.
2. Bekturov, E. A., Jumadilov, T. K., Korganbayeva, Zh. K. Remote effect in polymeric systems. Herald of Al-Farabi Kazakh National University, chemical series. 2010. №3 59. p. 108–110.
3. Jumadilov, T. K. Effect of remote interraction of polymeric hydrogels in innovative technology. Industry of Kazakhstan. 2011. № 2. p. 70–72.
4. Jumadilov, D. Shaltykova, I. Suleimenov. Anomalous ion exchange phenomenon. Austrian-slovenian polymer meeting. ASPM 2013. Slovenia, 2013. Abstr. S5. p. 51.

CHAPTER 11

EMISSION OF MULTI CHARGED IONS

GEORGE MESKHI

Samtskhe-Javakheti State University, Faculty of Engineering, Agrarian and Natural Sciences, Akhaltsikhe, 0800 Rustaveli, 106, Georgia, Email: george.meskhi@yahoo.com

CONTENTS

ABSTRACT

The interaction of the energetic ions with the surfaces plays an important role in the field of the surface science. Relevant topics are the ionization of atomic particles at surfaces, the modification of surfaces and surface analysis. In this work the production of multi charged escaping ions at ion-surface interactions are experimentally investigated. During the ion-surface bombardment some different manners of multi charged ions formation and neutralization at emission are also discussed. Quasimolecular approach of atomic particles interaction is used. Numerous experimental results depending on the various factors on formation and emission of the multi charged ions are presented. The emission of multi charged ions at bombarding of the NaCl, Mg, Al, Si, S surfaces by N^+, O^+, N_2^+, O_2^+, Ar^+ primary ions with energies E_0 = 0.5–50keV at different experimental conditions are investigated by using the SIMS method. At ion – surface interaction the inner shell vacancies can be formed and multi charged ions can be produced as a result of Auger-transitions either during Auger decay in solids or after ion emission in vacuum. The neutralization processes of the double and triple charged ions emitted from Al and Si surfaces were studied. The processes of formation and emission of the multi charged ions are rather sensitive to atomic and electronic structure of surfaces.

11.1 INTRODUCTION

During ion sputtering of solid surfaces the mass spectra of the emitted secondary ions contains the multi charged ions as well, its research is obviously important for determining the mechanism of their formation. So-called "kinetical" mechanism allows basic determination of the production of multi charged ions at ion-surface interaction. It is necessary to specify the occurrence of the multi charged ions, in order to get some information about the influence of the various factors on the formation and emission of the multi charged ions. It is known that [1] during atomic binary collisions in keV energy range inner shell vacancy can be produced and after that Auger and radiation transitions can occur. The Auger transitions lead to producing multi charged ions. Present experimental investigations of double and triple charged ion yields shows that Auger processes can take place either in solids or after emission in vacuum.

In the present work the multi charged ions (Mg^{n+}, Al^{n+}, Si^{n+}, S^{n+}) yields Y^{n+} (n=2,3) are measured during N^+, O^+, N_2^+, O_2^+, Ar^+ ions bombardment of NaCl, Mg, Al, Si, S surfaces in energy range Eo = 0.5÷50 keV under UHV (~10^{-10} torr) conditions. The yields of Al^{n+} are measured at various oxygen partial pressures in the interaction chamber. The yields Y^{n+} are measured also for Al and Si in dependence on angle of bombardment by Ar^+ ions, counted from normal to surface of sample. With the purpose of comparison of results the similar data of inner shell vacancy production at binary ion-atom collisions in gas, such as Ar^+ – Ar [1] are also discussed.

11.2 METHODS OF RESEARCH

Using the method of mass-spectrometry of secondary positive and negative ions (SIMS) multi charged ions yields are measured during primary ion sputtering of surfaces. The measurements of ions yields Y^{n+} at bombardment by N^+, O^+, N_2^+, O_2^+, Ar^+ ions of Al surfaces in energy range of the primary ions E_o = 5÷50 keV, and also research behavior of ions yields depending on pressure of oxygen in the chamber of interaction, are executed on experimental arrangement at the University of Salford (England) consisting of secondary ions mass-spectrometry method adding with energy analyzer [2]. Yields Y^{n+} at bombardment of NaCl, Mg, Al, Si, S surfaces by Ar^+ ions, in an energy range E_o = 0.5÷10 keV and depending on an angle of bombardment – α of primary ions Ar^+, counted from normal to a researched surface are measured on installation developed at Tbilisi State University for research of solid surfaces by a method SIMS on base to the UHV arrangement [3].

The N^+, O^+, N_2^+, O_2^+, Ar^+ ions are formed in a plasma type ions source with the high-frequency discharge, in energy range E_o = 5÷50 keV are accelerated, allocated on mass by mass-monochromator, collimated by slits and bombarded along normal researched surfaces. Emitting thus the secondary positive and negative ions get in the hemispherical energy-analyzer with radius 24 cm and further act in quadruple mass-spectrometer such as Finnigan -750 for the analysis of mass in a range 1–750 a.m.u. Using monochromator in visual spectra light emission is investigated during bombardment as well. Typical intensity of a primary ions beam is changing from some tens μA/cm² to a few mA/cm². For investigation of multi charged ions yields depending on a condition of researched surfaces in the interaction chamber the oxygen up to pressure $PO_2 = 10^{-6}$ torr was filled. The time of stabilization of pressure O_2 and time of establishment a condition of balance of multi charged ions yield at letting-to-oxygen gas in the interaction chamber were experimentally determined [2].

In the work [3], the primary ions beam of Ar^+ with energy E_o =0.5÷10 keV and density of a current a few mA/cm² collides with the researched sample surface along with the normal of the surface. The minimal diameter of etch pit by ion beam makes ~ 0.5 mm. Emitted during bombardment the secondary positive ions are analyzed by monopole mass-spectrometer directed under 45° angle to the surfaces of the sample. As a source of the primary ions duoplasmatron type ion source is applied, which represents a discharge plasma source with the cold cathode and double contraction of plasma. The source of ions is equipped with electronic optics, which provide focusing and scanning ion beam on the surface on the area 5×5 mm^2 for providing the balancing formation of the edges of a crater at the depth profile analysis. The secondary ions are analyzed by monopole mass-spectrometer MX7304A, which is complemented with electronic optics for the stretch of the secondary ions and is altered for a method of the SIMS. The range of registered mass is 1÷380 a.m.u. The technique also allows to investigate yields of secondary ions depending on an angle of incident of primary ions in range α = 0÷45°. The errors of relative measurements of multi charged ions yields in both methods are made ~ (10÷20)% and at low intensities of multi charged ions yields are determined within the factor 2.

11.3 RESULTS AND DISCUSSION

Results of measurements of multi-charged ions yield Y^{n+} (n = 2,3) during Ar^+-Mg, Al, Si, S interactions are given in a Figures 11.1–11.4 and in Table 11.1. The results from work [4] for Ar^+-Al, Si – are also specified, which as it is visible from figures, will be in line with our data satisfactorily. In a Figures 11.1–11.3 calculated value coefficients of sputtering S° for neutral atoms of Mg, Al, Si in a case Ar^+-Mg, Al, Si from [5] are also given.

11.3.1 MECHANISM OF MULTI-CHARGED IONS PRODUCTION

The investigations of Y^{n+} (n=2, 3) have shown their dependence on the initial energy of bombarding ions E_o, as against behavior of neutral, negative and positive ions, and the radiations, arising at decay of all observable exited states Al*, Al^{+*}, Al^{2+*} [11] from E_o have the pronounced, brightly expressed character. It is obvious, that the double and triple-charged secondary ions yield sharply increases with increase of E_o, that specifies

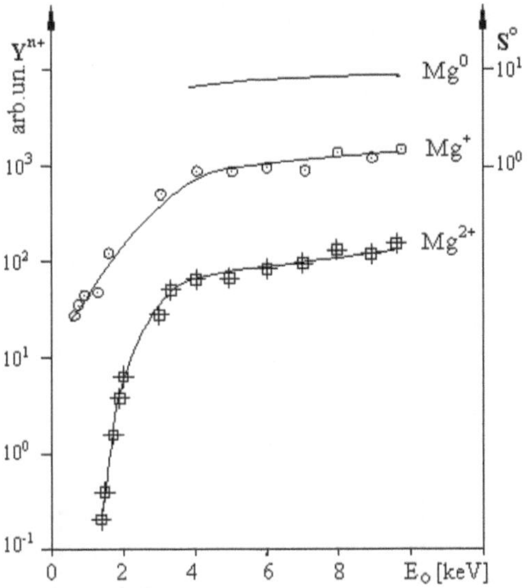

FIGURE 11.1 Secondary Mg^{+} and Mg^{2+} ions yield during $Ar^{+}-Mg$. Mg^{0} -coefficient of sputtering [5].

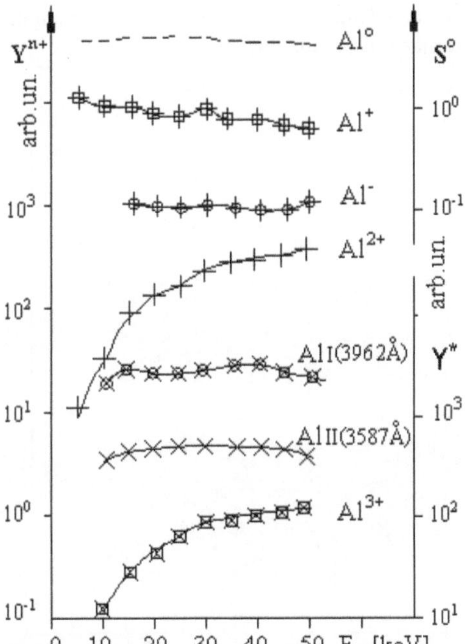

FIGURE 11.2A Secondary and ions emission in case. and – exited state for and *Al II* (3587 Å) – exited state for Al^{0} and Al^{+}.

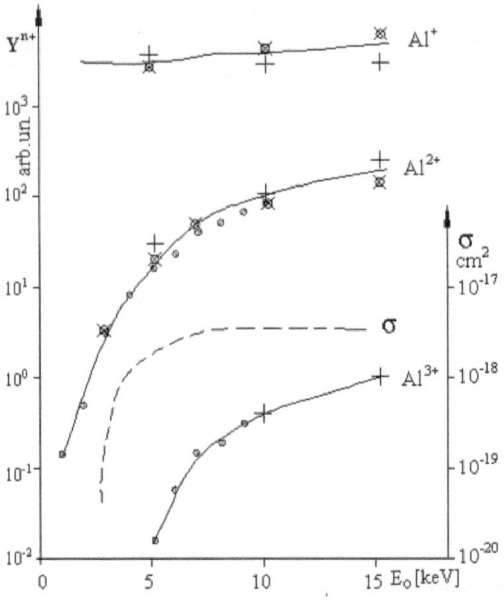

FIGURE 11.2B Secondary ions emission in $Ar^{+}-Al$ case. ⊗ – from [4]; +, ʘ – measurements in Salford and Tbilisi. Dashed lines – cross section of $2p(Al)$ production in $Ar^{+}-Al$, which was estimated using the formula (1).

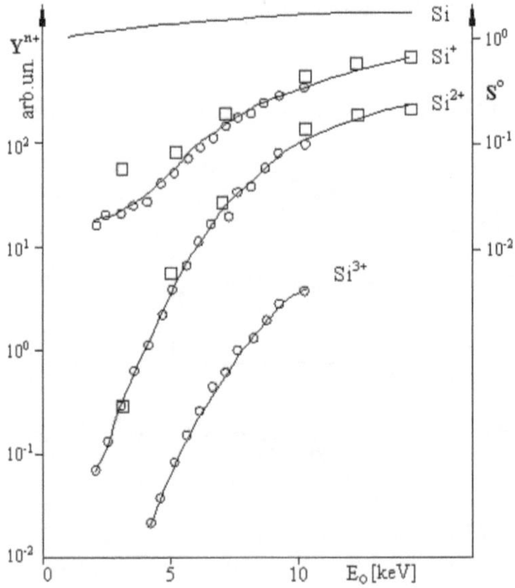

FIGURE 11.3 Secondary Si^{2+}, Si^{2+}, Si^{3+} ions emission in Ar^{+}-Si case. 4 form [4]. Si° – coefficient of spluttering [5].

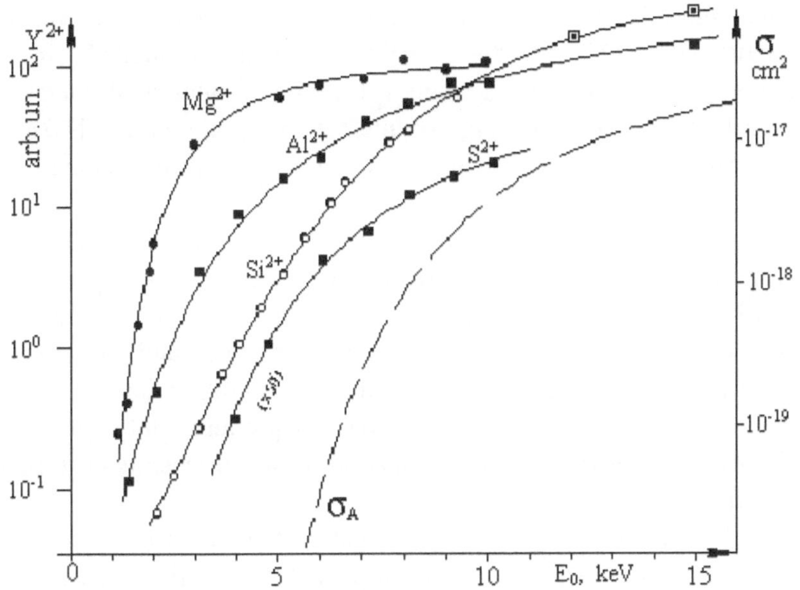

FIGURE 11.4 Double-charged ions Mg^{2+}, Al^{2+}, Si^{2+}, S^{2+} emission during Ar^{+}-Mg, Al, Si, S interaction. σ_A–cross section of 2p-Auger-electrons emission ar Ar^{+}-Ar collisions [1].

TABLE 11.1 Results of Measurements of Multi-Charged Ions Yield Y^{n+} (n = 2,3)

Collisions	Ar^+-Mg	Ar^+-Al	Ar^+-Si	Ar^+-S	Ar^+-Ar	N^+-Al	O^+-Al
Energy, E_0, keV	10	50	15	10	50	50	50
Y_{21}	0.1	3.3×10^{-2}	0.3	8×10^{-3}	5×10^{-2}	1.5×10^{-3}	2×10^{-3}
Y_{31}		3×10^{-3}	2×10^{-2}		1.8×10^{-4}		

difference between mechanisms of their formation. It is known [6], that at binary collisions of Ar^+ ions with atoms Mg, Al, Si, S in keV-energy range of collisions, with probability close to 1, at achievement critical inter nuclear distance r_K (appropriate overlapping of inner shells of interacting atomic particles), responding to drastic promotion 4fσ orbitals, one or two inner 2p-vacancies in the light partner of collision are formed, that is explained on the basis of "level shifts" model of molecular orbitals [6]. The basic channel of decay of 2p-vacancy in these cases is the Auger-transition, as a result after that the atom remains without two electrons.

In a Figure 11.2b the dashed line gives the cross section of 2p-vacancy formation in Al at binary Ar^+ – Al collisions, which is estimated using the formula $\sigma \sim \pi r_к^2(1-U/E_0)$ (1), where $r_к$ the sum of radiuses of interacting 2p-subshells, U-potential energy of interaction. From a Figure 11.2b follows, that the yield of multi charged Al^{2+} and Al^{3+} ions depending on E_0 have similarity σ, threshold character. In a Figure 11.4 the measured yield of double charged ions of the Mg^{2+}, Al^{2+}, Si^{2+}, and also cross section of Auger-electrons formation σ are given as a result of 2p-vacancy decay in Ar during Ar^+-Ar collisions from work [1]. It is visible, that in the investigated energy range E_0 of double charged ions yield and cross section of Auger-electrons σ behave in approximately similar manner. Different behavior of yields Al^{2+}, Al^{3+} and σ at the sharply threshold region of E_0 indicates a role of the collisional cascade in the surface, against binary interaction in gas.

Due to atomic collisional cascade interaction in surfaces, behavior of Y^{n+} (n=2,3) against E_0 is not as sharp as at binary Ar^+ – Ar collisions (Figures 11.1–11.4) and they indicate difference between binary and cascade interaction. Radiuses 2p-subshells of the investigated elements correspond as r(Mg)>r(Al)>r(Si)>r(S)>r(Ar).

Therefore, the threshold energy E_0, necessary for achievement the critical inter nuclear distance $r_к$ should obey to a correlation E(Ar)>E(S)>E(Si)>E(Al)>E(Mg). It is visible, that alleged appropriateness is observed in experiment (Figure 11.4). From a Figures 11.1–11.4 follows, that with increase of E_0 yields Y^{n+} (n=2,3) increase and approach to saturation at large E_0. The value of a ratio of yields double-charged ion to single-charged ion Y_{21} (Y_{21}=Y(A^{2+})/Y (A^+)) and triple-charged ion to single-charged ion Y_{31} (Y_{31}=Y (A^{3+})/Y(A^+)) is increased with the increase of E_0 and appears on saturation. The ratio of the yield of the triple-charged ions to double-charged ions Y_{32} (Y_{32}=Y(A^{3+})/Y(A^{2+})) at the saturation areas practically do not depend on E_0 which are given in the Table 11.1. At the thresholds region of E_0 triple-charged ions yields decrease more rapidly, then double-charged for Al and Si basically for Al, where neutralization processes for highly charged ions are more probable.

In our investigations [7, 10] in a case N^+, O^+ – Al 2p-vacancy in Al is formed as a result of promotion 3dσ – orbital, which is formed from 2p-level of Al and also will get mixed up at rather small inter nuclear distance r_{κ} with smaller number of free orbitals in comparison with 4fσ orbital, corresponding with formation of 2p-vacancy in the light collisional partner during interaction of two L-shell Ar^{+}- Mg, Al, Si, S, Ar. Consequently, thresholds energy of formation of 2p-vacancy in a case N^{+}, O^{+}- Al are at large E_0 and besides this case the cross section of formation of 2p-vacancy in Al is not big enough (Figure 11.5). Because of chemical activity of projectiles measurements of N^{+}, O^{+}- Al were carried out at small intensity of a primary beam of N^{+} and O^{+} (~ 50 nA/cm^2) also it is experimentally proved, that implanted small concentration of N and O in Al does not change yield of ions Al^{2+}.

Formation of inner shell vacancy is sometimes discussed in frame of the cascade of collision. For example, in the first act of impact N^{+}, O^{+}, Ar^{+} – Al the atom Al receives significant kinetic energy for formation further 2p-vacancies in a case Al -Al. Thus, the threshold energy of formation 2p(Al) in all cases should coincide, that is not proved to be true from our results (Figure 11.5).

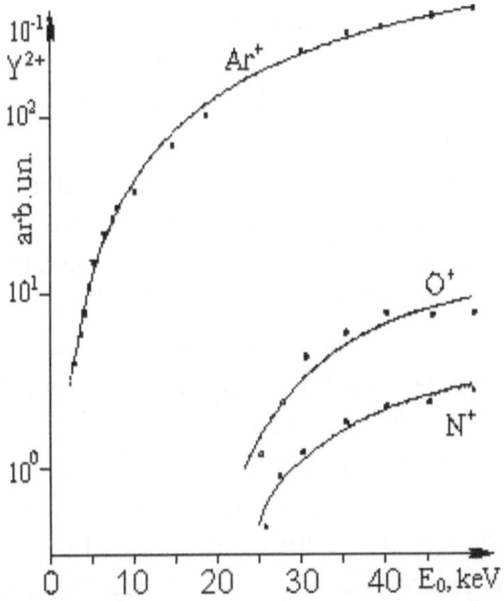

FIGURE 11.5 *Al^{2+} ions emission at N^{+}, O^{+}, Ar^{+}, –Al.*

Atom with a vacancy in an inner shell – high excited states for $t = 10^{-14}–10^{-15}$ sec is relaxed by emission of X-ray or Auger electron. It is known, that by the basic channel of decay of 2p-vacancy in the above-mentioned atoms is Auger-transitions, since fluorescence yield $\omega_x \sim 10^{-3}–10^{-4}$ [1] ($\omega_x = \sigma_x/\sigma_x + \sigma_A = \sigma_x/\sigma_{vac}$) where σ_x and σ_A are cross sections of X-ray and Auger-transitions, and – σ_{vac} cross section of inner shell vacancy production.

The presence of Auger-electrons in the secondary electrons energy spectra testifies the production of inner shell vacancies in atoms during bombardment by ions of surfaces in keV-energy range [8]. Observation on the corresponding X-ray photons is difficult because ω_x is small and also because of the inconvenience for the analysis of weak signals for such energy range of radiation (tens of eV).

In the work [3] during bombardment of surfaces Mg, Al, Si, S with Ar^+ ions yield of the multi charged ions is explained by formation of 2p-vacancy at binary collision in solids and subsequent Auger-transition, as a result of which the atom remains without two electrons.

Triple-charged ions can be produced: as a result of "shake" effect at Auger decay of inner vacancy [9]; as a result of simultaneous production of vacancies in outer and inner shells and subsequent Auger-transition; at simultaneous formation of two vacancies at once (molecular orbital contain two electrons with counter-oriented spins) in inner shell of atom and subsequent Auger-transitions. The most probable effect represents to "shake" (is especial for K-LL).

It is known, that the ions Al^{2+} can also be formed in result dissociation of a molecular ion $(AlO)^+$; at stripping of ion Al^+ in a case of interaction with adsorbed on the surface of aluminum oxygen and in case of tunneling electron from $(Al^+)^*$ in solid. However, probability of formation Al^{2+} in such cases will depend on a condition of aluminum surface, from presence impurity of atoms in participating in the process of production Al^{2+}.

In the present work the yield of ions Al^{n+} (n=1, 2, 3) and oxygen containing ions (such as $(AlO)^+$, $(Al_2O)^+$, $(AlO)^-$, $(AlO_2)^-$ etc.) are measured in dependence on pressure of oxygen in the chamber of interaction in a range $Po_2 = 5\cdot10^{-10} – 10^{-6}$ torr and at current intensity of ions $Ar^+ – I_o \sim \mu A/cm^2$ (Figure 11.6). Observable sharp increase of yield Al^+ and all oxygen containing ions with increase of Po_2 indicates adsorption process of oxygen

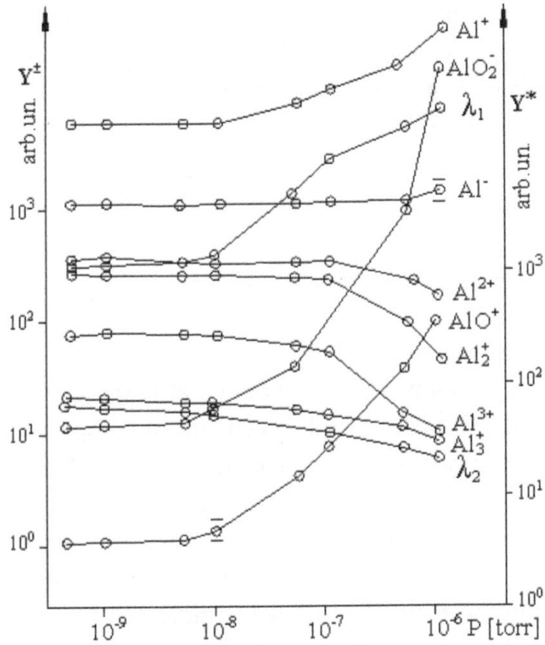

FIGURE 11.6 Secondary ions and photons emissions at various oxygen pressure (λ_1, λ_2 - excited states of $(Al^+)^*$).

on surface Al [10]. Considering that, yield of multi charged ions Al^{2+} and Al^{3+} is equally weak (about 2–3 times) and decreases with increase of Po_2, as a Al_2^+, Al_3^+ clusters. It appears in the research [11], that the yield of the radiation arising at decaying of exited states $(Al^+)^*$ also poorly decreases with increase of Po_2.Thus, contribution in formation of ions Al^{2+} in above specified processes is insignificant in comparison to the process of formation inner shell vacancy and subsequent Auger-transitions.

11.3.2 WHERE DO THE AUGER-TRANSITIONS OCCUR—IN SOLIDS OR AFTER EMISSION, IN VACUUM?

Assuming that emitted atomic particle Mg, Al, Si or S with 2p-vacancy has energy ~10 eV, and considering that time of life is defined as ~10^{-14} sec, parts of atoms with vacancy can pass < 1Å. In such case formed multi charged ions move in solid and before emission in vacuum tests a lot of interactions with atoms of solid, which leads to neutralization in process of capture by multi charged ion. At the close atomic collisions, when there is an ionization of inner electronic shells of atoms the partners of interaction receive significant kinetic energy. In the work [12] energy distribution of ions.

Mg^{n+}, Al^{n+}, Si^{n+} (n = 1, 2, 3), emitted at irradiation by Ar^{+} ions with energy E_0 =10 keV of Mg, Al, Si surfaces were investigated. The yield of Mg^{2+} ions have a single maximum at energy $E_i \sim 300$eV, and ions Si^{2+} – two maxima, first at energy ~8 eV, and second – at E_i ~500 eV with intensity of 60% from the first basic peak. The ions Al^{2+} have a maximum at E_i ~5 eV and insignificant intensity <10% a shoulder at large E_i. Thus, certain part of ions Mg^{2+} and Si^{2+} has the high speed and in comparison to Al^{2+} they, with inner 2p-vacancy "Live" up to exit in vacuum, where Auger-process takes place in isolated atom. The data given in the Table 11.1 also testifies it. Besides that, it is possible to explain suppression of a yield Al^{2+} to their high neutralization in aluminum (in comparison to Mg and Si).

Transitions in the isolated particles is indicated by narrow, atom-like line in spectra of Auger-electrons, which were measured in work [13] for a case Ar^{+} – Mg, Al, Si at E_0 = 9 keV, where the peaks in spectra of electrons are clearly visible, appropriate to Auger-transitions in atoms Mg, Al, Si. In spectra of secondary electrons their wide distribution, some "shoulder" at higher E_e, is visible, which was named by the authors [13] as "chemical shift."

Observed in [13], apart from atom-like lines, the wide distribution in spectra of the secondary electrons can be connected with so-called Auger-transitions in quasimolecule [14]. Because of smallness of lifetimes of inner vacancies at the binary collision of atomic particles, the Auger-transitions can occur "in quasimolecule" up to scatter of pushing atoms together – partners of collisions and the energy of Auger-transitions is determined at the inter nuclear distance where was an Auger-transition. In case of bombardment of solids the Auger-decay can occur in immediate proximity of partner of collision (length of free path of atom with vacancy <1Å), or under influence of electronic system of solid surface (electrons of surface can take part in Auger-process, similar to quasimolecule). In the latter case probability of Auger-transitions will be determined by electronic structure of solid surface and investigation of Auger-electrons spectra (analysis of the "shoulder") or multi charged ions will allow to receive information directly on the electronic structure of the surface.

Measured in Ref. [13] spectra of the secondary electrons in a case Ar^+ – Mg, Al, Si indicate about Auger-transitions in Mg, Al, Si as in vacuum – after emission, as well in solids – up to exit in vacuum.

Assume that the Auger-transitions occur in solids, the average depth of formation emitted multi charged ions will determine the probability of their neutralization during emission [15]. Depth of formation of multi charged ions is possible to be varied by the change of an angle of bombardment α.

The yield of ions Al^{n+} and Si^{n+} (n = 1, 2, 3) are measured at angles of bombardment $\alpha = 0°$ and $\alpha = 45°$ by ions Ar^+ in a energy range $E_0 = 5 - 9$ keV (Figure 11.7). It is visible, that the values of ratio Y_{21} and Y_{31} at the angles $\alpha = 45°$ are above than at $\alpha = 0°$. From the data on yields of various ions follow the weak dependence of yield on α in a range of angles (0–45°). Increase of yield of ions Si^{n+} (n = 1, 2, 3) at increase of an angle α can be connected to reduction of average depth of exit, and that with reduction of probability of their neutralization [7].

11.3.3 NEUTRALIZATION AT EMISSION

The additional ionization of outer shell in the emitting ion with the 2p-vacancy should occur as a result of the exchange interaction between the electronic structures the emitted ion + surface, that is not probable.

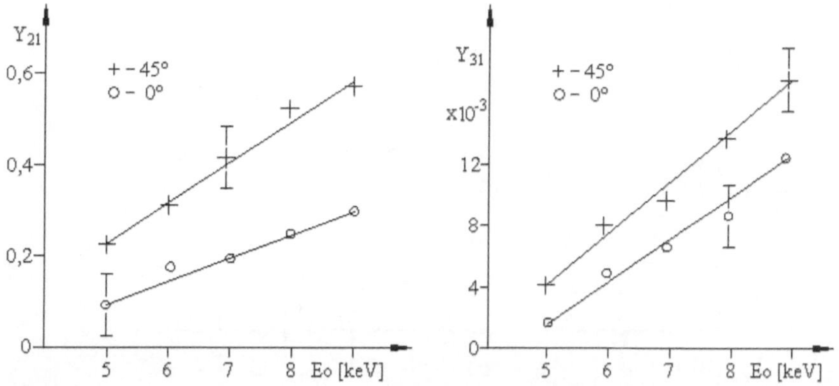

FIGURE 11.7 Ratios of secondary ions Y_{21} and Y_{31} at various angles of bombardment.

In this case the processes of Auger neutralization are more probable: with the participation of surface plasmons and the neutralization of ions as a result of resonance-tunnel capture of electrons from the semiconductor of metallic surface.

Numerous studies consecutively conform with the model of asymmetry, which considers the higher degree of the neutralization of ions.

Number of multiple charged ions show the difference between probability of neutralization ions at emission either during Auger decay in solids or after ion emission in vacuum [15, 16]. On the Figure 11.8 are presented ratio R of Al^{2+}/Al^{3+} and Si^{2+}/Si^{3+} against E_o. The yields of double and triple charged ions are linkages at the initial energy E_o=15 keV. At low initial energy of E_o partner of collision (with 2p-vacancy) can take corresponding low kinetic energy at recoil and therefore probability of Auger decay in solids increases. Therefore neutralization is more probable for A^{3+} than for A^{2+}. Influence ratio of R from E_o(from Figure 11.8) shows, that some part of Auger-transitions takes place in solids. Besides that, from Figure 11.2b and Figure 11.3 it follows, that at high E_o ratioY^{3+}/Y^{2+} for Ar^+ – Si is about one

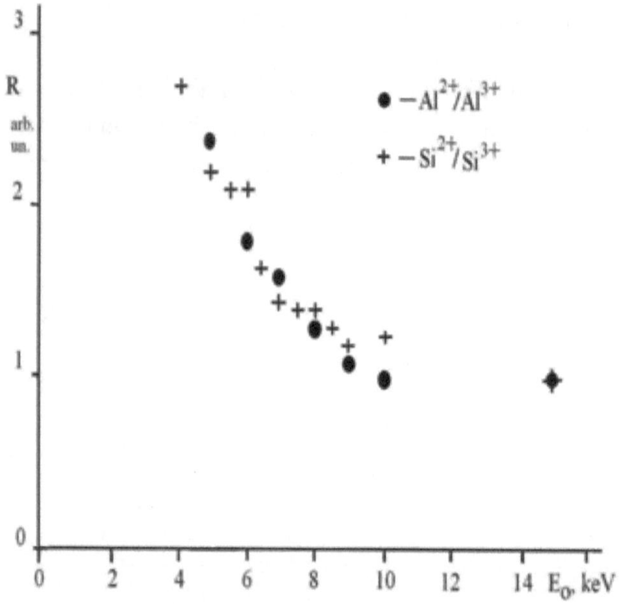

FIGURE 11.8 Ratio R of Al^{2+}/Al^{3+} and Si^{2+}/Si^{3+} against E_0.

order of magnitude more than Ar^+–Al case. Consequently, neutralization processes of multi charged ions at emission from solids are more probable in Al than in Si [16, 17].

11.4 CONCLUSIONS

The numerous experimental results confirm, that during ions bombardment of surfaces inner shell vacancies can be formed in atoms and as a result Auger-transitions emits Auger- electrons and multi charged ions. The experimental data shows, that the Auger-transitions take place both in solids, and in vacuum – after emission as well. The processes of formation and yield multi charged ions are rather sensitive to atomic and electronic structure of surfaces.

KEYWORDS

- **Auger-transitions**
- **ion emission**
- **ions yield**
- **multi charged ions**
- **quasimolecule**
- **sputtering**

REFERENCES

1. Afrosimov, V. V., Gordeev Yu. S., Zinoviev, A. N., Meskhi, G. G., Shergin, A. P. Abstracts of X International Conference on Phys. Electronic and Atomic Collisions, Paris, 202, 1977.
2. Armour, D. G., Jimenez-Rodriguez, J. J., Barber, C. H., Snowdon, K. S. Hedbavny, P. Vacuum 34, 217, 1984.
3. Armour, D. G., Kikiani, B. I., Meskhi, G. G., Chrelashvili, G. K. Mater. X Intern. Conf. on Ion Surface Interaction. Moscow, 1, 206, 1991. D. G. Armour, B. I. Kikiani, G. G. Meskhi, Bulletin of the Academy of Science of Georgia, 150, 3, 429–437, 1994.

4. Wittmack, K. Surface Sci. 53, 626, 1975.
5. R. Behrisch. "Sputtering by Particle Bombardment I", New York, 1981.
6. Garcia, J. D., Fortner, R. J., Kavanagh, T. M. Rev. Mod. Phys., 45,111, 1973.
7. Armour, D. G., Kikiani, B. I., Meskhi, G. G., Chrelashvili, G. K. Materials of XI International Conference on Ion Surface Interaction. Moscow, 2, 33, 1993.
8. Hennequin, J. F., Viaris de Lesegno, P. Surface Sci., 42, 50, 1974.
9. Parilis, E. S. et al. Pisma Zh. E. T. P., 2, 710, 1976.
10. Armour, D. G., Meskhi, G. G. Materials XII International Conference on Ion Surface Interaction. Moscow, 1, 298, 1995.
11. Armour, D. G., Meskhi, G. G. Abstracts of VI Conference Secondary Ion and Photon Emission, Kharkov (Ukraine), 98, 1991.
12. Hennequin, J. F., Inglebert, R. L., Viaris de Lesegno, P. V. International Conference on SIMS, Washington, 60, 1985.
13. Saiki, K., Rittaporn, I., Tanaka, S. Proceedings of 10th Symposium on ISIAT '86, Tokyo, 307, 1986.
14. Afrosimov, V. V., Meskhi, G. G., Tsarev, N. N., Shergin, A. P. Zh. E. T. P, 84, 2, 454–465, 1983.
15. Armour, D. G., Gorgiladze, B. G., Meskhi, G. G., Sichinava, A. V. Materials XVII Intern. Conf. on Ion Surface Interaction. Moscow, Russia, 2, 203, 2005.
16. Meskhi, G. G. Materials of XIX International Conference on "Ion Surface Interaction, ISI-2009." Moscow, Russia, 2, 127, 2009.
17. Meskhi, G. G. Materials of 3rd International Caucasian Symposium on Polymers and Advanced Materials, Tbilisi, Georgia, p. 54, 2013.

CHAPTER 12

GRADUALLY ORIENTED STATE OF THE LINEAR POLYMERS

L. NADAREISHVILI, R. G. BAKURADZE, N. S. TOPURIDZE, L. K. SHARASHIDZE, and I. S. PAVLENISHVILI

Georgian Technical University, Department of Cybernetics, 5 S. Euli St. Tbilisi, 0186, Georgia, E-mail: levannadar@yahoo.com

CONTENTS

ABSTRACT

The mathematical model of formation of the new structural state of the linear polymers – the gradually oriented/stretched state (GOS) – is discussed. The model allows to regulate the quantitative parameters of the gradually oriented/stretched polymers (GOPs). On the possibility of functionally graded materials' (FGMs) creation by graded orientation/stretching method is indicated.

12.1 INTRODUCTION

The simplest and most common kind of orientation of polymers is the uniaxial oriented stretching action of a uniform mechanical field on the polymer sample. Throughout the volume of the specific oriented polymer sample relative elongation $\Delta l/l$ (Δl is the real elongation and l is the initial length of the sample) practically is the same.

Uniaxial oriented stretching can be carried out in a different mode. Previously we have developed the landmark decision of uniaxial oriented stretching – graded oriented stretching and introduced the notion – gradient of relative elongation/orientation degree and established the conception about new structural state of the thermoplastic polymers – about GOS.

Quantitative parameters of GOPs are: range of change of relative elongation/orientation degree, its length and profile (linear, hyperbolic, parabolic, logarithmic, etc.) [1–17].

GOS points to untapped resources existing in the nature of the polymeric substance. According to this conception as a result of transformation of isotropic polymers and its composites to GOS (in other words, by graded oriented stretching) in materials is generated gradient of all the properties that depend on the value of relative elongation/orientation degree. These properties are: optical, electrical, magnetic, acoustic, thermal, mechanical, sorption, etc. Consequently the graded oriented/stretching method may be of interest to obtain a new type of functionally graded materials (FGMs) [18, 19].

In this paper mathematical model of controlled uniaxial graded oriented stretching of linear polymer films is discussed. Based on this model the configuration of the inhomogeneous mechanical field is defined, which provides transition of the isotropic polymer films into GOS with pre-determined values of parameters.

12.2 MATHEMATICAL MODEL OF CONTROLLED GRADED ORIENTED STRETCHING

Let's assume that the test sample has a form of curvilinear isosceles trapezoid *ABCD*, where the large base of trapezoid is *AD* and *BC* is a small base. Graded stretching of trapezoidal *ABCD*-sample is discussed.

Let's consider rectangular Cartesian co-ordinates *XOY*, where the abscissa is parallel to the large base of *ABCD*-trapezoid and ordinate is an axis of symmetry (Figure 12.1).

Introduce the notations: $AD \equiv 2a$, $BC \equiv 2b$, $NM \equiv H$. $NP_1 \equiv h$, $P_1P_2 \equiv l$, elongation of P_1P_2 segment $P_2P_3 = \Delta l$.

Let's assume that the side of trapezoid is $y = f(x)\ x \in (0, a]$ function. *MN* is an axis of symmetry. $P_1 \in MN$. Then we get:

$$h = f(l) \tag{1}$$

$$l = f^{-1}(h) \tag{2}$$

Two cases of graded oriented stretching are discussed:

1. After stretching the sample takes a form of curvilinear rectangle AB_1C_1D (design for this type of deformation is described in Ref. [6]). Quantitative parameters of obtained gradually oriented AB_1C_1D-sample are: range of change of relative elongation $0 - \frac{\Delta l}{MC}$; length of change of relative elongation – *MN*. Creating of different profiles of change of relative elongation (the third quantitative parameter) is discussed below.

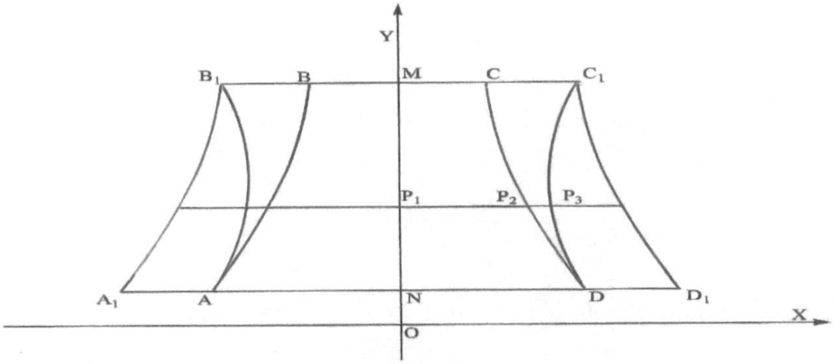

FIGURE 12.1 Rectangular Cartesian co-ordinates *XOY*, where the abscissa is parallel to the large base of *ABCD*-trapezoid and ordinate is an axis of symmetry.

Let's consider CDC_1 and P_2DP_3 curvilinear triangle. $\Delta CDC_1 \sim \Delta P_2DP_3$ Based on the conformal mapping, we may assume that

$$\frac{P_2P_3}{CC_1}=\frac{NP_1}{MN}\Rightarrow P_2P_3=CC_1\cdot\frac{NP_1}{MN}$$

Taking into account the above denotations this formula becomes:

$$\Delta l=P_2P_3=(a-b)\cdot\frac{h}{H}=w\cdot h \tag{3}$$

where $w\equiv\frac{a-b}{H}$. From Eqs. (2) and (3) we get:

$$\frac{\Delta l}{l}=w\cdot\frac{h}{f^{-1}(h)} \tag{4}$$

Let's consider different cases of *f(x)* function:

a) linear function: $f(x)=kx+c\Rightarrow f^{-1}(h)=\frac{h-c}{k}$

$$\frac{\Delta l}{l}=\frac{wkh}{h-c}=\frac{wkh-wkc+wkc}{h-c}=wk+\frac{wk}{h-c}$$

After introduction of the notation $m = wk$, we get

$$\frac{\Delta l}{l}=\frac{m}{h-c}+mk \tag{4_1}$$

b) quadratic function: $f(x)=x^2\Rightarrow f^{-1}(h)=h^{\frac{1}{2}}$, then

$$\frac{\Delta l}{l}=wh^{-\frac{1}{2}} \tag{4_2}$$

c) logarithmic function: $f(x)=\ln x\Rightarrow f^{-1}(h)=e^h$ then

$$\frac{\Delta l}{l}=wh\cdot e^{-h} \tag{4_3}$$

d) hyperbolic function: $f(x)=\frac{k}{x}\Rightarrow f^{-1}(h)=\frac{k}{h}$, then $\frac{\Delta l}{l}=\frac{w\cdot h^2}{k}$,

After introduction of the notation $m=\frac{w}{k}$ we get

$$\frac{\Delta l}{l}=m\cdot h^2 \qquad (4_4)$$

Similar approach can also be applied to other functions.

2. During stretching the sides of the trapezoid moved parallel to its initial position by Δl distance. After stretching the sample takes a form of curvilinear trapezoid $A_1B_1C_1D_1$. Unlike the first case $P_3\in C_1D_1$; In this case elongation of the segment is $P_2P_3\equiv\Delta l=\text{cost}$. Quantitative parameters of gradually oriented $A_1B_1C_1D_1$-sample are: range of change of relative elongation $\frac{\Delta l}{ND}-\frac{\Delta l}{MC}$; length of change of relative elongation – *MN*. Creating of different profiles of change of relative elongation (the third quantitative parameter) is discussed below.

$$\frac{\Delta l}{l}=\frac{\Delta l}{f^{-1}(h)} \qquad (5)$$

Let's consider different cases of *f*(*x*) function:

a) linear function: $f(x)=kx+c\Rightarrow f^{-1}(h)=\frac{h-c}{k}$, we get:

$$\frac{\Delta l}{l}=\frac{k\cdot\Delta l}{h-c}$$

After introduction of the notation $m=\Delta l\cdot k$, we get

$$\frac{\Delta l}{l}=\frac{m}{h-c} \qquad (5_1)$$

b) quadratic function: $f(x)=x^2\Rightarrow f^{-1}(h)=h^{\frac{1}{2}}$, then

$$\frac{\Delta l}{l}=\Delta l\cdot h^{-\frac{1}{2}} \qquad (5_2)$$

c) logarithmic function: $f(x) = \ln x \Rightarrow f^{-1}(h) = e^h$, then we get:

$$\frac{\Delta l}{l} = \Delta l \cdot e^{-h} \tag{5$_3$}$$

d) hyperbolic function: $f(x) = \frac{k}{x} \Rightarrow f^{-1}(h) = \frac{k}{h}$, then we get:

$$\frac{\Delta l}{l} = \frac{\Delta l \cdot h}{k}$$

Introduce the notation $m = \frac{\Delta l}{k}$, then we get

$$\frac{\Delta l}{l} = m \cdot h \tag{5$_4$}$$

Similar approach may be applied also to other functions. In the case, when $f(x)$ or $\frac{\Delta l}{l}(h)$ are complex functions, to calculate appropriate profile of the clamps quantitative technique can be applied.

12.3 CONCLUSIONS

The regularities of formation of GOS – the new specific structural state of linear polymers has been studied. Transition of the isotropic polymers to GOS occurs by action of inhomogeneous mechanical field on the polymeric sample.

Mathematical model of controlled uniaxial graded oriented stretching of linear polymer films is developed. Based on this model the configuration of the inhomogeneous mechanical field is defined which provides transition of isotropic polymer films to GOS with pre-determined values of parameters. These parameters are: range of changing of relative elongation/orientation degree, length and profile (linear, hyperbolic, parabolic, logarithmic, etc.).

Uniaxial graded oriented stretching method can be considered as a scientific and technological innovation for the creation of a new type FGMs/elements on the base of linear polymers/copolymers and its composites with gradient of all properties which are depended on the elongation/orientation degree.

KEYWORDS

- **functionally graded materials**
- **gradually oriented/stretched polymers**
- **gradually oriented/stretched state**
- **quantitative parameters**

REFERENCES

1. Nadareishvili, L. Fabrication method and investigation of polymer films with a specified gradient of birefringence. Georgian Engineering News, vol. 2, 73–77, 2001. (In Georgian).
2. Nadareishvili, L., Gvatua Sh., Lekishvili, N. Investigation of Optical Properties of Polymers Subject to Solid Phase Transformation. International Conference of Polymer Characterization, POLYCHAR-8, Denton, USA, January, 2000.
3. Nadareishvili, L., Gvatua Sh., Formation of Polymeric Macrosurface media with specified gradient refractive index and birefringence. Proceedings of the International Conference Applied Optics. 2000, vol. 1, 34–35, St. Petersburg, October, 2000. (In Russian).
4. Lekishvili, N., Nadareishvili, L., Zaikov, G. E., Khananashvili, L. New Concepts in Polymer Science. Polymers and Polymeric Materials for Fiber and Gradient Optics, (Eds. J. S. Vygodsky, Sh. A, Samsonia), VSP, Utrecht, Boston, Totyo, Koln, p. 222, 2002.
5. Nadareishvili, L., Gvatua Sh., Blagidze, Y., Zaikov, G. E. GB-optics – a new direction of gradient optics, J. of Apll. Polym. Science, 91, 489–493, 2004.
6. Wardosanidze, Z., Nadareishvili, L., Lekishvili, N., Gvatua Sh., The Possibilities of Application of the Gradient Birefringence Polymer Elements, Georcia Chemical Journal, 4, 270–273, 2004. (In Georgian).
7. Lekishvili, N., Nadareishvili, L., Zaikov, G. Polymer medium with gradient optical properties – advanced materials for next-generation optical instrument, in: Panorama of modern chemistry of Russia. Advances in the field of physical chemistry of polymers, (Ed. G. E. Zaikov), "Chemistry", Moscow, Russia, 624–675, 2004 (In Russian).

8. Nadareishvili, L., Lekishvili, N., Zaikov, C. E, Polymer Medias with Gradient of the Optical Properties, in: Modern Advanced in Organic and Inorganic Chemistry (Ed. G. E. Zaikov), Nova Science Publ., New York, USA, 31–134, 2005.
9. Nadareishvili, L., Gvatua, Sh., Topuridze, N., Japaridze, K., Polarization properties of polymer films with a gradient of birefringence, Optical Journal, St. Petersburg, 10, 12–18, 2005. (In Russian).
10. Nadareishvili, L., Bakuradze, R., Topuridze, N., Sharashidze, L. Gradient oriented state of polymers, APMAS-2011, Turkey, Antalya, May, 2011.
11. Lekishvili, N., Nadareishvili, L., G. E. Zaikov, Polymer Materials with the Structural Inhomogeneity for Modern Optical Devices (Review), in: New Trends in Natural and Synthetic Polymer Science, (Eds. G. Vasile, G. E. Zaikov), Nova Science Publishers, New York, 165–201, 2006.
12. Nadareishvili, L., Bakuradze, R., Topuridze, N., Some Regularities of Polymer Gradient Orientation in Inhomogeneous Mechanical Field, Proceedings of the Georgian National Academy of Science, Chemical Series, 36, 197–200, 2010. (In Georgian).
13. Nadareishvili, L., Bakuradze, R., Topuridze, N., Nakaidze, T., Pavlenishvili, I., Sharashidze, L., Some Regularities of Gradient Orientation of Polymers in Heterogeneous Mechanical Field, Georgia Chemical Journal, 11. 281–283, 2011 (in Russian).
14. Nadareishvili, L., Wardosanide, Z., Lekishvili, N., Topuridze, N., Zaikov, G., Kozlowski, R., Gradient Oriented State of Linear Polymers: Formation and Investigation, Molecular Crystals and Liquid Crystals, 556, 52–56, 2012.
15. Nadareishvili, L., Akhobadze, V., Gvatua Sh., Topuridze, N., Sharashidze, L., Pavlenishvili, I., Blagidze, Y., Skirtladze, I., Japaridze, K., Device for stretching polymer films. Georgian Patent P2992, 2003.
16. Nadareishvili, L., Wardosanidze, Z., Chelidze, G., Polymeric films deformation method. Georgian Patent P4182, 2007.
17. Nadareishvili, L., Wardosanidze, Z., Skirtladze, I. Chelidze, G., Akhobadze, V., Topuridze, N., Lekishvili, N., Pavlenishvili, I., Sharashidze, L., Japaridze, K., Device for stretching polymer films. Georgian Patent P4398, 2008.
18. Functionally Graded Materials, (Ed. Nathan, J. Reinolds), Nova Science Publ., New York, 2011.
19. Sobczak, J., Drenchev, L., Functionally Graded Materials. Processing and Modeling, Warsaw, 2009.

CHAPTER 13

METHOD OF OBTAINING OF GRADUALLY ORIENTED POLYMERIC FILMS

L. NADAREISHVILI, R. BAKURADZE, N. TOPURIDZE, L. SHARASHIDZE, and I. PAVLENISHVILI

Georgian Technical University, Department of Cybernetics, 5 S. Euli St. Tbilisi, 0186, Georgia, E-mail: levannadar@yahoo.com

CONTENTS

ABSTRACT

An algorithm for the formation of gradually oriented/stretched rectangular polymeric films with specified parameters (range of changes in elongation and length) in heterogeneous mechanical field is given. It is shown that the elongation of the films having curvilinear trapezoid form in the parallel clamps causes hyperbolic distribution of elongations.

13.1 INTRODUCTION

Widespread method of the structural modification of the linear polymers is the uniaxial oriented stretching. In result of stretching isotropic polymer passes into the oriented state, which is characterized by predominant location of the structural elements in the stretching direction. Throughout the volume of the specific oriented polymer sample relative elongation $\Delta l / l$ (Δl is the real elongation and l is the initial length of the sample) practically is the same.

Previously, we have developed a new conception on the structural state of linear polymers – about gradually oriented state (GOS) [1–14]. Quantitatively COS is described by three parameters: the range of change in relative elongation/degree of orientation, length of this change and the profile (linear, hyperbolic, parabolic, logarithmic, sinusoidal, etc.). Transition of the isotropic polymer into COS is realized by superimpose of the inhomogeneous mechanical field on the isotropic polymer sample. Several managed graded stretching methods. In these methods on the basis of mathematical models the configuration of the mechanical field is determined, which provides obtaining of the graded stretched/oriented film with specified parameters.

Development of the new graded oriented stretching methods is an important part in the study of scientific and technological aspects of the COS and the various perspective applications.

In this paper we discuss a matter of getting the rectangular blanks with a hyperbolic distribution of elongation perpendicular to stretching's direction.

13.2 ALGORITHM FOR GRADED ORIENTED STRETCHING

Let's say we want to produce graded oriented rectangular *ACDB* film. The films such as these will be obtained by uniaxial stretching of *ACDB* – curvilinear trapezoid (Figure 13.1). The arrow indicates the direction of stretching.

To calculate the geometric sizes of ***ACDB*** – curvilinear trapezoid let's consider sector ***AOB*** (Figure 13.2) the part of which is the curvilinear trapezoid ***ACDB***.

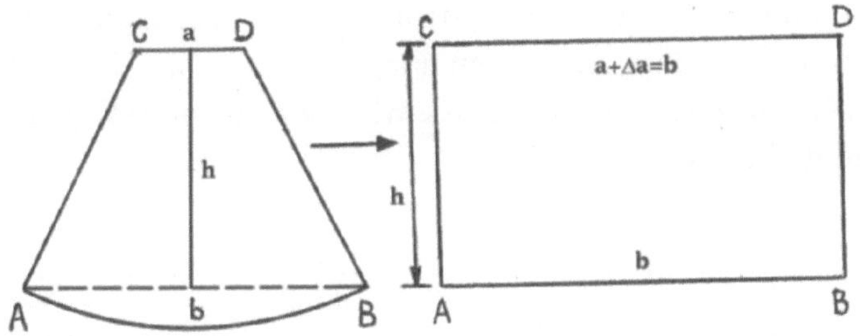

FIGURE 13.1 Explanations are in the text.

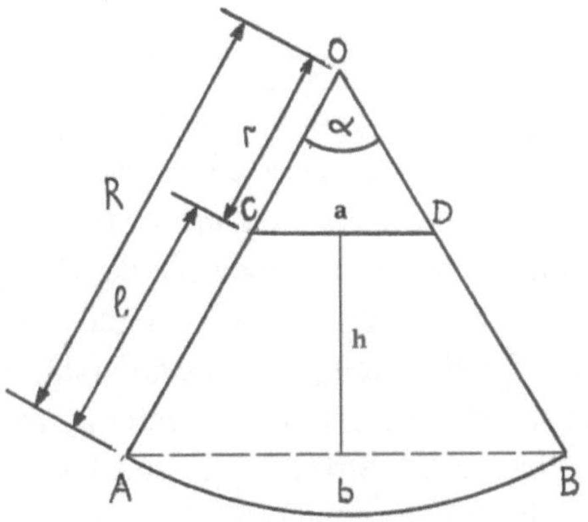

FIGURE 13.2 Explanations are in the text.

We introduce the following notation: ***OA=OB=R, OC=OD*** = ***r, OC+CA=OA, OD=DB=OB***; the length of the chord ***CD*** through – ***a***; the length of the ***AB*** – arc through – ***b***; ∠ ***AOB*** is denoted by 2α (in radians); the distance between the chords ***CD*** and AB through – ***h***; ***b***/***a*** = ***n***.

For the fixed ***a, l***, and ***n*** significances the parameters ***R***, ***r*** and α will be calculated. After simply transformation we obtain the system of equations:

$$2r \sin \alpha = a$$

$$R = n / \alpha$$

$$R - r = l$$

For solution of the equations system on the basis of chords method the algorithm and program on programming language Pascal is elaborated. The results of calculations for the case when ***a*** = 20mm and *l* = 40 mm are presented in the table:

n	∠ ***AOB*** **(in radians)**	***R*, mm**	***r*, mm**
2	28	81.85	41.34
3	56	61.4	21.3
4	85	56.6	15.4
5	110	52.1	12.2
6	135	51.3	10.9
7	160	50.13	10.15

Figure 13.3 illustrates the graded stretching's technique. The curvilinear trapezoid ***ACDB*** is fixed between the clamps of the stretching device disposed mutually parallel. The distance between the clamps equals to small base of trapezoid – ***a***. The test trapezoid – sample is fixed in the clamps along the sides ***AC*** and ***DB***. If you move the clip (shift's direction is indicated by the arrow) stretching front moves smoothly from a small base of the trapezoid ***(a)*** to the arc ***(b).*** Stretching of specimen was stopped when ***a*** + Δ***a*** (Δ***a*** increase the length of the small base of the trapezoid – ***a***) will be equaled to the length of arc ***b***, which at this point becomes a straight line.

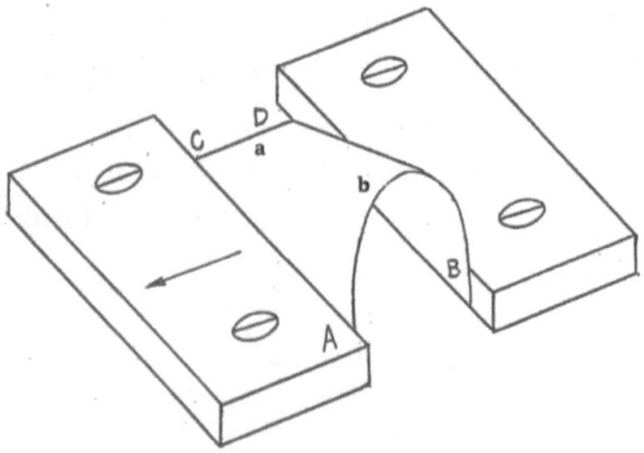

FIGURE 13.3 Graded stretching's technique.

As a result of stretching curvilinear trapezoid transforms to a rectangle with sides ***b*** and ***h***. In the described mode of graded stretching the lower limit of change in elongation is almost zero and the upper limit is determined by the ratio ***b*/*a*** = ***n***.

In order to visualize the distribution of elongations on the original sample the square grid is applied. Figure 13.4a shows the original curvilinear trapezoid-sample for which ***b*/*n*** = 30 mm: 10 mm = 3, and ***h*** = 35 mm. After a three-fold extension (relatively to small base of the trapezoid) the sample is converted into a rectangle (Figure 13.4b).

Changing of topographic pattern of a square grid clearly indicates the existence of a gradient elongation perpendicularly to the stretching direction.

Quantitative parameters of the resulting gradually oriented film (for its middle part) are shown in Figure 13.5. According to the Figure 13.5 range of change in relative elongation (**Δ*a*/*a***) 0.225 – 2; length of change in relative elongation ***h*** = 30 mm; profile of distribution of relative elongation's change – hyperbola.

Let us briefly consider the possibilities and limitations of the proposed method. In gradually oriented films the minimum lower limit of **Δ*a*/*a*** may be equal to zero, and the upper limit depends on the mechanical properties of the polymer and the deformation mode (stretching speed and temperature); the maximum length (***h***) of distribution of **Δ*a*/*a*** dictated by practical expediency, and the minimum value amounts to a few millimeters; this method provides a hyperbolic distribution of elongation.

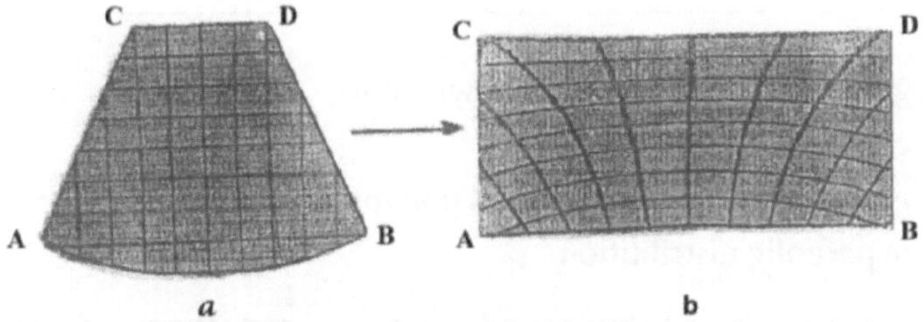

FIGURE 13.4 Curvilinear trapezoid-sample.

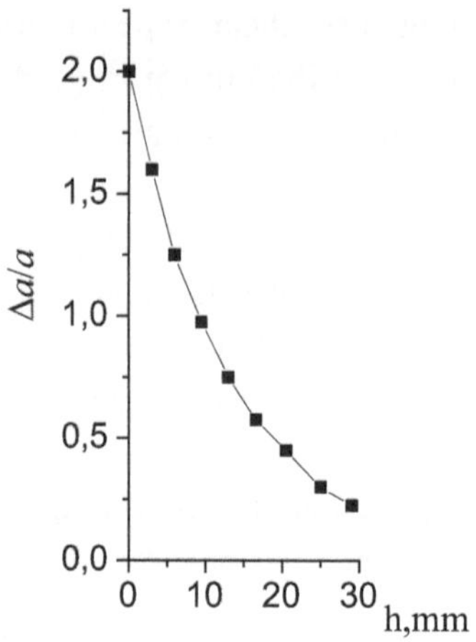

FIGURE 13.5 Distribution of relative elongation ($\Delta a/a$) in height (h) at the middle part of the gradually oriented polymer film.

13.3 CONCLUSIONS

Method of controlled graded stretching of the linear polymers is proposed. On the basis of mathematical model the configuration of the mechanical field is determined, which provides obtaining of the graded stretched/oriented film with specified parameters such as the range of change in relative elongation and length of this change. Method provides a hyperbolic distribution of elongation.

KEYWORDS

- **graded oriented stretching algorithm**
- **gradually oriented state**
- **gradually oriented/stretched polymers**
- **hyperbolic distribution**

REFERENCES

1. Nadareishvili, L., Fabrication method and investigation of polymer films with a specified gradient of birefringence. Georgian Engineering News, 2, 73–77, 2001. (In Georgian).
2. Nadareishvili, L., Gvatua, Sh., Lekishvili, N., Investigation of Optical Properties of Polymers Subject to Solid Phase Transformation. International Conference of Polymer Characterization, POLYCHAR-8, Denton, USA, January, 2000.
3. Nadareishvili, L., Gvatua, Sh., Formation of Polymeric Macrosurface media with specified gradient refractive index and birefringence. Proceedings of the International Conference Applied Optics–2000, 1, 34–35, St. Petersburg, October, 2000. (In Russian).
4. Lekishvili, N., Nadareishvili, L., Zaikov, G., Khananashvili, L., New Concepts in Polymer Science. Polymers and Polymeric Materials for Fiber and Gradient Optics, (Eds. Vygodsky, J.S, Samsonia Sh. A.), VSP, Utrecht, Boston, Totyo, Koln, p. 222, 2002.
5. Nadareishvili, L., Gvatua Sh., Blagidze, Y., Zaikov, G. GB-optics – a new direction of gradient optics, J. of Apll. Polym. Science, 91, 489–493, 2004.
6. Wardosanidze, Z., Nadareishvili, L., Lekishvili, N., Gvatua Sh., The Possibilities of Application of the Gradient Birefringence Polymer Elements, Georcia Chemical Journal, 4. 270–273, 2004. (in Georgian).
7. Lekishvili, N., Nadareishvili, L., Zaikov, G., Polymer medium with gradient optical properties – advanced materials for next-generation optical instrument, in: Panorama of modern chemistry of Russia. Advances in the field of physical chemistry of polymers, (Ed. G. E. Zaikov), "Chemistry", Moscow, Russia, 624–675, 2004. (in Russian).
8. Nadareishvili, L., Lekishvili, N., Zaikov, G., Polymer Medias with Gradient of the Optical Properties, in: Modern Advanced in Organic and Inorganic Chemistry (Ed. G. E. Zaikov), Nova Science Publ, New York, USA, 31–134, 2005.
9. Nadareishvili, L., Gvatua Sh., Topuridze, N., Japaridze, K., Polarization properties of polymer films with a gradient of birefringence, Optical Journal, St. Petersburg, 10, 12–18, 2005 (in Russian).
10. Nadareishvili, L., Bakuradze, R., Topuridze, N., Sharashidze, L., Gradient oriented state of polymers, APMAS-2011, Turkey, Antalya, May, 2011.
11. Lekishvili, N., Nadareishvili, L., Zaikov, G., Polymer Materials with the Structural Inhomogeneity for Modern Optical Devices (Review), in: New Trends in Natural and Synthetic Polymer Science, (Eds. G. Vasile, G. Zaikov), Nova Science Publishers, New York, 165–201, 2006.
12. Nadareishvili, L., Bakuradze, R., Topuridze, N., Some Regularities of Polymer Gradient Orientation in Inhomogeneous Mechanical Field, Proceedings of the Georgian National Academy of Science, Chemical Series, 36, 197–200, 2010. (In Georgian).
13. Nadareishvili, L., Bakuradze R, Topuridze, N., Nakaidze, T., Pavlenishvili, I., Sharashidze, L., Some Regularities of Gradient Orientation of Polymers in Heterogeneous Mechanical Field, Georgia Chemical Journal, 11. 281–283, 2011 (in Russian).
14. Nadareishvili, L., Wardosanide, Z., Lekishvili, N., Topuridze, N., Zaikov, G., Kozlowski, R., Gradient Oriented State of Linear Polymers: Formation and Investigation, Molecular Crystals and Liquid Crystals, 556, 52–56, 2012.

CHAPTER 14

METHOD OF PRODUCTION OF MICROCAPSULES

YEVGENIYA NIKOLAYEVA, SAULE KOKHMETOVA, ANDREY KURBATOV, ALINA GALEYEVA, and OLEG KHOLKIN

Centre of Physical-Chemical Methods of Research and Analysis, Al-Farabi Kazakh National University, 96a, Tole bi str., Almaty, Kazakhstan, E-mail: alinex@bk.ru

CONTENTS

ABSTRACT

The results of influence of the gelatin – multinuclear alcohol ratio on dispersivity of the synthesized microcapsules are presented. It has been established that at the 1/3 gelatin-multinuclear alcohol ratio the most finely dispersed microcapsules are obtained.

14.1 INTRODUCTION

Microencapsulation is a process of cladding fine particles of a substance (solid, liquid or gaseous) into a shell made of a film-forming material. Nowadays microencapsulation is a well-known, intensively developed technology, which is widely used in different branches of industry.

Microencapsulated systems are characterized by various properties, such as controllable and selective mass transfer of microcapsules through the shell, prolonged action of encapsulated substances, opportunity to control their reactivity, etc. This makes microencapsulation a perspective and high tech method, which could be used for development of absolutely new materials.

Besides, the basic component of microcapsules – a capsulated substance – may be in any aggregate state – liquid, solid or gaseous. Microcapsules' content can be released both as a result of mechanical destruction of shells due to exposure to pressure, friction, melting, and ultrasound, and diffusion caused by swelling walls of capsules in a surrounding liquid. In any case, a complete content release during its use should be combined with its reliable protection from environment during storage.

There are many ways of production of microcapsules, which can be carried out by various methods: chemical, physical and physical-chemical [1–7]. Some technologies use the combined methods of shell formation, and the choice of the most effective methods or their combination is determined by both the properties of the capsulated product, requirements to the received microcapsules and by the cost of the technological process. For each specific case the effectiveness of the method is defined by a set of final product properties.

Coacervation is stratification of a polymer solution into two liquid phases; one of them is enriched with a dissolved substance (coacervate), and the other – with water. This phase division occurs due to the decrease of polymer solubility in a medium, which is caused by various factors.

Microencapsulation based on coacervation is widely used for production of microcapsules containing a nonpolar liquid (oil) or a solid substance with a low-energy surface as the basic capsulated substance.

As a material for shells water-soluble high-molecular compounds (HMC), which are capable to dissociate in aqueous solution to ions, that is, polyelectrolytes (PE), and those ones which are capable of coacervation being exposed to external action on the solution are used.

Polyelectrolyte macromolecules in water solutions possess specific electric, conformational and hydrodynamic properties distinguishing them from ordinary non-dissociating polymers. In colloid chemistry polyelectrolytes (and other well-soluble in water polymers) are named hydrophilic colloid substances or hydrophilic colloids. Colloid properties of these substances are caused by the presence of big kinetic units having the sizes of 10^{-5}–10^{-7} cm [8] in their solutions.

14.2 EXPERIMENTAL PART

14.2.1 MATERIALS

Food gelatin, polyatomic alcohol, carbon tetrachloride, 5% hydrochloric acid solution, 5% caustic soda solution.

14.2.2 SYNTHESIS OF MICROCAPSULES

Synthesis of microcapsules from the real concentrated solution of gelatin has been carried out according to the well-known technique of emulsification of a hydrophobic substance in a water solution and included the following stages:

- modification of the gelatin water solution;
- production of the emulsion;
- formation of the microcapsules;
- dehydration of the shells of the gelatin microcapsules;
- hardening the shells.

The emulsion that represents itself the drops of capsulated liquid stabilized with structure layers of gelatin has been produced by trickling the capsulated substance into the concentrated solution of gelatin at the temperature of 40–50°C and constant stirring.

The size of emulsion drops and, consequently, the future size of capsules are defined by surface tension at gelatin water solution-capsulated liquid interface and also by the intensity of stirring during the emulsion formation. Emulsification of the capsulated liquid has been carried out with a magnetic stirrer.

The gelatin microcapsules produced by the coacervation method usually have the sizes of 300–1500 micrometers, which makes their use in our research unacceptable, for in our case the size of microcapsules should not exceed 50 micrometers. To reduce the size of the microcapsules the modification of the gelatin water solution by polyatomic alcohol has been carried out.

Then the capsules have been separated from the gelatin water solution, and the shells have been solidified on the capsulated liquid by dilution of the produced emulsion with water at the simultaneous sharp temperature decrease.

Dehydration of the gelatin microcapsules shells has been carried out with 20% sodium sulfate solution at 10°C, and is based on the property of gelatin to change its water solubility with a change of the medium pH. For hardening the shells the microcapsules have been treated with the formaldehyde water solution.

14.3 RESULTS AND DISCUSSION

Modification of 10% gelatin water solution has been carried out by polyatomic alcohol in the amount that promotes the best binding of gelatin. Besides, it was necessary to establish an optimal gelatin – polyatomic alcohol ratio. For this purpose a series of experiments in which gelatin – polyatomic alcohol ratio varied 2:1, 1:1, 1:3 and 1:5 has been carried out. One of the basic characteristics of the produced microcapsules is their size and size distribution (Figure 14.1: ***1*** – 2:1 gelatin – polyatomic alcohol ratio; ***2*** – 1:1 gelatin – polyatomic alcohol ratio; ***3*** – 1:3 gelatin – polyatomic alcohol ratio; ***4*** – 1:5 gelatin – polyatomic alcohol ratio).

The analysis of the particle sizes has been carried out by means of optical microscopy (Figure 14.1). In Figure 14.2 the curves of distribution of the microcapsule sizes are presented for different gelatin – polyatomic alcohol ratios.

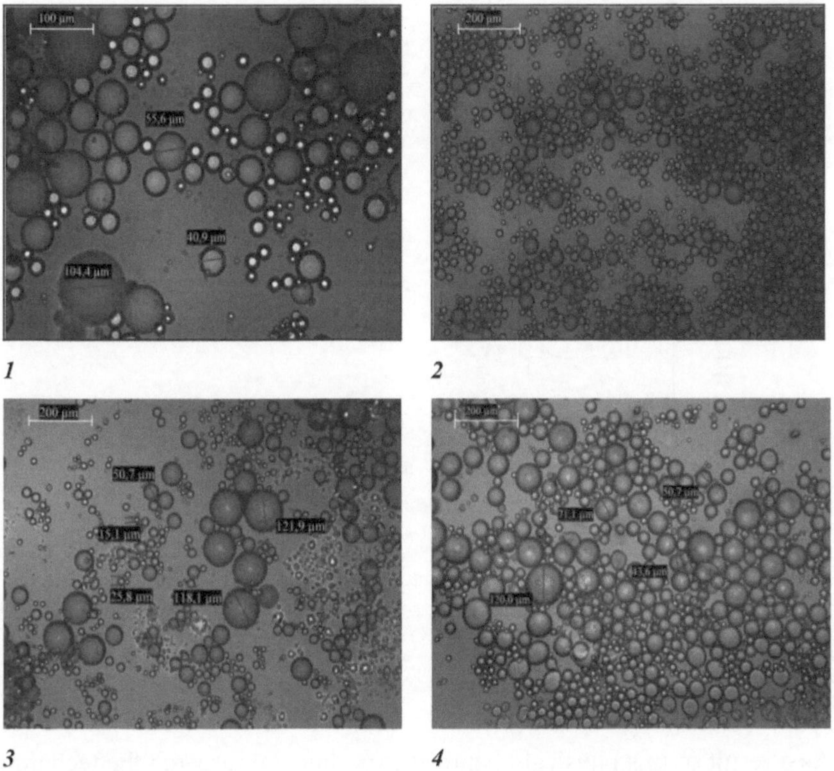

1 – 2 : 1 gelatin - polyatomic alcohol ratio; *2* – 1 : 1 gelatin - polyatomic alcohol ratio; *3* – 1 : 3 gelatin - polyatomic alcohol ratio; *4* – 1 : 5 gelatin - polyatomic alcohol ratio

FIGURE 14.1 Optical micro photos of the synthesized microcapsules on the basis of gelatin modified by polyatomic alcohol.

According to the data presented in Figure 14.2, at the 2:1 gelatin – polyatomic alcohol ratio the curve of distribution of the sizes of the produced microcapsules has a quite smooth transition from 20 up to 100 micrometers while at the 1:1 gelatin – polyatomic alcohol ratio the curves show clearly seen peak at 20–40 micrometers. At the 1:5 gelatin – polyatomic alcohol ratio the peak of distribution of the sizes of the synthesized microcapsules is observed at 40–70 micrometers. Therefore, the 1:1 and 1:3 gelatin – polyatomic alcohol ratios prove to be optimal for our purposes.

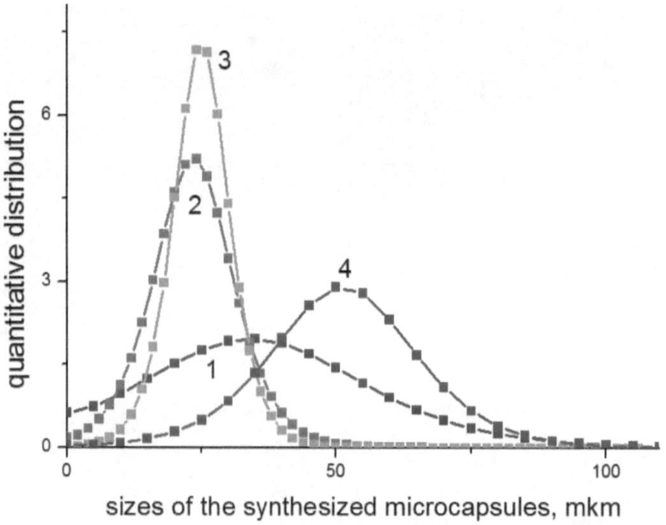

FIGURE 14.2 Distribution of the sizes of the synthesized microcapsules on the modified gelatin basis. The gelatin – polyatomic ratios: *1* – 2:1; *2* –1:1; *3* –1:3; *4* –1:5, correspondingly.

14.4 CONCLUSION

As a result of this physical-chemical experimental research the technique of modifying gelatin microcapsules has been improved allowing to lower their size up to 20–40 micrometers.

This work has been carried out within the framework of the project 063 "Development of the methods of obtaining self-healing composition coatings" under the program of Groups of junior researchers of the project "Commercialization of technologies" of Ministry of Education and Science of Republic of Kazakhstan and the World Bank.

KEYWORDS

- **coacervation**
- **emulsification**
- **gelatin**
- **microencapsulation**

REFERENCES

1. Tatsiana, G. Shutava, Shantanu, S. Balkundi, Yuri, M. Lvov. (−)-Epigallocatechin gallate/gelatin layer-by-layer assembled films and microcapsules. Journal of Colloid and Interface Science, 330, 276–283, 2009.
2. Amanda, K. Andriola Silva Brun-Graeppi, Cyrille Richard, Michel Bessodes, Daniel Scherman, Otto-Wilhelm Merten. Cell microcarriers and microcapsules of stimuli-responsive polymers. Journal of Controlled Release, 149, 209–224, 2011.
3. Weigang Li, Gang Wu, Hongzheng Chen, Mang Wang. Preparation and characterization of gelatin/SDS/NaCMC microcapsules with compact wall structure by complex coacervation. Colloids and Surfaces A: Physicochemical and Engineering Aspects, 333, 133–137, 2009.
4. Dai Runying, Wu Gang, Li Weigang, Zhou Qiang, Li Xihua, Chen Hongzheng, Gelatin/carboxymethylcellulose/dioctyl sulfosuccinate sodium microcapsule by complex coacervation and its application for electrophoretic display. Colloids and Surfaces A: Physicochemical and Engineering Aspects, 362, 84–89, 2010.
5. Liu Jiayi, Liu Chaohong, Liu Yingju, Chen Minjie, Hu Yang, Yang Zhuohong. Study on the grafting of chitosan–gelatin microcapsules onto cotton fabrics and its antibacterial effect. Colloids and Surfaces B: Biointerfaces, 109, 103–108, 2013.
6. Sakai Shinji, Ito Sho, Kawakami Koei. Calcium alginate microcapsules with spherical liquid cores templated by gelatin microparticles for mass production of multicellular spheroids. Acta Biomaterialia, 6, 3132–3137, 2010.
7. Nakagawa Kyuya, Nagao Hiromistu. Microencapsulation of oil droplets using freezing-induced gelatin–acacia complex coacervation. Colloids and Surfaces A: Physicochemical and Engineering Aspects, 411, 129–139, 2012.
8. Solodovnik, V. D. Microencapsulation. M.: Chemistry, 1980. 216 p.

CHAPTER 15

INVESTIGATION OF FULVIC ACIDS ISOLATED FROM NATURAL WATERS BY THE THERMAL ANALYZE

GIORGI MAKHARADZE, NAZI GOLIADZE, and TAMAR MAKHARADZE

Ivane Javakhishvili Tbilisi State University Chavchavadze Avenue 1, Tbilisi 0128, Georgia, E-mail: giorgi.makharadze@yahoo.com

CONTENTS

ABSTRACT

Fulvic acids, isolated from surface waters, have been studied for the first time by the thermal analyze. Fulvic acids investigated by thermogravimetric and differential thermal analysis are characterized by one endothermal and four exothermal effects. The loss of mass in low temperature range for fulvic acids isolated from the bottom sediments and water sample makes 30 and 38 percent and over the high temperature range it equals 62 and 48 percent respectively. It is shown that the degree of aromatization of fulvic acids isolated from the bottom sediments was higher than that of fulvic acids isolated from the water sample.

15.1 INTRODUCTION

Unlike soil humic acids [1–3], the thermal characteristics of humin and fulvic acids isolated from the surface waters are not studied. By means of this method it is possible to evaluate the thermal stability of fulvic acids, as well as to calculate the variants of their mass depending on temperature and to determine the quantitative correlation between central and peripheral parts in the molecule.

The aim of this work was to study the fulvic acids isolated from the bottom sediments and from the water sample of the river Mtkvari by the methods of thermogravimetric and thermal analysis.

15.2 EXPERIMENTAL PART

To obtain pure samples of fulvic acids we concentrated the water of the river Mtkvari by the frozen method. Filtered water samples were acidified to pH 2 and was put for 2 hours on water bath at 60° C, for coagulation of humin acids. Then the solution was centrifuged for 10 min at 8000 rpm. To isolate FA from centrifugate the adsorption-chromotographic method was used. The charcoal (BAU, Russia) was used as a sorbent. Desorption of amino acids and carbohydrates were performed by means of 0.1N HCl. For the desorption of polyphenols the 90% acetone water solution was used. The elution of FA fraction was performed with 0.1N NaOH solution. The obtained alkali solution of FA for the purification was passed through a cation-exchanger (KU-2, Russia) and dried under the vacuum until the constant mass was obtained [4, 5]. The extraction of the fulvic acids from the bottom sediments was performed with 0.1 M NaOH.

The thermal analysis was carried out on the Paulic-Paulic-Erday system derivatograph Q-1500 (Hungary); in the air, standard sample-Al_2O_3, over heating rate 5°C/min, temperature range 20–800°C. Differential thermal analysis (DTA), thermogravimetric (TG) and differential thermo-gravimetric (DTG) curves were recorded simultaneously. Sensibility of the balance – 100 mg, sensibility of DTA galvanometer – 250 μv and that of DTG – 500μv. Thermal effect were interpreted according to Refs. [1–3].

The thermograms of both samples of both samples were characterized by one endothermal effect, which was mainly caused by dehydration of the fulvic acids (Figure 15.1; Table 15.1). The effect was registered at

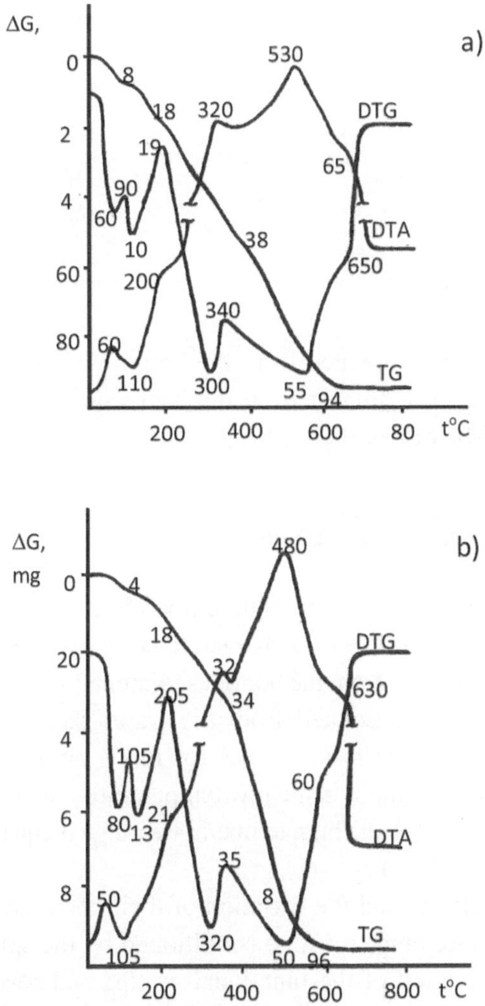

FIGURE 15.1 The thermogram of fulvic acids isolated from the sample of the river Mtkvari: (a) water, and (b) bottom sediment.

TABLE 15.1 Thermogravimetric Characterization of Fulvic Acids Isolated From the Lake Paravani

Fulvic Acids	Ash	DTA, T°C max		DTG T°C	TG %	Z
		Endothermal	Exothermal			
Water Sample	1.5	110	200	90	8	0.79
			320	190	10	
			530	340	28	
			650	550	39	
				650	9	
				105	4	
			210	205	9	
Bottom Sediments	3.5	105	330	350	21	0.48
			480	500	52	
			630	600	20	

105°C for fulvic acids isolated from the bottom sediments and at 110°C in case of fulvic acids isolated from the water sample. The loss of mass equals 4% and 8%, respectively.

15.3 RESULTS AND DISCUSSIONS

Both in low (190–350°C) and in high temperature range (350–650°C) the thermograms of two samples are characterized by two exothermal effects. For fulvic acids, isolated from the bottom sediments: 200°C, 330°C, 480°C, 630°C.For fulvic acids isolated from the water sample: 200°C, 320°C, 530°C, 650°C. Thus, the loss of mass for fulvic acids isolated from the sediments and water sample in the low temperature ranges makes 30% and 38%, respectively, but in the high temperature range it equals 62% and 48% (Figure 15.1; Table 15.1).

Exothermal effects and the variation of their corresponding masses in the low temperature range must be conditioned by the splitting of the aliphatic chain, liberation of the functional groups and partial oxidation of the formed products or because of the destruction of the structural components of peripheral parts of fulvic acids molecules accompanied by dehydrogenation and decarboxylation.

Exothermal effects and their corresponding changes in the high temperature range must be conditioned both by the splitting of aliphatic chain and liberating of more stable functional groups, releasing of cycles and aromatic nuclear [1–3].

Quantitative estimation of the correlation between the central and peripheral parts in the molecule of fulvic acids is possible by the value z=m/M, where m denotes the loss of mass in the low temperature range and M is the loss of mass in the high temperature range. The less the ration, the more aromatized is the substance. In our case for fulvic acids isolated from the bottom sediments z=0.48 and for those of isolated from the water sample z=0.79.

15.4 CONCLUSIONS

Thus on the basis of thermal analysis we can conclude that the investigated fulvic acids are characterized by one endothermal and four exothermal effects. The degree of aromatization of fulvic acids isolated from the bottom sediments is higher than that of fulvic acids isolated from the water sample.

KEYWORDS

- **endothermal effect**
- **exothermal effect**
- **fulvic acids**
- **natural waters**
- **thermal analysis**

REFERENCES

1. Schnitzer, M. S., Hoffman, J. Thermogravimetry of soil humic acids. Geochim. et Cosmochim. Acta, 29(8), 205–218, 1965.
2. Schnitzen M.S. Reaction between fulvic acids, a soil humic compound and inorganic soil constituents. Soil Sci. Soc. Amer. Proc., 33(1), 5–18, 1969.
3. Thompson, G. Chesters. Infra- red spectra and differential thermograms of lignins and soil humic materials saturated with different cations. Soil. Sci., 21(2), 320–327, 1970.
4. Varshal G.M., Kosheeva I.Y, Sirotkina I.S., Velukhanova T.K., Intskirveli L.N., Zamokina N.C. The study of organic substances of surface waters and their interaction with metals ions. Geokhimia, (Geochemistry, Russ.), 4, 598–607, 1979.
5. Revia R., Makharadze G., Cloud-point preconcentration of fulvic and humic acids. Talanta, 48(1), 409–413, 1999.

CHAPTER 16

FULVIC AND HUMIN ACIDS IN SURFACE WATERS OF GEORGIA

GIORGI MAKHARADZE, NAZI GOLIADZE, ANNA KHAIAURI, TAMAR MAKHARADZE, and GURAM SUPATASHVILI

Ivane Javakhishvili Tbilisi State University Chavchavadze Avenue 1, Tbilisi 0128, Georgia, E-mail: giorgi.makharadze@yahoo.com

CONTENTS

ABSTRACT

We have studied the concentrations of humin and fulvic acids in surface waters of Georgia. The concentration of fulvic acids changes from 0.08 mg/L to 4.23 mg/L. The concentration of humin acids changes from 0.00 mg/L to 0.60 mg/L. We have studied acid-base properties of humin fulvic acids isolated from the river Mtkvari, using potentiometric titration method. It has been established that in the case of humic acids pK (COOH) = 4.12 pK (phen.-OH) = 10.52 and in the case of fulvic acids pK (COOH) = 4.19 pK (phen.-OH) = 10.46.

16.1 INTRODUCTION

Geopolymers (humic acids: humin and fulvic acids) belong to those natural ligands, which can form complex compounds both in solid and soluble state. The intensity of complexing is mainly conditioned by the existence of carboxylic and phenolic hydroxyl groups being in various states in molecule of humic acids and their ionization ability. Literature data on dissociation of humic acids and quantitative substation of acidic groups in them is not uniform. The data of dissociation constants of one and the same acid groups of humic acids differs in several lines from each other [1–10]. Due to this the real role of humic acids in complexing reactions proceeding in natural waters is not clear. Also there are not experimental data about the concentration of humic acids.

Our objective was to study the concentration of humic acids in surface waters of Georgia and determine dissociation constants.

16.2 EXPERIMENTAL PART

To separate the fulvic and humin acids, we concentrated the surface waters by the frozen method. Filtered water samples were acidified to pH 2 and was put for 2 hours on water bath at 60°C, for coagulation of humin acids. Then the solution was centrifuged for 10 min at 8000 rpm. To isolate fulvic acids from centrifugate the adsorption-chromotographic method was used. The charcoal (BAU, Russia) was used as a sorbent. Desorption of amino acids and carbohydrates were performed by means of 0.1N HCl. For the desorption of polyphenols the 90% acetone water solution was used. The elution of fulvic acids fraction was performed with 0.1 N NaOH solution. The concentrations of fulvic and humin acids, isolated from the natural surface waters were determined by the photometric method (pH=10, l=420 nm). The obtained alkali solution of fulvic acids and humin acids for the purification were passed through a cation-exchanger (KU-2, Russia) and were dried under the vacuum until the constant masses were obtained [11, 12].

For purification, isolated humin acids were dissolved in 0,1 M NaOH, then humin acids were reprecipitated. The purification procedure was repeated four times. Purified humin acids were washed with bidistillated water until the negative reaction on the Cl^- ions, then humin acids were dried under the vacuum until the constant masses were obtained. The obtained humin and fulvic acids were used as standards for the photometric determination.

To determine dissociation constants of humin acids we used potentiometric titration (direct and back titration). As objects for investigation were chosen the samples of fulvic acids separated and purified by precipitation from the river Mtkvari. For potentiometric titration 50.0 mg of fulvic acids was solved in 20 mL 0.05 M NaOH. The end volume of solution by bidistallated water was reduced to 25 mL (m=0.1 KCl). The titration of alkali solution of fulvic acids was performed by 0.106 M HCl. Just after finishing of inverse titration (pH=2.98) direct titration was started with 0.095 M NaOH. The accounting was made on pH meter, after adding of each portion of reagent (0.1 mL). Both direct and back titration was carried out in nitrate area.

As objects for investigation were chosen the samples of humin acids separated and purified by precipitation from the river Mtkvari. For potentiometric titration 50.0 mg of humin acids was solved in 20 mL 0.05 M NaOH. By means of solution pH 0.5 M HCl reduced to 11.5–11.8. The end volume of solution by bidistillated water was reduced to 25 mL (m=0.1 KCl). As previous studies show, humin acids solution (T_{HA}=2 mg/mL) of such concentration is optimal for potentiometric titration. The titration of alkali solution of humin acids was performed by 0.01 M HCl. Just after finishing of inverse titration (pH=3.5–3.6) direct titration was started with 0.095 M NaOH. The accounting was made on pH meter (pH-673) after adding of each portion of reagent (0.1 mL). Both direct and back titration was carried out in nitrate area.

16.3 RESULTS AND DISCUSSIONS

As the results show (Table 16.1), in surface waters of Georgia the concentration of fulvic acids changes from 0.08 mg/L to 4.23 mg/L, the concentration of humin acids changes from 0.00 mg/L to 0.60 mg/L.

TABLE 16.1 The Concentrations of Humin and Fulvic Acids (mg/L) in Surface Waters of Georgia

The object	**Humin acids**		**Fulvic acids**	
Rivers	***Min***	***Max***	***Min***	***Max***
Mtkvari	0.11	0.58	0.31	4.23
Farvani	0.11	0.31	0.60	2.58
Aragvi	0.00	0.04	0.08	0.40
Xrami	0.00	0.03	0.33	0.43
Rioni	0.00	0.07	0.18	0.78
Tskhenisckali	0.00	0.04	0.20	0.58
Lakes	***Min***	***Max***	***Min***	***Max***
Lisi	0.00	0.00	0.56	0.70
Bazaleti	0.00	0.02	0.19	0.48
Tmogvi	0.00	0.03	0.12	0.33
Faravani	0.18	0.38	0.65	1.90
Sagamo	0.10	0.60	0.66	1.98

Figure 16.1 illustrates differential curve constructed by potentiometric titration data of humin acids isolated from the paravani Lake water. If we take in consideration that humin acids are mainly represented by alitatic and aromatic carboxylic and phenolic hydroxyl groups it is clear that the data corresponding to the first group on the differential curve must be placed within pH=4–7 and those of the second group above pH=7.

One peak is well pronounced on the differential curve of the direct potentiometric titration. According to the literature it must be conditioned by titration of carboxylic and partially phenolic hydroxyl groups because the fixation of the latter by separate peak is not possible. If we take into consideration that only phenol hydroxyl groups, for example, pKa (phenol)=10.0 pka (2,4-dinitrophenol) = 4.08 dislocated near electron acceptor groups (e.g., nitro groups) have rather high acidity then we can prove that the given peak might be conditioned only by COOH groups titration. The validity of this statement is testified by the data obtained by back titration (Figure 16.1). Two peaks are clearly observed on differential curve, which

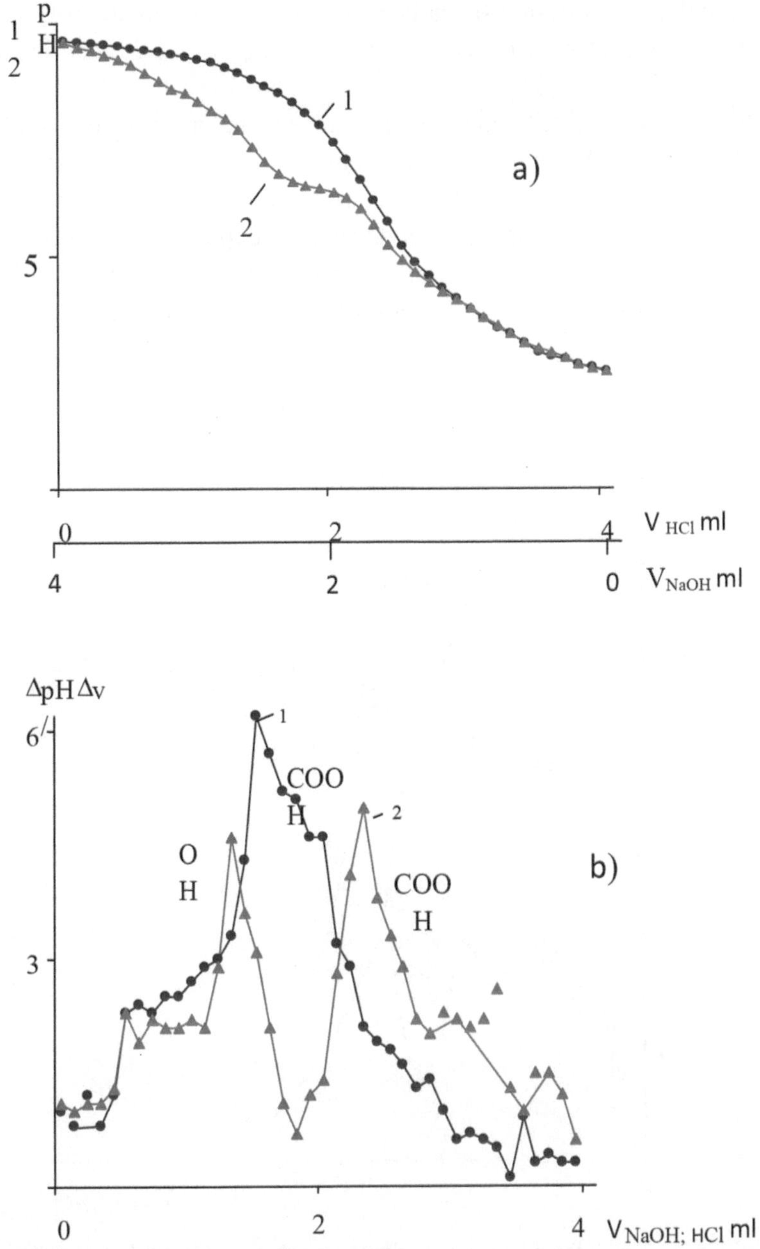

FIGURE 16.1 The integral (a) and differential (b) curves of potentiometric titration of humin acids isolated from the river Mtkvari (1 – the direct titration, 2 – back titration; V_1=25 mL; C_{NaOH}=0.095 M; C_{HCl}=0.106 M; μ=0.1 (KCl)).

correspond to the titration of carboxylic and phenolic hydroxyl groups. The similar character has differential curves in case of fulvic acids isolated from water (Figure 16.2):

This makes possible to regard humin acids in the given regions of pH as monofunctional substances and for calculation of dissociation constant to use the Henderson-Haselbach method. The link between pH solution, dissociation rate and acid dissociation constant is expressed by the equation pH=pKα–lg[(1–α)/α]. α = Ci/Cmax, where Ci-the concentration of base, mg.eqv, which was spent on titration of fulvic acids by the base before

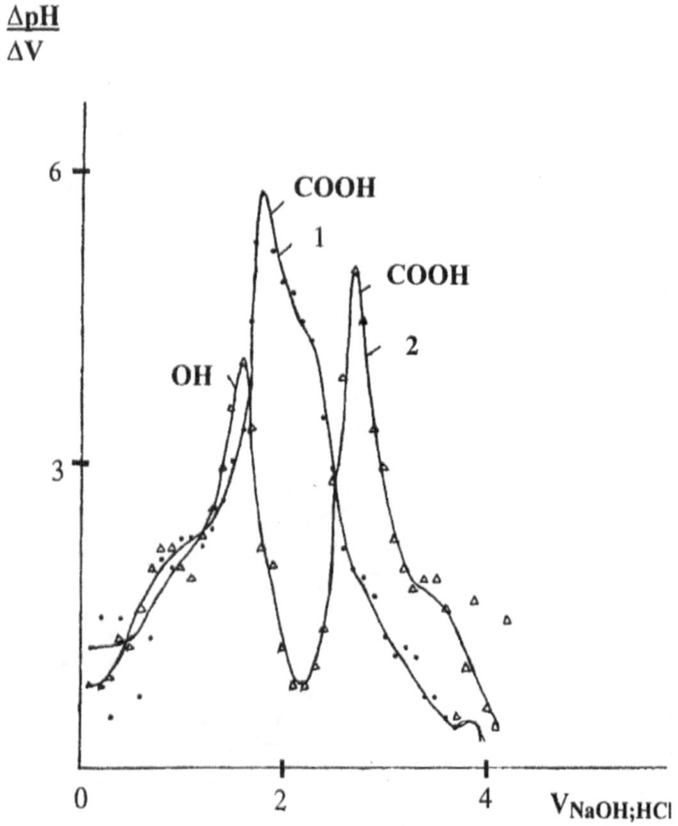

FIGURE 16.2 The differential curve of potentiometric titration of fulvic acids isolated from the river Mtkvari (1 – the direct titration, 2– back titration; V_1=25 mL; C_{NaOH}=0.095 M; C_{HCl}=0.106 M; μ=0.1 (KCl)).

given pH, and Cmax-the quantity of base (mg.eqv) which was spent on titration of fulvic acids, on potentiometric curve, before the pH of the corresponding the maximum of the peak. Each concrete value of pH corresponds to the determined dissociation rate of the given functional group, which can be calculated according to the additional reagent amount. When $\lg[(1-\alpha)/\alpha] = 0$, pH= pKα.

The data to calculate the dissociation constants of carboxylic and phenolic-OH groups of fulvic and humin acids are given in the Tables 16.2–16.5.

As the results show in the case of humin acids pK (COOH)=4.12 (Table. 16.2); pK (phen.-OH)=10.52 (Table 16.3). As the results show in the case of fulvic acids pK (COOH)=4.19 (Table 16.4); pK (phen.-OH)=10.46 (Table 16.5).

TABLE 16.2 The Experimental Data to Calculate the Meaning of Dissociation Constant of COOH Groups of Humin Acids Isolated From the River Mtkvari

pH	add. NaOH, mg.ekv $\times 10^{-3}$	α	(1–α)/α	lg[(1–α)/α]
3.12	9.5	0.062	15.13	1.1798
3.20	19.0	0.125	7.00	0.8451
3.32	28.5	0.187	4.35	0.6385
3.40	38.0	0.250	3.00	0.4771
3.52	47.5	0.312	2.20	0.3424
3.75	57.0	0.375	1.67	0.2227
3.99	66.5	0.437	1.29	0.1106
4.12	76.0	0.500	1.00	0.0000
4.37	85.5	0.562	0.78	–0.1079
4.62	95.0	0.625	0.60	–0.2218
4.89	104.5	0.687	0.46	–0.3372
5.18	114.0	0.750	0.33	–0.4815
5.48	123.5	0.812	0.23	–0.6383
5.81	133.0	0.875	0.14	–0.8539
6.24	142.5	0.938	0.07	–1.1549
6.86	152.0	1.000	0.0	–

TABLE 16.3 The Data to Calculate the Stability Constant of Phenolic-OH Groups of Humin Acids Isolated From the River Mtkvari

pH	add. HCl, mg.ekv ×10^{-3}	α	(1–α)/α	lg[(1–α)/α]
11.39	10.6	0.071	13.08	1.1166
11.29	21.2	0.143	5.99	0.7774
11.18	31.8	0.214	3.67	0.5647
11.07	42.4	0.286	2.50	0.3979
10.94	53.0	0.357	1.80	0.2553
10.71	63.6	0.429	1.33	0.1239
10.52	74.2	0.500	1.00	0.0000
10.30	84.8	0.571	0.75	–0.1249
10.19	95.4	0.643	0.55	–0.2596
9.98	106.0	0.714	0.40	–0.3979
9.76	116.6	0.786	0.27	–0.5686
9.55	127.2	0.857	0.17	–0.7696
9.26	137.8	0.929	0.08	–1.0969
8.80	148.4	1.000	0.00	–

TABLE 16.4 The data to Calculate the Dissociation Constant of COOH Groups of Fulvic Acids Isolated From the River Mtkvari

pH	add. NaOH, mg.ekv ×10^{-3}	α	(1–α)/α	lg[(1–α)/α]
3.10	9.5	0.056	16,86	1.2269
3.25	19.0	0.111	8,01	0.9036
3.30	28.5	0.167	4.99	0.6981
3.45	38.0	0.222	3.50	0.5441
3.58	47.5	0.278	2.60	0.4150
3.65	57.0	0.333	2,00	0.3010
3.99	66.5	0389	1.57	0.1959
4.19	85,5	0.500	1.00	0.0000
4.42	95.5	0.556	0.80	–0.0969
4.65	104,5	0.611	0.64	–0.1938
4.87	114.5	0.667	0.50	–0.3010

TABLE 16.4 Continued

pH	add. NaOH, mg.ekv ×10^{-3}	α	(1–α)/α	lg[(1–α)/α]
5.22	123,5	0.722	0.38	–0.4202
5.38	133,0	0.778	0.28	–0.5528
5.69	142,5	0.833	0.20	–0.6990
6.03	152,0	0.889	0.12	–0,9208
6.48	161,5	0944	0.06	–1,2218

TABLE 16.5 The Data to Calculate the Dissociation Constant of Phenolic-OH Groups of Fulvic Acids Isolated From the River Mtkvari

pH	add. HCl mg.ekv. ×10^{-3}	α	(1–α)/α	lg[(1–α)/α]
11.46	10.6	0.062	15.13	1.1798
11.38	21.2	0.125	7.00	0.8451
11.29	31.8	0.187	4.35	0.6385
11.16	42.4	0.250	3.00	0.4771
11.04	53.0	0.312	2.20	0.3424
10.88	63.6	0.375	1.67	0.2227
10.68	74.5	0.437	1.29	0.1106
10.46	84.8	0.500	1.00	0
10.24	95.4	0.562	0.78	–0.1079
10.04	106.0	0.625	0.60	–0.2218
9.85	116.6	0.687	0.46	–0.3372
9.62	127.2	0.750	0.33	–0.4815
9.36	138.8	0.812	0.23	–0.6383
9.06	148.4	0.875	0.14	–0.8539
8.70	159.0	0.937	0.17	–1.1546
8.30	169.6	1.000	0.00	–

16.4 CONCLUSIONS

The concentration of Fulvic acids changes from 0.08 mg/L to 4.23 mg/L. The concentration of humin acids changes from 0.00 mg/L to 0.60 mg/L in surface waters. Thus, we can make the conclusion, that fulvic and humin acids, isolated from the surface waters practically don't differ from each other by the acid–base properties. On pH, characteristic the surface water, Humic acid, which are dissolved in surface waters, are practically dissociated. As the results show in the case of humin acids pK (COOH)=4.12 (Table 16.2); pK (phen.-OH)=10.52 (Table 16.3). As the results show in the case of fulvic acids pK (COOH)=4.19 (Table 16.4); pK (phen.-OH)=10.46 (Table 16.5).

KEYWORDS

- **dissociation constant**
- **fulvic acids**
- **geopolymers**
- **humin acids**

REFERENCES

1. Ephreim, J. H., Boren, H., Arsenie, I., Pettersson, C., Allard, B. A combination of acid- base titrations and derivatization for functional group determinations of an aquatic fulvic acid. The Science of the Total Environment, 81/82, 615–624, 1989.
2. Leenheer, J. A., Wershaw, R. L., Reddy, M. M. Strong acid, carboxyl-group structures in fulvic acid from the Suwannee river. Georgia. 1. Minor structures. Environmental Science Technology, 29(2), 393–398, 1995.
3. Leenheer, J. A., Wershaw, R. L., Reddy, M. M. Strong acid, carboxyl-group structures in fulvic acid from the Suwannee river. Georgia. 2. Major structures. Environmental Science Technology, 29 (2), 399–405, 1995.
4. Ephraim, H., Pettersson, C., Norden, M., Allard, B. Potentiometric titrations of humic substances: do ionic strength effects depend on the molecular weight? Environmental Science Technology, 29(3), 622–628, 1995.
5. Bowles, E. C., Antweiler, R. C., MacCarthy, P. Acid-base titration and hydrolysis of fulvic acid from the Suwannee river. In: Humic substances in the Suwannee river, Georgia: U. S. Geological Survey Open-file Reports, 87–557 Denver, CO, 209–229, 1989.
6. Gamble, D. S. Potentiometric titration of fulvic acid: equivalence point calculations and acidic functional groups. Can. J. Chem., 50, 2680–2690, 1972.

7. Andres, J. M., Romero, C., Gavilan, J. M. Potentiometric titration of fulvic acids from lignite, in dimethylformamide and dimethilsulphoxide media. Talanta, 34(6), 583–585, 1987.
8. Mathuthu, A. S., Marinsky, J. A., Ephraim, J. H. Dissociation properties of Laurentide fulvic acid: identifying the predominant acidic sites. Talanta, 42(3), 441–447, 1995.
9. Sardessai Sugandha. Characteristics of humic and fulvic acids in Arabian Sea sediments. Indian Journal of Marine Sciences. 24, 119–127,1995
10. Lobartini, J. C., Tan, K. H., Asmussen, L. E., Leonard, R. A., Himmelsbach, D., Gingle, A. R. Chemical and spectral differences in humic matter from swamps, streams and soil in the southeastern United States. Geoderma, 49, 241–254, 1991.
11. Varshal, G. M., Kosheeva I. Y., Sirotkina, I. S., Velukhanova, T. K., Intskirveli, L. N., Zamokina, N. C. The study of organic substances of surface waters and their interaction with metals ions. Geokhimia, (Geochemistry, Russ.), 4, 598–607, 1979.
12. Revia, R., Makharadze, G. Cloud-point preconcentration of fulvic and humic acids. alanta, 48(1), 409–413, 1999.

CHAPTER 17

SIDE CHAINS AZOBENZENE MOIETIES IN POLYMETHACRYLATES FOR LIQUID CRYSTAL ALIGNMENT

V. TARASENKO, O. NADTOKA, and V. SYROMYATNIKOV

Taras Shevchenko National University of Kyiv, Volodymyrs'ka str., 64, 01033 Kyiv, Ukraine, E-mail: oksananadtoka@ukr.net

CONTENTS

ABSTRACT

Polymers based on the polymethacrylic acid with azobenzene-containing side-chains have been obtained and investigation of their orientation ability has been carried out. For this purpose the method of azopolymer synthesis by polymer analogues reaction of carboxylic groups of polymethacrylic acid with hydroxyazobenzenes has been proposed. In addition to functional carboxylic and photosensitive azobenzene groups the alkyl moieties of different length were involved into polymeric side chain. It was shown that "azobenzene-containing" polymers have ability to produce liquid crystal photoalignment. Exposed to actinic UV light photosensitive azobenzene fragments undergo orientation due to *trans-cis* isomerization. Liquid crystal (LC) covered on such polymer film can be oriented.

17.1 INTRODUCTION

In the field of up-to-date materials the creation light-controlled high-sensitivity polymeric systems is the most topical question to date. Amorphous polymer materials with side-chain azochromophore groups are attractive in this respect due to some their properties. So azobenzene fragments are able to involve the photoinduced dipole orientation in polymeric matrices, which can be used for creation reversible optical storage materials [1]. In addition azobenzene derivatives are dipole nonlinear optical chromophores, which are used for electro optical polymer materials synthesis indispensable to wide-band light modulator and integral optoelectronic systems based on it [1–4].

Efficiency and stability photoinduced processes in polymers depends on both chemical nature of system chromophore-polymer and the way of including chromophore in polymer. It is known that azochromophores can be put into polymeric matrices by the polycondensation or radical (co) polymerization ways as well as by polymer analogues reaction of functional groups with hydroxyazobenzenes.

Radical (co)polymerization of methacrylic azomonomers is widely applicable method of incorporating of chromophore group in side chain of macromolecule [5–9]. Nevertheless this method has some essential disadvantages due to the necessity of utilizing of hydrogen chloride and difficulties of chromophore purification from ionic impurities, which lead to film conductivity and dipole polarization efficiency reduction in electric field. In addition polymerization of methacrylic azomonomers doesn't lead to the high-molecular products as a result of low azomonomer activity, where azogroups play the role of free radical "traps" during polymerization.

To overcome these difficulties, we have used the method of polymer analogues reaction for obtaining light-sensitive polymers based on polymethacrylic acid [10]. So condensation of carboxylic groups with hydroxyazobenzenes leads to transformation part of them into photochromic moieties. As a result amorphous polymers containing photosensitive azobenzene fragments and functional carboxylic groups were synthesized. These materials attract the special interest due to their dual properties. The presence of the azofragments capable of forming ordering in combination with functional chemically responsive groups allows to substantially enhancing the ability of new photosensitive materials creation. Functional carboxylic groups can not only to participate in the chemical reactions but also to form noncovalent binding with different low-molecular dopants such as chiral substances, dyes, biological active fragments and liquid crystals (LC) molecules in particular. Such "binding" may be realized as due to ionic interaction of corresponding complementary pairs, and due to formation of hydrogen bonds between interacted groups.

In our paper the synthesis and the investigation of LC alignment ability of multifunctional azopolymers are presented. Illumination of these materials with exciting light stimulates reversible or irreversible *trans-cis* isomerization of azochromophores. When the exciting light is polarized, initially isotropic orientation distribution of photosensitive fragments transforms into anisotropic one characterized by some degree of orientational order [11]. The LC molecules, adjacent to the orientationally ordered photoaligning film, reproduce to certain extend this order due to anisotropic interaction with the molecules or anisotropic fragments from the surface of this film.

In addition to functional carboxylic and photosensitive azobenzene groups the alkyl moieties of different length were introduced into polymeric side chain to improve of liquid crystal molecules mobility.

17.2 EXPERIMENTAL PART

17.2.1 MATERIALS

All starting chemicals were purchased from commercial sources and used without further purification unless otherwise noted. The structures of all the precursors and final products were confirmed by ^{1}H NMR spectroscopy. The ^{1}H NMR spectra were measured on samples dissolved in $CDCl_3$. Obtained results are in a good agreement with proposed structure.

17.2.2 HYDROXY-AZOBENZENE SYNTHESIS (STAGE 1)

4-*methyl-4'-hydroxyazobenzene (Azo1).* 4-methylaniline 6.8 g (0.05 mol) was dissolved in 100 mL hydrochloric acid (2M). Solution was cooled by ice bath up to 0°C and solution of 3.3 g (0.05 mol) sodium nitrate in minimal volume of water was added slowly. To the cooled solution of phenol 4.4 g (0.05 mol) in water the diazonium salt was added under stirring. Yellow azodye precipitate was filtered, washed by cool water and dried. Substance purification was made by recrystallization from isopropanol.

Yield 87 %. T_g = 155°C. Rf = 0.7 (eluent–ethyl acetate, toluene = 1:1). NMR ^{1}H (400 MHz, DMSO-d_6), ppm: 9.92 (s, 1H, OH), 7.71 (m, 4H, Ar), 7.27 (d, 2H, Ar), 6.86 (d, 2H, Ar), 2.42 (s, 3H, CH_3).

17.2.3 COPOLYMERS SYNTHESIS (STAGE 2)

Methacrylic acid and alkyl methacrylate (methylmethacrylate (MMA), buthylmethacrylate (BMA), hexylmethacrylate (HMA), octylmethacrylate (OMA), decylmethacrylate (DMA)) at different molar ratios (2:1, 1:1, 1:2) were dissolved in dimethylformamide (10 ml) to obtain total monomers concentration 6 mol/L. Thermo induced radical polymerization was carried out at 70°C in sealed ampoules about 2 hours. 2–2'-azobisisobutyronitrile (AIBN) was used as free radical initiator. After reaction the contents of the ampules were transferred into cooled methanol, and copolymers were isolated as a precipitate. The product was dried to constant weight in a vacuum drying oven.

The yields of the polymers were determined by gravimetry. After titrimetric analysis samples with similar molar ratios of functional groups were chosen for next synthetic stage.

17.2.4 AZOPOLYMERS SYNTHESIS (STAGE 3)

About 0.5 g (0.014 mol) of copolymer methylmethacrylate-co-methacrylic acid (MMA-co-MA) (at molar ratio 1:3) and 8.9 g (0.042 mol) of 4-methyl-4'-hydroxyazobenzene (equimolar amount to carboxylic groups) were dissolved in 5 mL DMF. 0,3 g (0.014 mol) dicyclohexylcarbodimide as catalyst and 0.06 g 4-dimethylaminopyridine (20 wt. % from catalyst) were added to the resulting solution. Reaction mixture was stirred for 5 hours and was left at overnight at room temperature. Then it was filtered.

Oxalic acid trace was added for removing of catalyst remainder and solution was left overnight once more. The solution was iteratively filtered and the filtrate was precipitated in alcohol. Polymers were purified by reprecipitation from DMF in alcohol. So azopolymers containing 4-methyl-4'-oxyazobenzene based on MMA-co-MA (P1–0), BMA-co-MA (P1–3), HMA-co-MA (P1–5), OMA-co-MA (P1–7) and DMA-co-MA (P1–9) were obtained. All of them are dissolved in acetone, dioxane, toluene, dichloroethane. The copolymers compositions were determined from elemental analysis data for nitrogen and ^{1}H NMR spectra data.

17.2.5 TECHNICS

Films preparation and samples irradiation were carried out in the following way. Selected polymers were dissolved in dichloroethane (3 wt. %) and the solution was spin-coated on quartz substrates. Polymer films were annealed at 70°C for an hour and left overnight at room temperature for solvent residue removing. Film thickness (d) was measured by profilometer. They varied from 300 to 500 nm. Absorption spectra in the films were measured in the spectral range of 250–600 nm by the spectrometer from Ocean Optics (USA).

The anisotropy in the polymer films was induced by UV light with λ_{exp} = 365 nm and I = 4.5 mW/cm^2. The light was linearly polarized by Glan-Thompson prism. Thus-obtained polarized monochromatic light beam was directed normally to polymer film so that the polarization of exciting light was parallel to the film axis *x* (Figure 17.1).

For testing of LC alignment we used symmetric LC cells made of two equally treated photoaligning layers withy parallel easy axis. The cell thickness was defined by spacers of 20 μm in diameter. The cells were filled with nematic LC E7 purchased from Merck. LC alignment quality

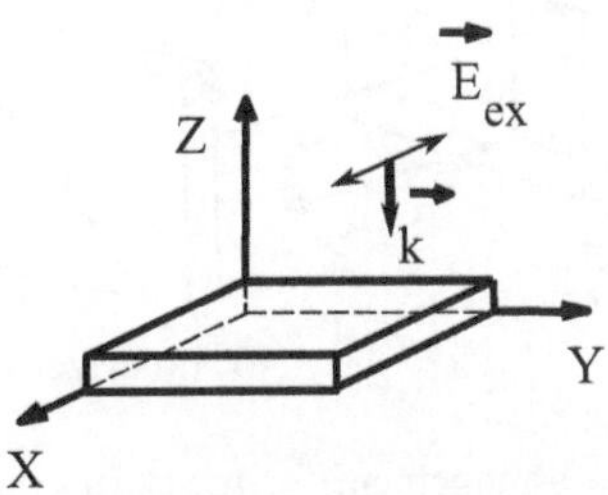

FIGURE 17.1 Sample irradiation geometry.

in the cells was observed with the naked eye and polarizing microscope. It was evaluated by the five-point scale: “excellent” (e), “good” (g) (singular orientation defects), “satisfactory” (s) (traces of defects), “bad” (b) (a lot of defects) and “no alignment” (n/a).

For *pretilt angle measurements* the substrates with covered polymers were irradiated in two steps. At the beginning linearly polarized light (I_{exp} = 4.5 mW/cm^2) was directed normally to polymer film for 15 minutes and anisotropy in film was detected as result. Thereafter these films were exposed to a nonpolarized exciting light at 45° to film surface for a minute. Obtained in this manner two equally treated photoaligning layers of each polymer were used for LC cells preparing.

The pretilt angle of a liquid crystal in the cell was measured by the commonly used crystal rotation method between the crossed polarizers [12] (Figure 17.2). A LC cell was adjusted between the crossed polarizers so that director *d* made an angle 45° to both their axes. The cell was rotated around the axis OY, perpendicular to the director. The dependence of the system transparency $T(\varphi)$ for a weak He-Ne laser beam on the angle φ between the beam and the cell normal was measured. In the case of a uniform director orientation through over a cell the value of a pretilt angle θ may be estimated by the expression:

$$\theta \approx \frac{\Delta\varphi}{(n^{o} + n^{e})} \qquad (1)$$

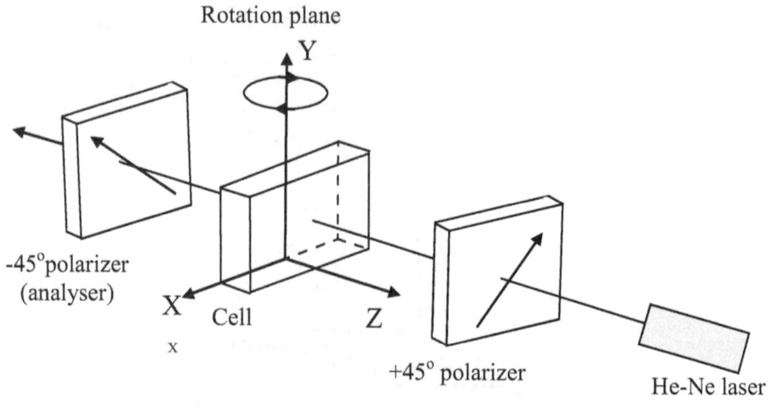

FIGURE 17.2 Experimental arrangement for measuring pretilt angles by cell rotation method.

Here $\Delta\varphi$ is the shift of the symmetry axis of the curve $T(\varphi)$ about the point $T(\varphi = 0)$, n^o and n^e are the refractive indexes of ordinary and extraordinary waves respectively, the angle θ is determined as the angle between the aligning surface and the direction of d near it.

17.3 RESULT AND DISCUSSIONS

The attempts to copolymerize methacrylic acid with methacrylic azomonomers results in the formation of polymethacrylic acid (PMMA) with isolated azobenzene moieties. Apparently, at high temperature required for radical copolymerization and at presence of active functional carboxylic groups in reaction mixture the acidolysis of ester groups occurs. As a result, the blend of polymethacrylic acid and separate covalent unbound azodye molecules was obtained. Therefore we tried to use the post-coupling reaction for incorporation azobenzene moieties in polymethacrylates. As polymethacrylate base were used copolymers of methacrylic acid with alkylmethacrylates containing alkyl substituents of different size. Synthesized azopolymers containing functional carboxylic and photosensitive azobenzene groups as well as the alkyl moieties is shown below:

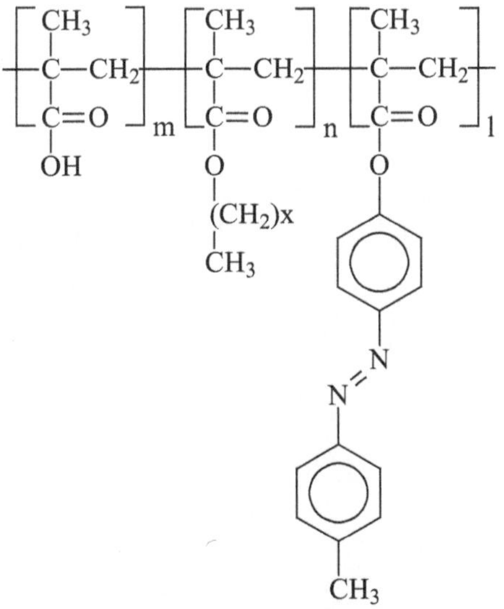

where x = 0 (P1-0), 3 (P1-3), 5 (P1-5), 7 (P1-7), 9 (P1-9)

Azopolymers synthesis by post-coupling reaction of alkylmethacrylate-co-methacrylic acid with azobenzene derivative was carried out in three steps.

On the *first stage* of synthesis (Figure 17.3) model 4-hydroxyazobenzene containing electrodonor methyl substituent was obtained. Azo-compound was synthesized according to the classical scheme of diazotation of aniline derivatives (4-methylaniline) and their further azo-coupling with phenol [6]. Purity of corresponding hydroxy-azobenzene

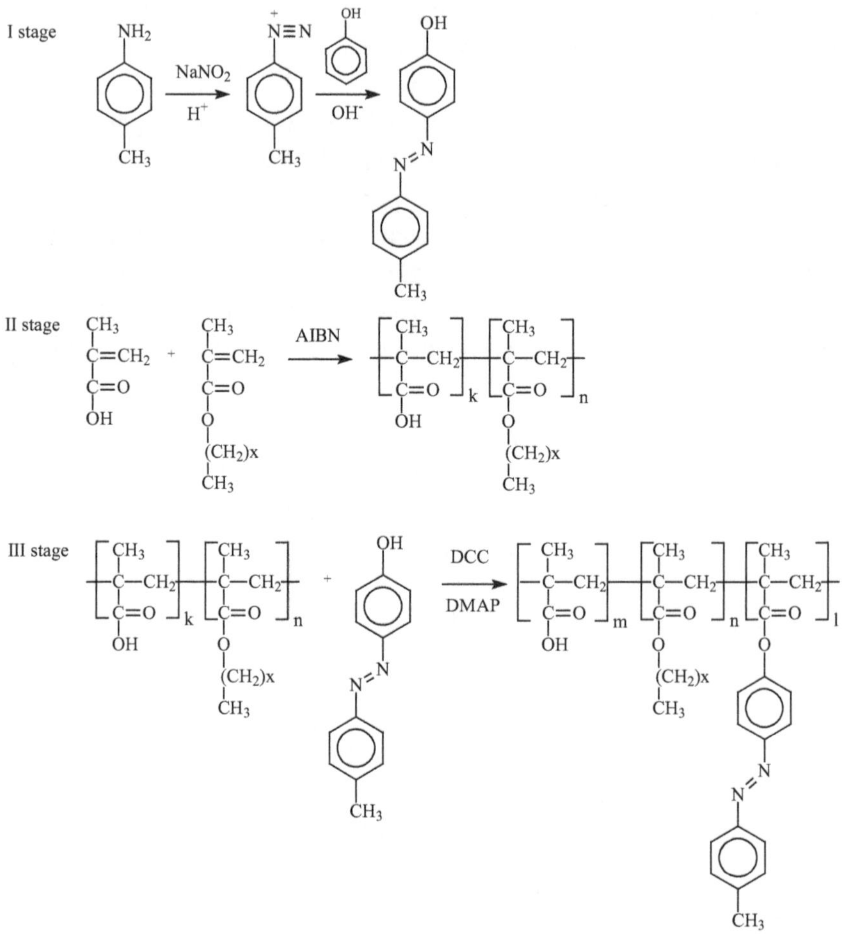

FIGURE 17.3 Polymer synthesis scheme.

derivative was controlled by thin-layer chromatography. The substance was identified accordingly to NMR-spectroscopy data. Synthesized material is well-solved in majority of organic solvents; they possess yellow color as well as stability to light and air.

On the *second stage* the radical copolymerization of methacrylic acid with alkyl methacrylate at different ratios was carried out. After titrimetric analysis samples with similar molar ratios of functional groups were chosen for next synthetic stage.

On the *third stage* the condensation of hydroxyazocompound with polymethacrylic copolymers was carried out. This reaction was occurred at dicyclohexylcarbodimide as catalyst and 4-dimethylaminopyridine as cocatalyst in dry DMF at room temperature.

Scheme of synthesis of these stages is represented on Figure 17.3 and synthetic details are described in experimental part.

Concentrations of azofragments covalently bonded to carboxylic groups of methacrylic acid chains were determined from NMR-spectroscopy data. For further LC alignment investigation it was used samples containing an equal side-chain proportion (Table 17.1).

In our study, first of all, we measured the UV/Vis absorption spectra of azopolypolymers containing azobenzene groups, one of them are represented on Figure 17.4. All investigated polymers strongly absorb in the visible region of the electronic spectrum. These UV/Vis spectra display high-intensity $\pi\pi^*$ bands in the UV (at about 360 nm) and low-intensity $n\pi^*$-bands in the visible region (at about 450 nm). Considering the absorption spectra of trans- and cis-azobenzene [13], the band at nearly 350 nm corresponds essentially to the absorption of the trans-isomers while the

TABLE 17.1 Azopolymers and Their Properties

polymer	x	Alkyl fragment % mol.	Azo fragment % mol.	λ_{max}, nm	Photo orientation quality	Pretilt angle Θ, °
P1–0	0	25	30	360	good	0.5–2
P1–3	3	25	30	360	g	1–1.5
P1–5	5	24	30	360	s	0.8–1.2
P1–7	7	24	30	360	s	0.7–1.3
P1–9	9	23	30	360	b	0.7–1.1°

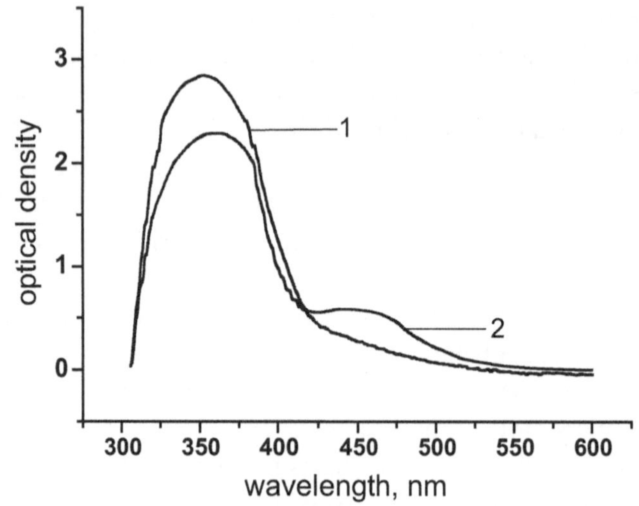

FIGURE 17.4 The absorption spectra of P1–0 polymer films before (1) and after (2) irradiation. The light parameters are λ_{exl} = 365 nm, I = 5 mW/cm^2, linear polarization along the x axis.

band at about 450 nm is mainly due to the absorption of the cis-isomers. The exact positions of the maximum of the $\pi\pi^*$ absorption bands, λmax, are presented in Table 17.1.

It should be noted, that polymers P1–3, P1–5, P1–7, P1–9 have identical spectra to P1–0 as far as all they have the same azochromophore in side chain. The presence of alkyl side-chain groups of different length in azopolymers practically has no effect on the maximum of the $\pi\pi^*$ absorption bands.

The anisotropy in the polymer films was obtained under UV polarized light due to reversible or irreversible *trans-cis* isomerization of azochromophores [11]. It was detected in crossed polarizers and appropriate spectra changes were recorded on the UV/Vis absorption spectra.

According to LC alignment tests, all obtained polymers show photoalignment effect (Figure 17.5). The azopolymer P1–0, having the smallest alkyl fragment in the side chain demonstrates the best LC alignment capability. Alignment characteristics of other polymers, having bigger alkyl groups are worse, that can be explained by steric effect (Table 17.1). Preferable direction of LC alignment determined by pretilt angle θ indicates tilted alignment.

FIGURE 17.5 Photoalignment of nematic liquid crystal E7 in the cells based on azopolymers P1–0, P1–3 tilted alignment. Cells are in crossed polarizers.

17.4 CONCLUSIONS

The polymethacrylates with alkyl-, carboxyl- and azobenzene-containing side-chains were synthesized by copolymerization and further polymer analogues reaction. It was shown, that their films has ability to orientation under polarized light. Oriented films have important property to orient liquid crystal (LC), covered them. It was shown that large alkyl group in side chain makes steric difficulties and don't improve LC alignment quality. It should be noted that azopolymer having photosensitive methylazobenzene groups, functional carboxylic groups and the methyl moieties possess good photoaligning properties. This peculiarity can be used for creation of photosensitive layers for LC displays and other devices.

KEYWORDS

- **azopolymer**
- **liquid crystal**
- **photoalignment**

REFERENCES

1. Natansohn, A., Rochon, P. Photoinduced Motions in Azo-Containing Polymers. Chem. Rev., 102 (11), 4139–4145, 2002.
2. Burland, D. M., Miller, R. D., Waish, C. A. Second-order nonlinearity in poled-polymer systems. Chem. Rev., 94 (1), 31–40, 1994.
3. Shibaev, V. P., et al. Polymers as Electrooptical and Photooptical Active Media – Springer: New York, 1996.
4. Savchenko, I., Davidenko, N., Davidenko, I., Popenaka, A., Syromyatnikov, V. Syntesis and electrooptical properties of metal-containing azopolymers. Mol. Cryst. Liq. Cryst., 467, 203–213, 2007.
5. Xie, S., Natansohn, A., Rochon, P. Recent developments in aromatic azo polymers research. Chem. Mater., 5, 403–412, 1993.
6. Freiberg, S., Labarthet, F., Rochon, P., Natansohn, A. Synthesis and characterization of a series of azobenzene-containing side-chain liquid crystalline polymers. Macromolecules, 36, 2680–2688, 2003.
7. Nadtoka, O., Syromyatnikov, V., Olkhovik, L. New photochromic polymers based on methacrylic azoesters. Mol. Cryst. Liq. Cryst., 427, 259–262, 2005.
8. Cojoariu, C., Rochon, P. Light-induced motions in azobenzene-containing polymers. Pure Appl. Chem., 76. (7–8), 1479–1484, 2004.
9. Meng, X., Natansohn, A., Barrett, C., Rochon, P. Azo Polymers for Reversible Optical Storage. 10. Cooperative Motion of Polar Side Groups in Amorphous Polymers. Macromolecules. 29. (3), 946–954, 1996.
10. Vretik, L., Syromyatnikov, V., Zagniy, V., Paskal' L., Yaroshchuk, O., Dolgov, L., Kyrychenko, V., Lee, C.-D. Polymethacryloylarylmethacrylates: New concept of photoalignment materials for liquid crystals. Mol. Cryst. Liq. Cryst., 479, 121. 2007
11. Nadtoka, O. N., Yaroshchuk, O. V., Bednaya, T. V., Ol'khovik, L. A., Syromyatnikov, V. G. Photoinduced Orientational Ordering in the Series of Methacrylic Azopolymers. Polymer Science. 52(3), 261–271, 2010.
12. Cuminal, M.-P., Brunet, M. A technique for measurement of pretilt angles arising from alignment layers. Liq. Cryst., 22 (2), 185–192, 1997.
13. Sekkat, Z., Wood, J., Knoll, W. Reorientation mechanism of azobenzenes within trans-cis photoisomerization. J. Phys., Chem., 99, 17226–17234, 1995.

PART 2:

ENGINEERED-BASED COMPOSITES AND MODELS

CHAPTER 18

PREPARATION OF NANOPOLYANILINE AND ITS POLYMER-POLYMER NANOCOMPOSITIONS WITH HIGH AND STABLE ELECTRIC CONDUCTIVITY

B. A. MAMEDOV, A. YA. VALIPOUR, S. S. MASHAEVA, and A. M. GULIYEV

Institute of Polymer Materials of Azerbaijan National Academy of Sciences, Sumgait, S. Vurgun Str.124, Azerbaijan,
E-mail: ipoma@science.az

CONTENTS

ABSTRACT

By the oxidative polycondensation of 4-N-methylamine phenol the polyfunctional polyconjugated oligo-4-N-methylamine phenols have been synthesized. Polyaniline in nanosizes and its nanocompositions with new matrix polymers have been prepared. It has been shown that they possess solubility and meltability and also high and stable electric conductivity.

18.1 INTRODUCTION

In recent years a use of polyfunctional reactive high-molecular compounds as the active additives to the industrially-produced polymers for improvement of their operational indices, and also for giving to them a complex of necessary and useful properties is of great interest. These compounds are polyfunctional aromatic polyconjugated homo- and co-oligomers. They show semiconductivity, paramagnetism, thermal stability, catalytic activity and in most cases possess solubility and meltability. As a result such high-molecular compounds and their compositions with fibers, thermoplasts, resins and elastomers are widely used in creation of sensors, transformers and "clever materials" of various purpose and also in textile industry [1–3]. Purposeful change and control of the electrical properties of such materials while maintaining the solubility and meltability is an actual and important problem [3–5].

18.2 EXPERIMENTAL PART

During carrying out of the experiments were used: 4-N-methylamine phenol aniline, methanol, ethanol ("rectificate"), 30% aqueous solution H_2O_2 and NaOCl, ammonium persulfate, maleic anhydride (German company MERCK) and styrene (Tabriz petrocompany), isopropylamine (German company MERCK), polyethylene glycol with Mn−500 (company ACROSS USA). Carried out: oxidative polycondensation of 4-N-methylamine phenol in the presence of oxidizers O_2+OH^-, H_2O_2 and NaOCl, aniline in the presence of ammonium persulfate in aqueous dispersion under the impact of ultrasound waves and template oxidative polycondensation of aniline in the presence of matrix polymers on methodologies cited in works [6–8]. For study of composition, structure and properties of the synthesized high-molecular compounds and polymer compositions the methods of elemental, chemical and spectral (IR-, UV-, EPR- and PMR-) analyzes, XRD and SEM have been used. The values of electrical conductivity of these substances were measured by a method of four points. The corresponding measurements were carried out in devices pH meter-Behineh model 2000, IQ spectrometer-Tbermo Nicolet Nexus–670, UB-spectrometer-T80PG Instruments Ltd., ^{1}HNMR-Spectrospin Advance (300 MHZ) and Scanning Electron Microscopic Bruker Company.

18.3 RESULTS AND DISCUSSION

Usually the polymers on nature are good isolators. A number of polymers possessing relatively high own electric conductivity is comparatively small. They are mainly high-molecular compounds, macromolecules of which include developed system of polyconjugated bonds, that is, polyacetylene, polyphenylenes, polymers of aniline, pyrrol, thiphene and their derivatives. However, these polymers, in its turn, have the serious lacks: practical absence of solubility in usual solvents and melting temperatures, very low adhesion, deformation and strength indices. In addition, the values of the electric conductivity of such polymers are changed depending on conditions of their synthesis in the very wide range and usually are not high. For elimination of such lacks the numerous investigations were carried out. For improvement of solubility and meltability and also adhesion and strength properties the functional derivatives of above-mentioned polymers have been synthesized. An essential increase of the electric conductivity, that is, an increase of quantity of solitons, polarons and bipolarons in their composition has been reached by the methods of thermal and chemical influence and in wider scale alloying. The alloying can be realized by various ways and with use of compounds of various natures. We have synthesized new representatives of polyfunctional polyconjugated polymers synchronically possessing semiconducting properties, solubility and meltability by oxidative of polycondensation of 4-N-methylamine phenol in the presence of various oxidizers (NaOCl, H_2O_2 and O_2+OH^-). It has been established that in this case oligo-4-N-methylamine phenols which consist of 1-hydroxy-4-N-methylamine-2,6-phenylene links are formed. Oligo-4-N-methylamine phenols in influence of various oxidizers (for ex. O_2+OH^-) and at high temperature (523÷–673K) in atmosphere O_2 form the stable microradicals of aroxyl type. Both the initial oligomers and their macroradicals show paramagnetic and semiconducting properties. By change of conditions of synthesis one can purposefully vary the concentrations of paramagnetic centers (PMC) in composition of these compounds. The results of the investigations showed that the electric conductivity of the samples both in constant and in alternating electric fields is essentially (~3÷5 order) increased with concentration growth of PMC (~1÷2 order) in their composition. The carried out corresponding calculations by means of Pollack equation showed that the concentrations of paramagnetic centers in composition of these compounds determine a density of the localized states near the Fermi level on which a transfer of electric charges is realized. By an introduction of the determined quantities of these compounds into composition of thermoplasts, thermoelastoplasts, elastomers and resins the antistatic polymer composition materials, including rubbers with high heat-physical and physical-mechanical indices have been prepared. They show more high (~10–100 times) and stable electric conductivity in comparison with the known analogs.

However, the preparation of electro-conductive polymers in nanosizes, optimization and stabilization of their electric properties and also creation of their nanocompositions with matrix polymers are more interesting and tempting, as they have more wide possibilities for development of polymer materials with high and stable electric conductivity.

Polyaniline (PANI) in nanosizes with more high electric conductivity has been synthesized by oxidative polycondensation of aniline in the presence of ammonium persulfate (APS) in a medium of HCl under action of ultrasound waves. For revealing of optimal parameters of synthesis of PANI in nanosizes the process has been carried out in atmospheres of air and nitrogen, at various values of pH medium and ratios monomer: oxidizer (mol) and measurement of the electric conductivity of each synthesized sample by a method of four points. It has been established that the samples of PANI prepared in a medium N_2 show more high electric conductivity than the samples synthesized on air. The electric conductivity of PANI prepared under action of ultrasound in the presence of N_2 for 4 h becomes higher. During using of various concentrations of hydrochloric acid the more high results are reached at 1.5 mol/L concentration of HCl. At the same time, it was known that a molar ratio of APS/monomer during preparation of various polymers are differed: for polyaniline, polypyrrole and poly-3,4-ethylenedioxythiophene a use of the values 1.5; 0.5 and 1.0, respectively is recommended. However, the results of the carried out investigations showed that for samples of PANI synthesized under action of ultrasound waves its optimal value is1.25 (mol). In this case, the electric conductivity of the prepared PANI is higher and reached to 23.5 S/cm. The structural changes occurring in the samples of PANI at alloying have been investigated by the methods of thermogravimetric analysis, IR- and UV-spectroscopy.

It becomes clear from XRD diagram (Figure 18.1) and figure SEM (Figure 18.2) of the synthesized samples of PANI that the particles of PANI have the sizes approximately 50–100 nm, that is, the samples of PANI synthesized under action of ultrasound waves consist of nanoparticles.

A comparison of value of the electric conductivity of the synthesized samples of PANI after treatment by various alloying compounds shows that a nature of the alloying agent also influences on value of the electric conductivity of polyaniline. As, the values of the electric conductivity of the samples of PANI prepared under action of ultrasound and treated with HCl and CSA are 23.5b and 16.0 S/cm, respectively

We have also carried out the template oxidative polycondensation of aniline in the presence of APS and matrix polymer under action of ultrasound waves in a medium of nitrogen. As a matrix polymers were used: poly(styrene-alt-maleic anhydride) (PSMA), products of its hydrolysis

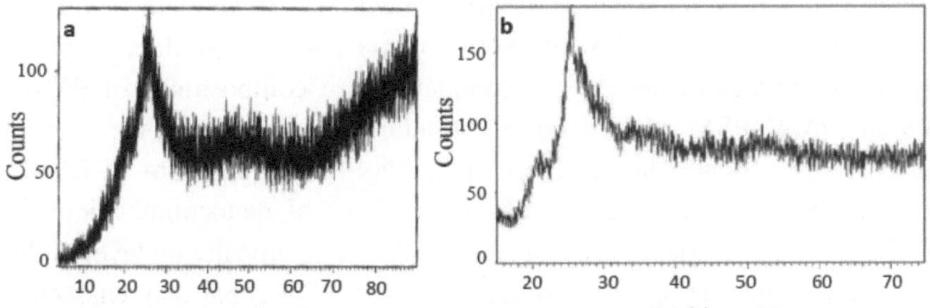

FIGURE 18.1 Diagrams XRD, synthesized samples of PANI: (a) with use of magnetic mixer, (b) under action of ultrasound.

a b

FIGURE 18.2 Figures SEM, synthesized samples of PANI (a) 50,000 KX and (b) 1000 KX).

(PSMT) in interaction with isopropylamine and polyethyleneglycol and also aniline sulfoacid (PSMA-SAT). As a result, a number of nanocompositions of polyaniline with high and stable electric conductivity have been developed.

The water-soluble polymers, which can be stable as nanosized particles, attracting surface of molecules of other monomers create a possibility for carrying out of their oxidative polycondensation. As a result, the newly obtained polymer particles are formed in nano-sizes. In addition, the polymer stabilizers acting by mechanism of phase stabilization can directly be used as a matrix polymer for prevention of accumulation of polymer nanoparticles. The compositions consisting of nanoparticles of spherical form in sizes about 100 nm have been prepared by the oxidative polycondensation in dispersions PSMT–PS and PSMT–SAT–PS in acidic aqueous medium. The electrical properties of these compositions have been studied for their thin films. The process has been carried out under action of ultrasound waves in a medium of nitrogen with use of APS as an oxidizer, HCl – alloying agent (37%) and PSMT – phase stabilizer, product of hydrolysis. The values of the electric conductivity of compositions of PANI–PS and PANI–PSMT–PS of various structures are changed in the interval of 0.17÷2.39 S/cm. The electric conductivity of compositions of PANI–PSMT–PS is higher. The electric conductivity of nanocompositions of PANI–PSMT (PEQ) and PANI–PSMT (IPA) are practically on level of the electric conductivity of compositions of PANI–PSMT, but they characterized by more high deformation and physical-mechanical indices. An introduction of PS into structure of these compositions leads to the decrease of their electric conductivity approximately in 20 times, however the prepared template of PANI–PSMT–PS is sufficiently homogeneous and elastic. A growth of quantity of PANI in structures of compositions and also a quantity of PSMT in a mixture of matrix polymers is accompanied by value growth of the electric conductivity of film. This has been connected with the fact that PANI is the basic electro-conductive component and acidic groups of PSMT, apparently, fulfill the role of the alloying agent.

The synthesized new matrix polymer of PSMA–SU includes in composition of macromolecule simultaneously carboxyl aniline sulfoacidic groups ($pk_a = 4$, $pk_a = 0.7$; and for aniline – $pk = 4.63$). In media of $pH < 4.5$ PSMA–SU becomes negatively charged and aniline positively charged. PSMA–SU fulfills the function of electrostatic template which creates a favorable condition for combination of molecules of aniline orderly through para-position and the polyconjugated macromolecules of linear structure are formed. A high concentration of cations of aniline near surface of template creates a favorable condition for proceeding of the oxidative polycondensation. As a result, it is formed PANI with more high molecular mass. Thus, the prepared nanocompositions of PANI possess considerably higher and stable electric conductivity than its known compositions made by other methods, for example, by a method of displacement. The nanocomposition of PANI–PSMA–SU prepared at ratio PANI: PSMA–SU= 0.06:1.0 (mol) is characterized by sufficiently high electric conductivity. A value of its specific volume electric conductivity is 0.41 S/cm and with content growth of PSMA–SU in comparison with PANI the electric conductivity of compositions is decreased (Figure 18.3). At the same time, the

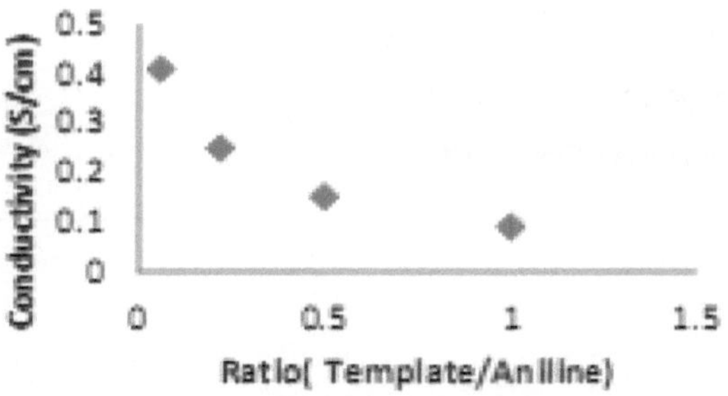

FIGURE 18.3 Influence of ratio template/aniline (mol) on electrical conductivity of nanocomposition.

electric conductivity of compositions of PANI prepared in the presence of various template polymers is noticeably differed.

The distribution and sizes of components in nanocomposition have been determined with use of methods XRD and SEM.

18.4 CONCLUSIONS

By the oxidative polycondensation of 4-N-methyamine phenol the oligo-4-N-methylamine phenols macromolecules of which consist of 1-hydroxy-4-N-methyamine-2,6-phenylene links have been synthesized. The oligo-4-N-methyamine phenols under the impact of various oxidizers and high temperature (523÷673K) in oxygen atmosphere form the stable macroradicals of aroxyl type.

Under the impact of ultrasound waves from aniline by template polycondensation of aniline in the presence of matrix polymer the polyaniline in nanosizes and polymer nanocompositions of polyaniline have been prepared.

It has been established that oligo-4-N-methylamine phenols, nanopolyaniline and its nanocompositions synchronically possess solubility and meltability and also high and stable electrical conductivity. The values of their specific volume electro-conductivity reach fields close to metallic conductivity.

KEYWORDS

- **nanopolyaniline**
- **oligo-4-N-methylamine phenols**
- **polymer nanocompositions**
- **template polycondensation**

REFERENCES

1. Hu, W. L., Chen, S. Y., Yang, Z. H. J. Phys. Chem. B, №115, 8453–8457, 2011.
2. Khastgir, D., Nikhil, K., Singha, J., Progress in Polymer Science, №34, 783–810, 2009.

3. Pud, A., Ogurtsov, N., Korzhenko, A., Shapovol, Q. Proq. Polym. Sci., vol.28, 1701–1758, 2003.
4. Valipour, A. Ya., Moghaddam, P. N., Mammedov, B. A. Archives Des Science Journal, Switzerland, Geneva. vol. 65 (7), 14–20, 2012.
5. Valipour, A. Ya, Moghaddam, P. N., Mammedov, B. A., Life Sci. J., vol.9 (4), 409–421, 2012.
6. Valipour, A. Ya., Moghaddam, P. N., Mamedov, B. A. et al. Azer. Chem. J., № 1, 39–43, 2011.
7. Valipour, A. Ya., Moghaddam, P. N., Mamedov, B. A., Processes of Petro-Chemistry and Oil-Refining, vol.12, № 1(45), 20–32, 2011.
8. Valipour, A. Ya., Moghaddam, P. N., Mammedov, B. A. Jokull Journal Reykjavik, Iceland, issue July, 211–251, 2013.

CHAPTER 19

SYNTHESIS OF BENTONITE AND DIATOMITE-CONTAINING POLYMER NANOCOMPOSITES AND THEIR CHARACTERISTICS

A. O. TONOYAN, D. S. DAVTYAN, A. Z. VARDERESYAN, M. G. HAMAMCHYAN, and S. P. DAVTYAN

State Engineering University of Armenia 105 Teryana Str., Yerevan, 0009, Armenia, E-mail: davtyans@seua.am

CONTENTS

ABSTRACT

Frontal polymerization of acrylamide filled by bentonite, diatomite and fine-dispersed chalk was investigated. Steady state propagation of heat waves was determined both for descending and ascending polymerization waves (from top to bottom as well as from bottom to top). Front velocity dependence on the initiator concentration (benzoic peroxide, azoisobutyric acid dinitrile) and filling degree was studied.

19.1 INTRODUCTION

There are many papers [1–6] devoted to the research of influence of various kinetic factors on the velocity of polymerization front propagation [1–10], conversion [1, 4, 6, 11, 12], molecular mass characteristics [13–16] of generated polymers. However, the majority of the specified research papers refer to the frontal polymerization of liquid monomers [1, 4–8] in relation to which the gravitational convective mass transfer [8, 14, 17] as well as natural convection [18] can be observed when the reaction front propagates vertically from top to bottom or vice versa (descending and ascending). It should be noted that there is extremely small quantity of analogous research for the crystal monomers [9,19] despite the fact that this kind of monomers is of great interest both scientific and practical point of view. Earlier, we studied and explained some features of the frontal polymerization of solid-state crystal monomers by the example of acrylamide (AAm) and acrylamide complexes with transition metal nitrates [9]. Specifically, it was shown that the difference of velocities of descending and ascending thermal waves upon the AAm frontal polymerization [19] is associated with the gravitational convective mass transfer of melted polymer from the reaction region to the monomer medium. It should be noted that in much of publications there is no information relative to the research in the area of frontal polymerization of filled systems based on the solid-state crystal monomers. The purpose of the presented paper is to fill this gap to the extent possible.

19.2 EXPERIMENTAL PART

Acrylamide and initiating agents of radical polymerization (benzoyl peroxide – BP and azoisobutyric acid dinitrile – AIBN) of Sigma-Aldrich brand were used without preliminary cleaning.

Polymerization initiating agents in certain proportions were introduced into acrylamide (AAm) from solutions in acetone, and further sample were dried in vacuum cabinet at the room temperature till they obtained constant weight. Dried AAm with initiating agent in certain proportions was mixed with particulate filler until homogeneous mass was obtained. Then obtained mass was introduced into the cylindrical glass ampoules (with inner diameter of 5.0 mm) in small portions with their sequential compaction. Packing density of AAm filled mixtures was determined gravimetrically calculating the sample volume according to the height of its column in the ampoule with known diameter. Frontal polymerization of AAm filled mixtures was studied in the vertically established glass ampoules depending on various parameters. Methods of frontal polymerization realisation are described in the paper [9] in details. Bentonite, diatomite and in certain cases fine-dispersed chalk were used as filling agents for the AAm frontal polymerization. Average dimensions of these filling agents (bentonite, diatomite and chalk) were ~5 μm. Herewith, bentonite has lamellar structure and according to different estimations [20–22] the thickness of individual layers is 1 to 2 nm. Also it should be mentioned that the microporosity of diatomite is 90–92% of its total surface area and it has density of 0.27–0.3 $g\cdot cm^2$.

19.3 INFLUENCE OF AAM FILLED MIXTURES DENSITY ON THE FRONTAL POLYMERIZATION VELOSITY

Under the conditions of frontal polymerization of crystal monomers (especially for the filled systems) the formation of stationary modes of thermal polymerization waves depends quite significantly on the density of initial reaction mixture [9]. Indeed, as is seen from Figure 19.1, for AAm mixtures with bentonite (initiating agent – BP with quantity of AAm 0.5% wt) with low packing densities of reaction mixture the degeneration

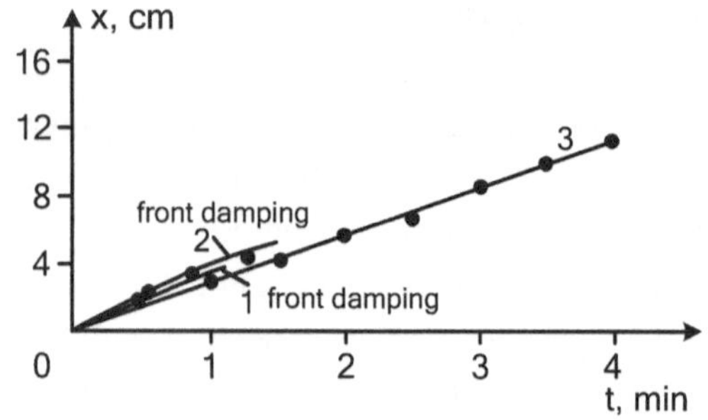

FIGURE 19.1 Dependence of polymerization front coordinate for the descending waves on time for AAm-bentonite nanocomposites (filler content of 30% wt) with various packing densities (in g/cm^3): 0.76 (1); 0.90 (2); 1.10 (3). Initiating agent – BP (0.5% wt of AAm).

and attenuation of the frontal modes (curves 1 and 2) are observed both for ascending and descending thermal waves.

Therefore, determination of the range where the steady frontal modes exist depending on the degree of filling and density of reaction mixture is of practical interest. Influence of the reactive medium packing density on the mass velocity of polymerization front propagation depending on filler amount is shown in Figure 19.2.

Three regions (I, II, III) of the frontal polymerization thermal modes stand out on curves in Figure 2a, 2b. In region I which corresponds to the packing densities of (0.7–0.85) g·cm^{-3} of initial reactive medium the formation of frontal modes is not observed. In this case application of high

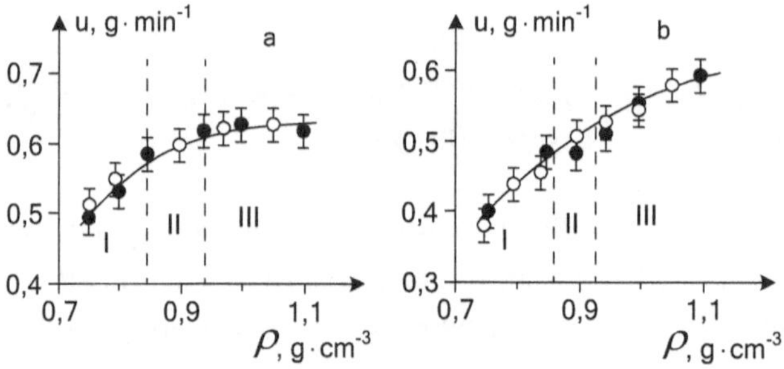

FIGURE 19.2 Dependence of mass velocity of frontal polymerization of AAm-bentonite mixtures on packing density upon the front propagation from top to bottom (●) and vice versa (○). Content of filler – 20 (a) and 30 (b) % wt. Polymerization initiating agent – BP (0.5% wt of AAm).

temperatures (~200°C – initiation) to the reactive ampoule edge causes the formation of polymerization frontal mode in adjacent layers and the degeneration of thermal polymerization wave occurs at the distance of ~1–4 cm. Most probably, the attenuation of frontal modes is caused [9, 19] by the gravitational convective mass transfer of the polymer generated in reaction zone to monomer medium.

When we increase the packing density of initial reactive medium (region II) up to $0.85 \leq \rho \approx 0.92$ g·cm^{-3}, the formation of unstable frontal polymerization modes is observed with occurrence of oscillatory and spin modes of thermal polymerization waves. Because unstable frontal polymerization modes were studied in details in the paper [9] we do not represent it here. Formation and propagation of the stable thermal modes of frontal polymerization is accomplished in region III which corresponds to the packing densities of initial reactive medium $\rho \geq 0.92$ g·cm^{-3}.

It should be noted that similar results were obtained when studying the AAm frontal polymerization with diatomite additives in different quantities. Also, the boundaries, which separate the regions of degeneration, unstable and stable stationary frontal modes are quite close to the results shown in Figure 19.2; therefore, this data is not specified here.

Further study of regularities of the frontal polymerization of Aam with various filling agents was carried out within the range of densities $\rho \geq 0.92$ g·cm^{-3} which provide the stable thermal modes of thermal polymerization waves propagation.

19.4 TEMPERATURE PROFILES OF THE FRONTAL POLYMERIZATION OF AAM MIXTURES WITH BENTONITE AND DIATOMITE

Temperature profiles of the frontal polymerization of AAm and its mixtures with bentonite (with different contents of filler in mixtures) are shown in Figure 19.3.

As is seen from the curves 1–4 in Figure 19.3, the temperature profiles of the frontal polymerization with different filling degrees are homotypic. As it should be expected, the increase of bentonite content in AAm mixtures results in the decrease of maximum heating-up

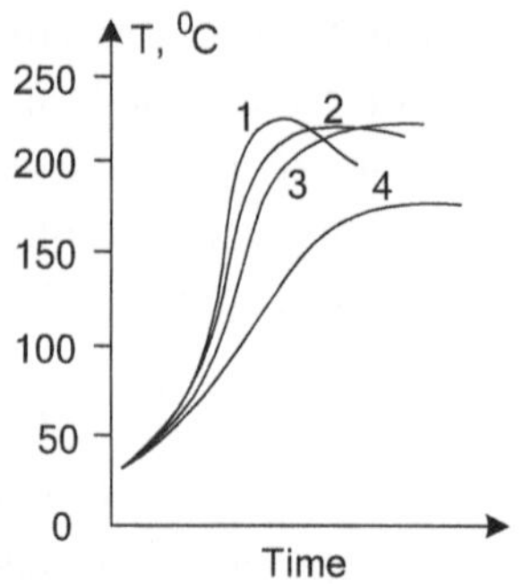

FIGURE 19.3 Temperature profiles of the frontal polymerization of unfilled AAm (1) and AAm-bentonite nanocomposites. Filler content in mixtures (in % wt): 20 (2); 30 (3) and 40 (4). Polymerization initiating agent – AIBN (0.5% wt of AAm).

temperature of reaction zone. However, it should be specifically noted that this decrease is not proportional to the filling degree but considerably lower. Perhaps, similar phenomenon is connected with the specific processes, which occur at the boundary of AAm-bentonite phases. The point is that the bentonite has lamellar structure and during its polymerization the delamination occurs [21, 22] which assists to the process of exothermic interaction [23–25] of polyacrylamide macromolecules with individual bentonite layers.

Non-proportional decrease of the limiting temperature of thermal polymerization waves from the degree of bentonite filling confirms this assumption.

Indeed, as is seen from Figure 19.4, the limiting heating-up temperature in reaction zone decreases considerably more slowly than it could have been expected in case of adding of inert fillers to the same system.

It should be specifically noted that frontal polymerization of acrylamide up to the temperatures 100–120°C is generating only linear macromolecules [26] and the imidization process starts at the temperatures >100–120°C and results in the formation of polyamide with spatially cross-linked structure.

Behavior of the temperature profiles of the AAm frontal polymerization in the presence of diatomite additives with different amounts (Figure 19.5) is analogous to the data given in Figure 19.3; the only difference consists in the following: when we increase the amount of filler the limiting temperature of thermal polymerization waves decreases to the greater degree than in case with bentonite adding (Figure 19.6).

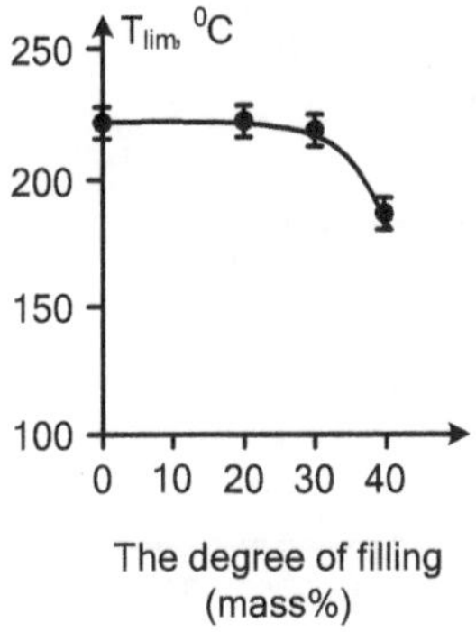

FIGURE 19.4 Dependence of limiting heating-up temperature of reaction zone on the degree of filling with AAm bentonite.

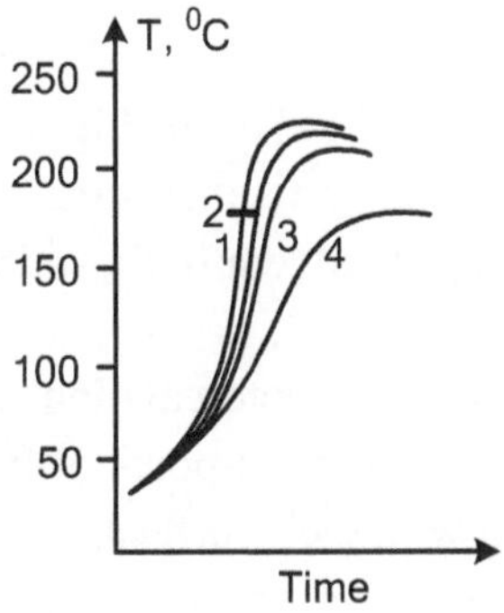

FIGURE 19.5 Temperature profiles of the frontal polymerization of unfilled AAm (I) and AAm-diatomite nanocomposites. Filler content in mixtures (in % wt): 0 (1); 20 (2); 30 (3); 40(4). Polymerization initiating agent – AIBN (0.5% wt of AAm).

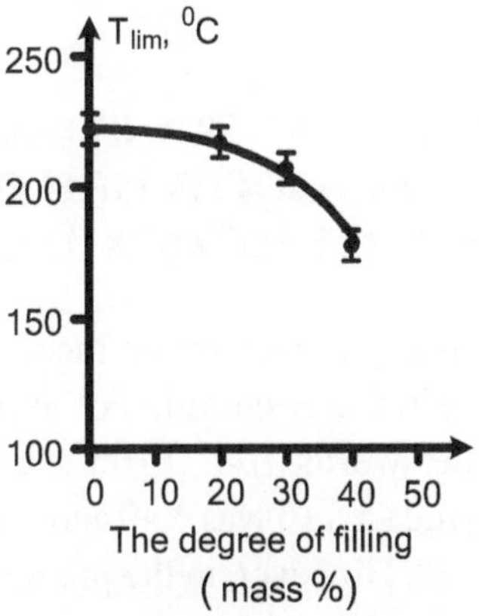

FIGURE 19.6 Dependence of the limiting heating-up.

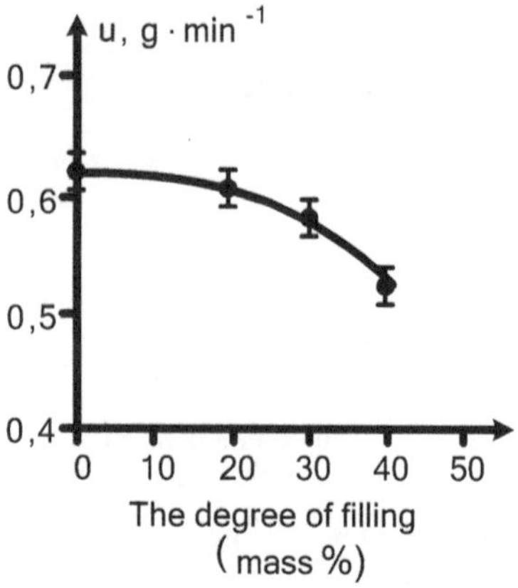

FIGURE 19.7 Dependence of mass front velocity on diatomite filling degree.

However, as is seen from the data in Figure 19.6, in this case non-additive decrease of the limiting temperature of thermal waves caused by the corresponding amount of diatomite additives is also observed (Figure 19.6).

This fact can be the result of intercalation of linear polyacrylamide macromolecules into micro- and nano-pores of filling agent which causes the exothermic interaction with their surface.

It is interesting that the change of mass front velocity caused by the diatomite amount (Figure 19.7) has the nature analogous to the curve in Figure 19.6 which is connected with the influence of limiting heating-up temperature of thermal waves on the velocity of polymerization front propagation.

19.5 INFLUENCE OF CHARACTER AND CONCENTRATION JF INITIATING AGENT ON VELOCITY OF FRONTAL POLYMERIZATION OF AAM FILLED MIXTURES

It is known from the frontal polymerization theory [27, 28] that the front velocity depends on the initial concentration of initiating agents according to the power law, in other words, $u \sim I^n_0$. And according to the results of numeric calculation [27] the value n was 0.40 and according to the results of analytical study [28] – 0.48. However, in the previous experimental papers

[1] it was shown that upon the frontal polymerization of 3-(oxyethylene)-γ, ω-dimethacrylate (OEDMA) and methyl-methacrylate under high pressures (up to 5 KBr) the value n depends on the character of initiating agent and monomer. Thus, in the condition of frontal polymerization of OEDMA initiated by peroxides: ditretbuthyl (t-BP), benzoyl (BP), dicyclohexylperoxydicarbonate (DCPC) the value *n* was 0.22, 0.32 and 0.34, respectively, and for methyl-methacrylate (initiating agent – BP) – 0.36. Relying only on these results we could assume that similar change of the value n is caused by the specific effect of high pressures on the efficiency of initiation, chain-breaking process etc. However, in further study [29] of the frontal polymerization of methacrylic acid and triethylene-glycol-dimethacrylate (TEGDM) without high pressures affected by AIBN, peroxides: cumil (PC), lauryl (PL), t-BP and AIBN, BP, LP the following values n were obtained: 0.24, 0.25, 0.27, 0.26 and 0.2, 0.23, 0.31. Also the data associated with the influence of character and concentration of initiating agent on the value n for the frontal radical polymerization of acrylamide should be considered. Influence of concentration of BP and AIBN on the velocity of acrylamide frontal polymerization as it is shown at the paper [19] indicates that when the packing density of initial reactive medium for descending and ascending waves is 0.95–1.0 g·cm^{-3} the obtained dependence thereof is practically identical and described through the following expression: u~Io$^{0.43\pm0.02}$. Thus, analysis of the represented data shows that upon the frontal polymerization of vinyl monomers the degree by initiating agent depends on the character of initiating agent, monomer and state of aggregation (liquid, crystal) of monomer. Until now similar dependence of the front velocity on the initial concentration of initiating agent has not been explained within the framework of current concepts of the frontal polymerization or in of the radical polymerization of vinyl monomers. Against the background of the considered data it should be investigated whether the specified regularities persist when introducing fillers, which are different by their character into AAm.

The dependence of the linear velocity of the AAm frontal polymerization on initiating agent concentration is given in Figure 19.8.

According to the data in Figure 19.9 the degree of front velocity is determined from the BP concentration which is ~0.6 with the filling degree of 30%. In the same way the degrees by initiating agent (for BP and AIBN) are

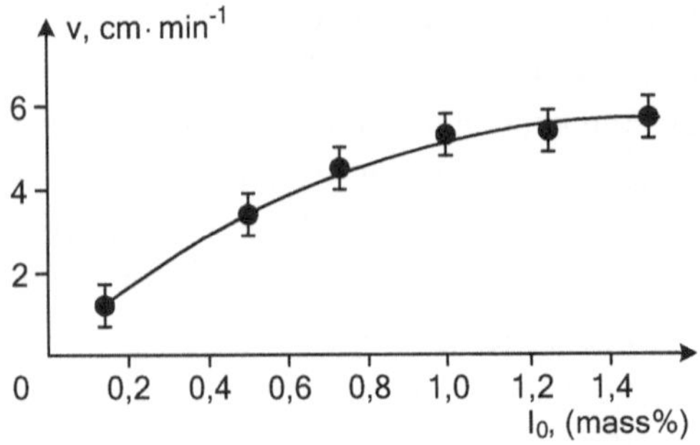

FIGURE 19.8 The influence of the initiator concentration (AIBN) on the linear velocity of frontal polymerization of mixture AAM – bentonite. Filler content in the mixture – 30%.

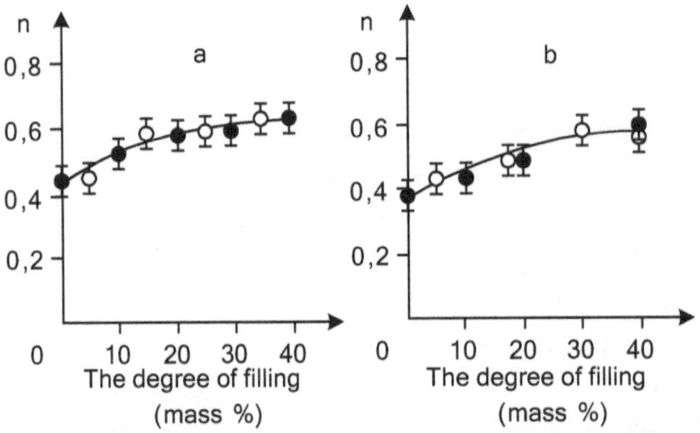

FIGURE 19.9 Dependence of the degree by initiating agent upon the frontal polymerization of AAm-bentonite mixtures on the filling degree for descending (●) and ascending (○) waves when using BP (a) and AIBN (b) as the initiating agents.

determined upon the polymerization of AAm mixtures with bentonite from the filling degree, which are given in Figure 19.9a and Figure 19.9b, respectively.

The obtained results indicate the complex nature of dependence of the degree by initiating agent on composition of mixtures and type of initiating agent. And as is seen from Figure 19.10, the degree does not depend on direction of the front propagating vertically from top to bottom or vice versa. For both initiating agents the increase of filler amount causes increase of the value *n*. Herewith, in case of initiation of the BP frontal polymerization the increase of filling degree leads to the growth of degree

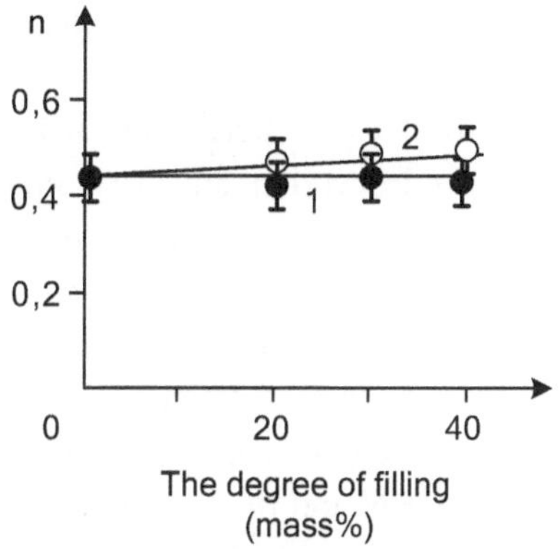

FIGURE 19.10 Dependence of the degree by initiating agent upon the frontal polymerization of AAm-chalk mixtures on the filling degree using BP (curve 1) and AIBN (curve 2) as the initiating agents.

by initiating agent right up to 0.61±0.02 but in case with AIBN the degree grows from 0.38 to 0.58±0.02. As it was mentioned before, such increase of the degree by initiating agent higher than 0.5 with growth of the filling degree is caused by the interaction of polyacrylamide macromolecules with the surface of individual layers of filler and occlusion of active polymerization centers in near-surface zone during the interaction process of macromolecules of bentonite which binds the surface of individual layers.

It is interesting that upon the frontal polymerization of AAm with diatomite additives the following results are obtained: n = 0.65±0.02 for BP and n = 0.6±0.02 for AIBN, respectively. Increase of the value n in case of diatomite adding can be explained through the intercalation of polyacrylamide macromolecules into micro- and nano-pores of diatomite and their interaction with the pore surface with the occlusion of active centers in filler pores.

On the basis of the obtained data, conclusion can be drawn that for the considered fillers the termination of growing macro-radicals is accomplished in accordance with both bimolecular and monomolecular mechanisms. Particularly this fact is the reason for notable increase of the degree by initiating agent under conditions of the AAm frontal polymerization in the presence of bentonite and diatomite.

In order to determine the role of lamellar or porous structure of filling agents analogous study was also carried out for AAm-fine-dispersed chalk mixtures using the same polymerization initiating agents. Data associated with the influence of the filling degree on the degree by initiating agent is given in Figure 19.10.

As well as upon the frontal polymerization of unfilled AAM [19] the degrees by BP and AIBN for descending and ascending thermal waves coincide in the whole range of filling degrees under study; however, their absolute values for each type of the initiating agent vary insignificantly. When using BP as the initiating agent the dependence of polymerization front velocity on the initial concentration of initiating agent is described through the equation $u \sim I_0^{0.43\pm0.02}$ for the composites filled with chalk (filler content is 20–40% wt). When using AIBN and filling AAm with chalk the degree by initiating agent grows to 0.48 (upon the filling degree of 50% wt).

Most probably, the difference in absolute values of degrees by initiating agents for the mixtures filled with bentonite, diatomite and chalk is associated with the constitution and structure of filling agents as well as character of chemical compounds on the surface layers of bentonite, diatomite and fine-dispersed chalk.

It should be noted that when using other fillers which are more coarsely-dispersed than chalk (expanded pearlite, potassium chloride) the dependence of the front velocity on BP and AIBN concentration is described through the regularities obtained for fine-dispersed chalk.

19.5.1 MORPHOLOGICAL CHARACTERISTICS OF POLYACRYLAMIDE/BENTONITE, DIATOMITE NANOCOMPOSITES

Textures of nanocomposite samples obtained via the frontal polymerization of acrylamide by bentonite (20% wt) and diatomite (20% wt) additives are shown in Figure 19.11. Data in Figure 19.11a shows that during the frontal polymerization the delamination of bentonite particles, which are evenly distributed in the volume of polymer binder is observed. According to Figure 19.11b the particles of added diatomite are also evenly distributed in the volume of polymer matrix.

In the article [25] we showed that the frontal mode benefits to the even distribution of nano-additives in the finite polymer matrix. Quite even distribution of individual bentonite layers as well as diatomite additives also can be explained through the frontal mode of

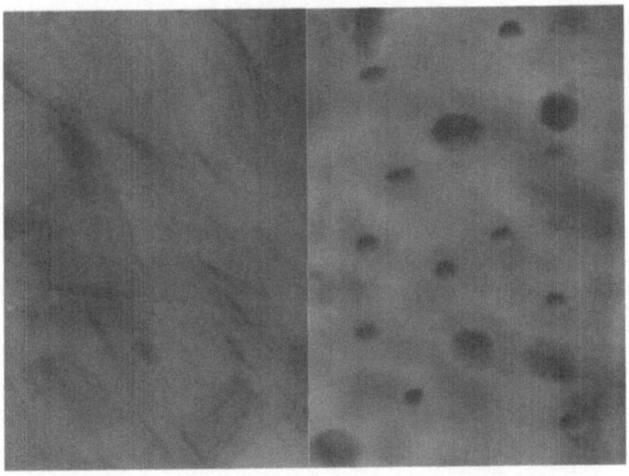

FIGURE 19.11 TEM photomicrographs of nanocomposites: polyacrylamide/bentonite (a) and polyacrylamide/diatomite (b).

polymerization. In this case thermal wave causes the even distribution of particles and the reaction zone leads to their fixation in polymer composite obtained.

Indeed, as is seen from Figure 19.11a, during the AAm frontal polymerization the delamination of bentonite particles is observed which is connected with the intercalation of acrylamide linear macromolecules into the interlaminar space with their further splitting to individual layers, as it was mentioned before. And the diatomite particles keep their initial form in the finite polymer matrix.

Delamination of bentonite particles (Figures 19.12a and 19.12b) during the AAm frontal polymerization as well as the exothermic interaction of polyacrylamide linear macromolecules with the surface of individual layers (Figure 19.12c) can be demonstrated in diagram form as follows.

It should also be noted that during the frontal polymerization a quite strong interaction of linear polyacrylamide macromolecules with the surface of individual bentonite layers takes place [23–25]. As a result, at the phase boundary the solid amorphous fraction [23–25] can be formed from the linear polyacrylamide macromolecules; it can cause notable change of relative thermal capacity, glass transition temperature and dynamic mechanical properties of nanocomposites depending on the filling degree.

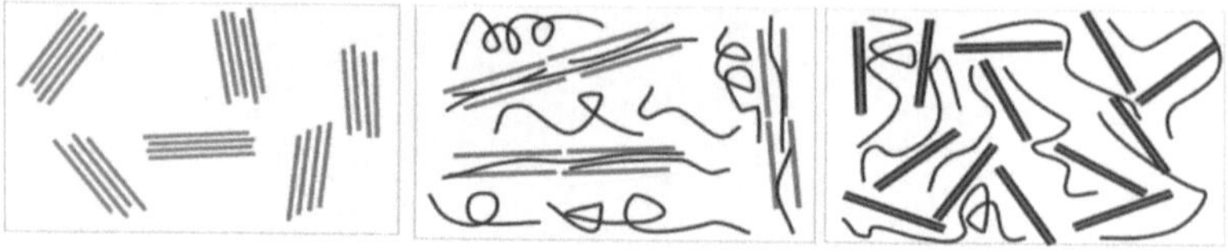

a b c

FIGURE 19.12 Schematic diagram of the process of bentonite delamination during the frontal polymerization.

KEYWORDS

- **bentonite**
- **diatomite**
- **front velocity**
- **frontal polymerization**
- **polymer nanocomposites**

REFERENCES

1. Chechilo, N. M., Enikolopyan, N. S. Structure of the front of the polymerization wave and the mechanism of propagation of polymerization reaction. Dokl. AN SSSR, 214,1131–1135, 1974.
2. Khanukaev, B. B., Kozhushner, M. A., Enikolopyan, N. S. Theory of the propagation of a polymerization front. Dokl. AN SSSR, 214, 625–629, 1974.
3. Aleksanyan, G. G., Arutyunyan Kh., Bodneva, V. L., Davtyan, S. P. Some rules governing the extension of the front of radical polymerization of vinyl monomers. Vysokomol. soedineniya. Ser. A, 17, 1052–1059, 1975.
4. Davtyan, S. P., Surkov, N. F., Rozenberg, B. A., Enikolopyan, N. S. Kinetics of radical polymerization under the conditions of the polymerization front propagation. Dokl. AN SSSR, 232, 379–387, 1977.
5. Davtyan, S. P., Zhirkov, V. P., Volfson, S. A. Problems of nonisothermal character in polymerization processes. Uspekhi Khimii, 539 (2), 251–274, 1984.
6. Pojman, J. A., Fortenberry, D. I., Ilyashenko, V. M. Frontal polymerization as an analog of SHS. International Journal of Self-propagating High-temperature Synthesis, 6, 355–366, 1997.
7. Chekanov, Y. A., Pojman, J. A. Preparation of functionally gradient materials via frontal polymerization. J. Appl. Polym. Sci., 78, 2398–2406, 2000.
8. Davtyan, D. S., Tonoyan, A. O., Bagdasaryan, A. E., Karapetyan, Z. A., Davtyan, S. P. To the contribution of thermal convective mass transfer for the reactive mixture

components to frontal curing of epoxydiane oligomers. Chem. Phys. Reports 19 (9), 1727–1765, 2001.

9. Davtyan, S. P., Hambartsumyan, A. F., Davtyan, D. S., Tonoyan, A. O., Hayrapetyan, S., Bagyan, S. H., H. S. Manukyan. The structure, rate and stability of autowaves during polymerization of co-metal-complexes with acrylamide. European polymer, J., 38, 2423–2431, 2002.
10. Davtyan, S. P., Tonoyan, A. O., Zakaryan, H. H. Quasi-stationary concentration principle for auto-wave processes. Polymer, 17, 5146–5154, 2007.
11. Pojman, J. A., Willis, J., Fortenberry, D., Ilyashenko, V. M., Khan, A. M. Factors affecting propagating fronts of addition polymerization: velocity, front curvature, temperature profile, conversion and molecular weight distribution. J. Polym. Sci., Part A: Polym. Chem., 33, 643–652, 1995.
12. Davtyan, S. P., Davtyan, D. S., Tonoyan, A. O., Radugina, A. A., Savchenko, V. I., Abrosimov, A. F. Frontal radical polymerization of methyl metakrylate in a cylindrical flow reaction. J. Polym. Sci., Part A: Polym. Chem., 41, 138–146, 1999.
13. Enikolopyan, N. S., Kozhushner, M. A., Khanukaev, B. B. Molecular weight distribution during isothermal and frontal polymerization. Dokl. AN SSSR, 217, 676–678, 1974.
14. Pojman, J. A., Greven, R., Khan, A. M., West, W. Convective instabilities induced by traveling fronts of addition polymerization. J. Phys. Chem., 1,2, 7466–7475, 1992.
15. Khachatryan, A. R., Tonoyan, A. O., Davtyan, D. S., Volpert, V. A., Davtyan, S. P. Control of the final conversion and MMD characteristics during propagation of polymerization heat waves in the tubular systems. Khimicheski, J. Armenii, 49(4), 39–46, 1996.
16. Davtyan, S. P., Davtyan, D. S., Tonoyan, A. O., Radugina, A. A., Savchenko, V. I. Control of conversion and molecular masses during frontal polymerization of MMA in cylindrical flow reactor. Polymer Sci, Part A: Polym. Chem., 41(2), 147–157, 1999.
17. Chechilo, N. M., Enikolopyan, N. S. Effect of pressure and initial temperature of the reaction mixture during propagation of a polymerization reaction. Dokl. AN SSSR, 260, 160, 1976.
18. Davtyan, D. S., Baghdasaryan, A. E., Tonoyan, A. O., Davtyan, S. P. To the contribution of thermal convective mass transfer of the reactive mixture components to frontal curing of epoxy diane oligomers. Khimicheskaya fizika, 19(9), 100–109, 2000.
19. Davtyan, D. S., Tonoyan, A. O., Hayrapetyan, S.M, Manukyan, L. S. Davtyan, S. P. Peculiarities of frontal initiated polymerization of acrylamide. Izvestia NAN RA i GIUA, 52(1), 38, 2003.
20. Tran, N. H., Wilson, M. A., Milev, A. S., Dennis, G. R., Kannangara, G. S. K., Lamb, R. N. Dispersion of silicate nano-plates within poly (acrylic acid) and their interfacial interactions. Science and technology of advanced Materials, 7(8), 786–791, 2006.
21. Kell, P., Akelah, A., A. Moet. Reduction of reccidual stress in montmorillonite/epoxy compounds. J. Mater. Sci. 29, 2274–2280, 1994.
22. Chvalun, S. N. Polymer nanocomposites, Priroda, 7, 22–30, 2000.
23. Sargsyan, A. G., Tonoyan, A. O., Davtyan, S. P., Schick, C. The amount of immobilized polymer in PMMA SiO_2 nanocomposites determined from calorimetric data. European Polymer Journal, 8, 3113–3129, 2007.
24. Sargsyan, A. G., Tonoyan, A. O., Davtyan, S. P., Schick, C. Rigid amorphous fraction in polymer nanocomposites. NATAS Notes, 39(4), 6–9, 2007.

25. Davtyan, S. P., Berlin, A. A., Tonoyan, A. O., Schick, C., Rogovina, S. Polymer nanocomposites with a uniform distribution of nanoparticles in a polymer matrix synthesized by the frontal polymerization. Rossiyskie nanotekhnologii, 4(7–8), 489–496, 2009.
26. Abramova, L. I., Bayburtov, T. A., Grigoryan, E. P., Zilberman, E. N., Kurenkov, V. F., Myagchenkov, V. F. Poliakrilamid (Polyacrylamide). Moskva, Izd. Khimia, 1992.
27. Khanukaev, B. B., Kozhushner, M. A., Enikolopyan, N. S. Theory of polymerization-front propagation. Combust. Explos. Shock Waves, 10(1), 562–568, 1974.
28. Pojman, J. A., Willis, J., Fortenberry, D., Ilyashenko, V., Khan, A. Factors affecting propagating fronts of addition polymerization: velocity, front curvature, temperature profile, conversion and molecular weight distribution. J. Polym. Sci., Part A, Polymer Chem., 33, 643–652, 1995.

CHAPTER 20

INFLUENCE OF SINGLE-WALL NANOTUBES ON THE STABILITY OF FRONTAL MODES AND PROPERTIES OF OBTAINED POLYMER NANOCOMPOSITES

D. S. DAVTYAN, A. O. TONOYAN, A. Z. VARDERESYAN, and S. P. DAVTYAN

State Engineering University of Armenia, 105 Teryana Str., Yerevan, 375009, Armenia, E-mail: atonoyan@mail.ru

CONTENTS

ABSTRACT

Characteristics of the frontal copolymerization of acrylamide with methyl-methacrylate in the presence of single-wall carbon nanotubes in different amounts are studied. It is shown that adding of bentonite which represents the natural lamellar nanomaterial with nanodimensional layers results in the formation of polyacrylamide-bentonite hydrogels. It is shown that the filling by nanotubes by more than 20% (of the initial weight of comonomers) causes the loss of stability of copolymerization thermal waves with occurrence of periodical, spin and chaotic modes. The mechanism of periodical modes formation is offered. Physical and mechanical, dynamic and mechanical and thermochemical properties of obtained polymer nanocomposites are studied.

On the basis of analysis of the data on the influence of amounts of single-wall nanotubes on the properties of copolymer nanocomposites the conclusion is drawn relative to the intercalation of copolymer macromolecules into the inner surface of nanotubes.

20.1 INTRODUCTION

Poor compatibility of carbon nanotubes with many polymer binders, organic and aqueous solutions considerably restricts their application as nanofillers. Therefore, there are many papers (for example, see Refs. [1–3] and cited references) devoted to the research of capabilities of considerable enhancement of interaction of single-wall (SWCNT) and multi-wall (MWCNT) carbon nanotubes surface with polymer macromolecules.

High physical and mechanical performance of carbon nanotubes (tensile strength ~100 GPa, modulus of elasticity ~1000 GPa and elongation up to ~0.4%) are good preconditions for the enhancement of properties of nanocomposites – polymer/carbon nanotubes. However, as it was mentioned in the paper [4], and as analysis of other papers shows [5–11], the data of physical and mechanical properties of nanocomposites (polymer/carbon nanotubes) is inconsistent. Most probably, first of all such status is connected with the uneven distribution of nanotubes in the polymer volume. Also the methods of nanocomposites generation [4, 12], which influence on the morphology of binder macromolecules, directly on the surfaces of phases of nanotube-polymer matrix are very important factors. Reliable comprehension of the results of many papers is complicated also due to the fact that often they do not give data on the thermal and temperature conditions of nanocomposites synthesis. Therefore, development of new methods of polymer nanocomposites synthesis which will provide the even distribution of carbon nanotubes in the binder volume as well as the enhancement of reliability and reproducibility of their generation process are the topical tasks for the obtaining of nanomaterials – polymer/carbon nanotubes.

The purpose of this chapter is to synthesize nanocomposites using the method of frontal copolymerization of acrylamide (AAM) with methylmethacrylate (MMA) in the presence of SWCNT and to distribute them evenly in the polymer matrix; to investigate their physical and mechanical, dynamic and mechanical, thermochemical and electroconductive properties; **to** determine the boundaries of stable frontal modes depending on the nanotubes filling degree, considering the direct dependence of obtained nanocomposite properties on the capability of setting the stationarity of frontal process thermal wave propagation. It is also interesting to investigate the geometric shapes and constitution of nonlinear structures, which are formed as a result of non-stationary front wave propagation.

20.2 EXPERIMENTAL PART

Sigma Aldrich AAMs and MMAs were used as co-monomers. MMAs were purified according to the methods [12]. AAMs were purified via double recrystallization from the saturated solutions of ethyl alcohol.

Initiating agent of copolymerization is dicyclohexylperoxydicarbonate (DCPC) which was used after the double recrystallization from ethyl alcohol and drying in vacuum cabinet at the room temperature until the constant weight was obtained. SWCNTs (of Sigma Aldrich brand) and aluminum nanopowder with the particles size of 40 nm (of Sigma Aldrich brand) were used as the nanofillers during the copolymerization process.

For the frontal copolymerization of AAM with MMA in the presence of SWCNT the initial mixtures were prepared as follows. In the beginning, the powdery AAM was carefully mixed with the necessary amount of nanotubes. Then, in order to ensure the stationary modes of frontal copolymerization [13, 14] the AAM mixture with nanoparticles was put into the reaction glass ampoules in individual portions and compacted. Further, MMA in quantity of 20% (of AAM weight) together with initiating agent was added to the prepared mixture. DCPC concentration in all experiments was 2% (of weight) of co-monomer amounts.

The frontal copolymerization of AAM with MMA adding nanotubes in the appropriate amounts was accomplished according to the methods, which were described by us before in the papers [14, 15]. The reaction was carried out in the vertically established glass ampoules with diameter of 5 mm, length of 100 mm. Polymerization front was initiated from the top of reaction ampoules by application of hot (~200°C) metallic surface to the edges of reaction ampoules [16]. Temperature profiles of the frontal copolymerization were determined through the copper-constantan thermocouples performance. Thermocouple junctions were located in the middle part of ampoules. And the velocity of front propagation was determined visually according to the dependence of front coordinate on time.

Physical and mechanical (under conditions of elongation), dynamic and mechanical properties of nanocomposite samples were determined on the Perkin-Elmer Diamond DSA device.

Thermal-oxidative degradation of polymer binders was investigated via derivatographic method on MOM device with the heating-up velocity of 3.2°/min. Electroconductive properties of nanocomposite samples (cross-section 0.2 cm^2, length 1cm) were determined via impedance measurements (frequency 1000 Hz, amplitude 5 mv) on Electrochemical workstation CHI 660D.

20.3 INFLUENCE OF AMOUNTS OF SWCNT ON THE CHARACTERISTICS OF FRONTAL COPOLYMERIZATION

Data on the influence of SWCNT amounts on the temperature profiles (Figure 20.1a) and propagation velocity (Figure 20.1b) of copolymerization front of AAM with MMA are given in Figure 20.1.

Comparison of the data in Figures 20.1a and 20.1b with analogous results obtained in the paper [12] displays their considerable difference. In this case, the limiting temperature (Figure 20.1a) of thermal waves and velocity of copolymerization front (Figure 20.1b) decrease

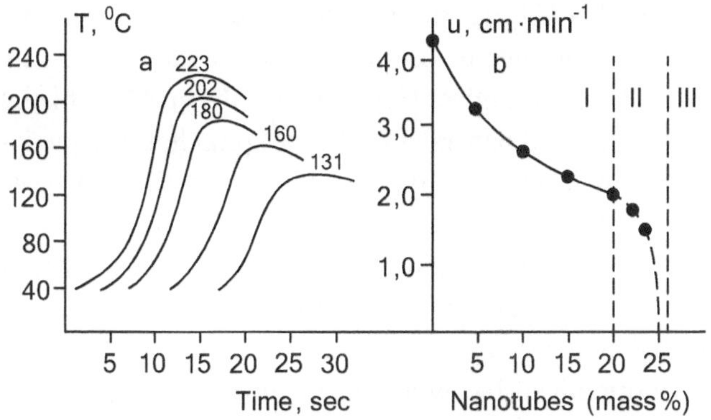

FIGURE 20.1 Influence of amounts of SWCNT on the behavior of temperature profiles (a) and front velocity (b). Ratio of AAM co-monomers: MMA= 80:20, amount of SWCNT (% of co-monomers weight): 1 – 0, 2 – 5, 3 – 10, 4 – 5 and 5 – 20.

practically simultaneously with adding of nanotubes. Herewith, as is seen in Figure 20.1a, the structure of temperature profiles also changes.

Observed changes of typical values of the AAM-MMA frontal copolymerization can be explained by two factors. From one hand, this is the joint effect of nanofiller quantities and intensity of thermal loss from the reaction zone to the environment on the thermal conditions of frontal copolymerization. But on the other hand, there is absence of chemical interaction of binder macromolecules with outer and inner surfaces of SWCNT. Proof of the first factor is found in such aspects as quite significant decrease of the limiting temperatures (Figure 20.1a) and data on the influence of nanotubes amounts on the stationary status of frontal modes and their stability. And indeed, as is seen from the data in Figure 20.1b, depending on the amounts of SWCNT added, three regions of copolymerization front velocity change with different nature are observed. Region I (amount of nanoadditives up to 20%) corresponds to the stationary stable statuses of thermal copolymerization waves. In region II (amount of nanotubes is 20– 25%) when amount of SWCNT grows the front velocity quite sharply reduces and the stability of frontal modes is lost. And in region III when amounts of nanotubes are higher than ~25–26% the frontal copolymerization modes do not exist. Let us consider unstable frontal modes in region II in details.

It is known [17–27] that the loss of frontal modes stability as a rule is followed by occurrence of oscillatory, periodical, single-, two-, three- and multiple-start spin modes. Herewith, on the surfaces of polymerized samples in the specified papers [17–27] the spiral hollows, which are typical for unstable modes of frontal polymerization are revealed. In this case, as is seen from the data in Figure 20.1b and Figure 20.2 (samples 1–6), the stability of co-polymerization thermal waves is lost when adding the nanotubes in amount of 20% and higher. Indeed, when filling the polymerized medium with nanotubes up to 19%, the frontal modes are stable and samples have smooth surface with black color. In Figure 20.2 the photo of one nanocomposite sample (sample 1) with 15% filling is displayed.

Stability loss (Figure 20.2) is followed by the formation of periodical (sample 2), single- (sample 3), two- (sample 4), three- (sample 5) start spin and chaotic (sample 6) modes. Formation of the specified nonlinear phenomena is displayed in the form of white colorings against the background of smooth surfaces of obtained samples.

In order to reveal the sequence of unstable modes occurrence with filling up to 20% and higher the amount of SWCNT additives was increased in small portions or by 1% of total weight of co-monomers.

For the processes of frontal copolymerization of AAM with MMA in the presence of SWCNT the loss of stability of stationary thermal waves starts from the occurrence of periodical modes (Figure 20.2, sample 2).

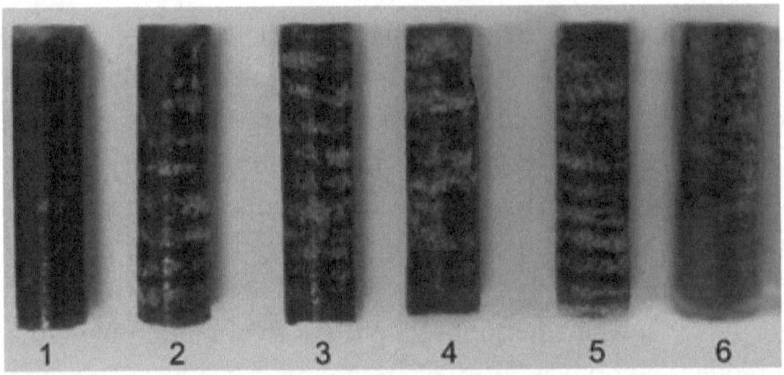

FIGURE 20.2 Samples of copolymer nanocomposites obtained under conditions of the frontal copolymerization of AAM with MMA. Filling degree (% wt of AAM and MMA amounts): 15–1, 20–2, 22–3, 23–4, 24–5, 25–6. Stationary stable – 1 and unstable modes: periodical – 2, single-start – 3, two-start – 4, three-start – 5, spin, chaotic – 6.

Periodical modes of the frontal copolymerization are characterized by the fluctuation [20] of front velocity about its stationary value (Figure 20.3).

Study of the mechanism of periodical modes formation and geometrical shape of front is of interest. For this purpose the polymerization was stopped by the freezing of reaction ampoules using liquid nitrogen in pre-assigned time intervals (points specified in Figure 20.3 by Figures 20.1–20.4), which correspond to the half of front velocity fluctuation period.

After, the reaction ampoules were independently heated up to the room temperature, without causing damage to the samples, then the glass housing was removed and photos of the obtained samples were taken (Figure 20.4).

It is seen in Figure 20.4 that in point 1 (Figure 20.3) the front shape is flat but then its incurvation occurs (Figure 20.4, sample 2) together with forming of "tongue" (point 2 in Figure 20.3). Then (Figure 20.4, sample 3), the formed "tongue" becomes longer (point 3 in Figure 20.3) and in point 4 (Figure 20.3) the front shape becomes flat again (Figure 20.4, sample 4). Dynamics of the front geometrical shape change shown in Figure 20.4 is connected with the heat loss from the reaction zone to environment and effect of inert SWCNT additives on the reaction mixture heating up (Figure 20.1a). Most probably, at the moment of maximum incurvation of the front geometrical shape (which corresponds to the minimum

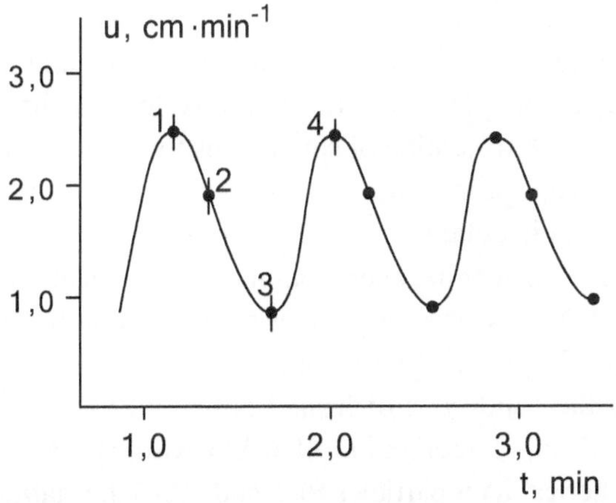

FIGURE 20.3 Oscillatory mode of the frontal copolymerization.

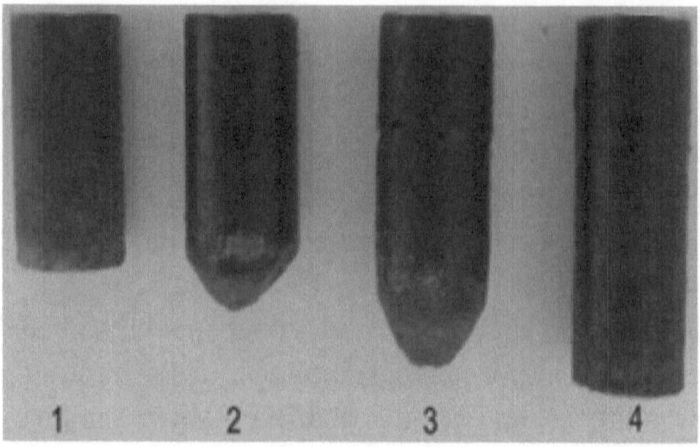

FIGURE 20.4 Change of geometrical shape of front caused by time. Time in sec. (time reference corresponds to the point 1 in Figure 20.3): 0 – 1, 20 – 2, 40 – 3 and 50 – 4.

temperature in reaction zone) the nucleation site occurs which closes on itself. Or by analogy with results of the paper [21], thermal wave propagates not only in axial but also in radial directions of reaction ampoules. Both considered mechanisms can result in the front shape alignment and increase of the temperature in reaction zone, respectively. Therefore, velocities of thermal copolymerization waves (Figure 20.3) have maximum values for the flat (Figure 20.4, sample 1 and 20.4, sample 4) and vice versa, minimum values (Figure 20.3) for the most curved front shapes (Figure 20.4, sample 3).

Further increase of amounts of nanotubes at first results in the formation of single-start (Figure 20.2, sample 3), then two-start (Figure 20.2, sample 4), multiple-start (Figure 20.2, sample 5) spin and at the end chaotic (Figure 20.2, sample 6) modes. Stability loss of thermal waves of chemical nature with formation of spin modes for the processes of burning and SHS (self-propagating high-temperature synthesis) is considered in the papers [25–27] in details.

It should be noted that degeneration of copolymerization frontal modes of AAm with MMA is observed when adding nanotubes in amounts of 26% and higher (Figure 20.1b, region III).

Also under the conditions of frontal polymerization of MMA [12] or upon the frontal copolymerization of AAM with MMA [24, 28] (in the presence of spherical nanoparticles SiO_2 and TiO_2) the stability loss of stationary frontal modes is observed when the filling degrees are 25–30%.

This phenomenon [12, 24, 28] is explained by the existence of additional heat generation source in reaction zone at the expense of exothermic interaction of binder macromolecules with nanoparticles surface.

20.4 PHYSICAL AND MECHANICAL, DYNAMIC AND MECHANICAL AND THERMOCHEMICAL PROPERTIES OF NANOCOMPOSITES

Influence of the filling degree on tensile strength (s), modulus of elasticity (E) and elongation (e) is displayed in the Table 20.1. Increase of amounts of SWCNT additives in nanocomposites leads to the increase of the values s and E and decrease of deformability of samples. 20% filling causes the growth of limiting tensile strength by ~30%, modulus of elasticity – by ~20% and the decrease of deformability by ~50%.

When the amounts of SWCNT additives increase, there is the notable growth of the tensile strength and modulus of elasticity, which indicates their even distribution in the copolymer binder volume. Even distribution is provided at the expense of deagglomeration of agglomerated nanoparticles [12] (nanotubes) under the influence of thermal copolymerization waves and fixation of this status in polymer binder.

Behavior of the dynamic module (E′) and tangent of angle of mechanical loss (tgδ) for copolymer nanocomposites which content different amount of SWCNT is illustrated in Figures 20.5a and 20.5b.

As should be expected (Figure 20.5a), the values E′ upon the same filling degrees keep constant and only at the temperatures ~220°C their decrease occurs.

TABLE 20.1 Influence of Amounts of SWCNT Additives on the Values σ, E and ε

SWCNT. % of binder weight	s. MPa, kgf/mm^2	E. MPa, kgf/mm^2	ε, %
0	84±5	136 ± 7	40
5	93± 5	140 ± 7	36
10	105± 5	150 ± 8	25
20	111 ± 5	170 ± 8	20

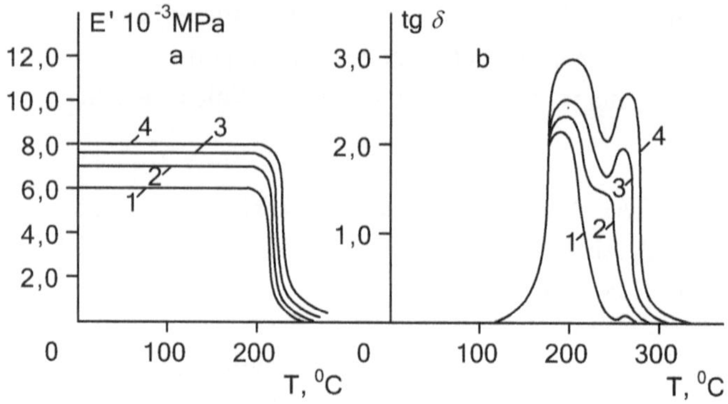

FIGURE 20.5 Change of the dynamic module (a) and tangent of angle of mechanical loss (b) caused by the temperature with different filling degrees. Filling degree corresponds to the data in Figure 20.1.

Obviously, this change of the dynamic module is associated with the increase of mobility of macromolecules and individual fragments of copolymer binder at the devitrification temperatures. Growth of the values E′ (curves 1–4, Figure 20.5a) is observed when amounts of nanotubes increase. Most probably, observed growth of the value E′ caused by the amounts of nanotubes occurs due to the intercalation of copolymer macromolecules or their fragments into the inner surface of nanotubes.

Behavior of tangent of mechanical loss angle (curves 1–4, Figure 20.5b) caused by the amounts of nanotubes has quite uncommon shape. Here two transitions are observed in the range of temperatures higher than 200°C, which correspond to the devitrification of nanocomposite copolymer samples. In the beginning quite intensive primary (devitrification) transition is observed and then the secondary transition occurs. Herewith, the intensity of the secondary transition grows (Figure 20.5b, curves 2–4) when the amounts of SWCNT increase. This fact, the growth of the secondary transition intensity with the increase of amounts of nanotubes, confirms the assumption that the intercalation of individual elements or binder macromolecules into the inner surface of nanotubes takes place.

Curves of weight loss depending on the temperature with different amounts of SWCNT are given in Figure 20.6.

It is seen from the data in Figure 20.6 that weight loss for the pure copolymer of AAm with MMA starts at the temperature ~300°C (curve 1). Adding

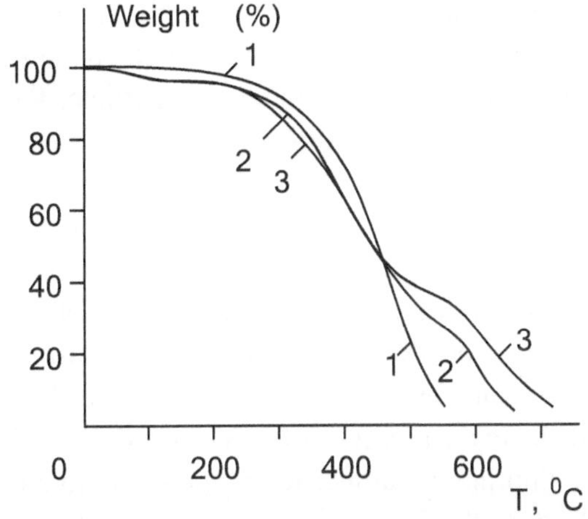

FIGURE 20.6 Weight loss of pure copolymer based on AAM with MMA and copolymer nanocomposites with different amounts of nanotubes (% wt of total amount of comonomers): 1–0, 2–10, 3–20.

SWCNT causes some decrease of the initial temperature of thermal-oxidative degradation and quite tangible change of character of weight loss curves.

Indeed, as is seen from the curve 2 and 3 in Figure 20.6 in the temperature range ~450–500°C small plateau occurs (curve 2) and only after this occurrence the second stage of weight loss starts. When the amounts of SWCNT grow the plateau value increases (curves 2, 3).

Most probably, observed two-stage character of nanocomposite weight loss curves is associated with the intercalation of AAM-MMA copolymers macromolecules into the inner surface of SWCNT, which slows down the thermal-oxidative degradation process to some extent. The following fact remains incomprehensible: increase of SWCNT amounts causes some decrease of initial temperature of nanocomposites thermal-oxidative degradation. Results given in Figure 20.5 correspond to the paper conclusions [29] quite well and clarify the data [30] on the influence of SWCNT and MWCNT on the process of nanocomposites thermal-oxidative degradation.

Samples of copolymer nanocomposites which content 15 to 25% of SWCNT have practically zero electrical conductivity. In order to generate the electroconductive nanocomposites we added 5% (of comonomers weight) of aluminum nanopowder to the initial reaction mixture. As it turned out, for copolymer nanocomposites which contain 18–19% of single-wall nanotubes and 5% of aluminum nanoparticles the electrical conductivity reaches up to ~95,000 $Si{\cdot}m^{-1}$.

Thus, the results which were obtained in this paper show that in the process of the frontal copolymerization of AAM with MMA in the presence of single-wall nanotubes, there takes place intercalation of fragments or binder macromolecules into the inner surface of nanotubes which causes the increase of tensile strength, modulus of elasticity and decrease of deformability of nanocomposite samples. The frontal copolymerization of AAM with MMA in the presence of ~18% of single-wall nanotube additives and 5% of aluminum nanoparticles (according to the total weight of comonomers) leads to the formation of nanocomposites with insignificant current-carrying properties.

KEYWORDS

- **frontal polymerization**
- **heat transfer**
- **nanocomposites**
- **nanotubes**
- **nonlinearity**

REFERENCES

1. O'Connell, M. J., Boul, P., Ericson, L. M., Huffman, C., Wang, Y. H., Haroz, E. et al. Reversible water-solubilization of single-walled carbon nanotubes by polymer wrapping. Chem. Phys. Lett., vol. 342, 265–271, 2001.
2. Hill, D., Lin, Y., Qu, L., Kitaygorodskiy, A., Connel, J. W., Allard, Sun, Y. P. Functionalization of carbon nanotubes with derivatized polyimide. Makromolecules. 38, 7670–7675, 2005.
3. Neira-Velázquez María Guadalupe, Ramos-de Valle Luis Francisco, Hernández-Hernández Ernesto, Zapata-González Ivan. Toward greener chemistry methods for preparation of hybrid polymer materials based on carbon nanotubes. e-Polymers, 162, 1618–7229, 2008.

4. Hobbie, E. K., Bauer, B. J., Stephens, J., Becker, M. L., McGuiggan, P. Colloidal particles coated and stabilized by DNA-wrapped carbon nanotubes. Langmuir. 21, 10284, 2005.
5. Wagner, H. D., Vaia, R. A. Carbon nanotube-based polymer composites: Outstanding Issues at the interface for mechanics. Materials Today, 11(7), 38–42, 2004.
6. Barber, A. H., Cohen, S. R., Wagner, H. D. Measurement of carbon nanotube-polymer interfacial strength. Appl. Phys. Lett., 82, 4140, 2003.
7. Barber, A. H., Cohen, S. R., Kenig, S., Wagner, H. D. Interfacial fracture energy measurements for multi-walled carbon nanotubes pulled from a polymer matrix. Compos. Sci. Technol., 64(15), 2283, 2004.
8. Barber, A. H., Cohen, S. R., Wagner, H. D. Static and dynamic wetting measurements of single carbon nanotubes. Phys. Rev. Lett., 92(18), Art No.186103, 2004.
9. Buchachenko, A. L. New horizons of chemistry: single molecules. Uspekhi khimii, 75(1), 3–26, 2006.
10. Chi-Yuan Huang, Ching-Shan Tsai, Keng-Yu Tsao, Po-Chian Hu. Carbon black nano composites for over voltage resistance temperature coefficient. Proceedings of the World Polymer Congress – Macro, 41st International Symposium on Macromolecules, p. 41, 2006.
11. Wagner, H. D., Lourie, O., Feldman, Y., Tenne, R. Stress-Induced Fragmentation of Multiwall Carbon Nanotubes in a Polymer Matrix. Appl. Phys. Lett., 72(2), 188–190, 1998.
12. Davtyan, S. P., Berlin, A. A., Tonoyan, A. O., Rogovina, S. Z., Schik, C. Polymer nanocomposites with a uniform distribution of nanoparticles in a polymer matrix synthesized by the frontal polymerization. Rossiyskie Nanotekhnologii, 4(7–8), 489–498, 2009.
13. Arutyunyan Kh.A., Davtyan, S. P., Rozenberg, B. A., Enikolopyan, N. S. Curing of epoxy resins of bis-phenol A by amines under conditions of reaction front propagation. Dokl. AN SSSR, 223(3), 657–660, 1975.
14. Davtyan, S. P., Hambartsumyan, A. F., Davtyan, D. S., Tonoyan, A. O., Hayrapetyan, S. M., Bagyan, S. H., Manukyan, L. S. The structure, rate and stability of autowaves during polymerrization of Co metal-complexes with Acrylamide. Eur. Polym. J., 38, 2423–2431, 2002.
15. Davtyan, D. S., Tonoyan, A. O., Hayrapetyan, S. M., Manukyan, L. S., Davtyan, S. P. Peculiarities of frontal initiated polymerization of acrylamide. Izvestia NAN RA i GIUA, 52(1), 38, 2003.
16. Davtyan, S. P., Zakaryan, H. H., Tonoyan, A. O. Steady state frontal polymerization of vinyl monomers: the peculiarities of. Chemical Engineering Journal, 155(1–2), 292–297, 2009.
17. Begishev, I. P., Volpert, Vit. A., Davtyan, S. P. Existence of the polymerization wave with crystallization of the initial substance. Dokl. AN SSSR, 273(5), 1155–1158, 1983.
18. Volpert Vit, A., Volpert, Vl. A., Davtyan, S. P., Megrabova, I. N., Surkov, N. F. Two-dimensional combustion modes in condensed flow. SIAM. J.Appl. Math., .52(2), 368–383, 1992.
19. Davtyan, S. P., Tonoyan, A. O., Davtyan, D. S., Savchenko, V. I. Geometric shape and stability of frontal regimes during radical polymerization of methyl metacrilate in a cylindrical flow reactor. Polymer Sci. Ser.A, 41(2), 153–162, 1999.

20. Davtyan, D. S., Baghdasaryan, A. E., Tonoyan, A. O., Karapetyan, Z. A., Davtyan, S. P. The mechanism of convective mass trasfer during the frontal (radical) polymerization of methyl metacrilate. Polymer Sci. Ser. A, 42(11), 1197–1216, 2000.
21. Davtyan, D. S., Baghdasaryan, A. E., Tonoyan, A. O., Davtyan, S. P. To the contribution of thermal convective mass transfer of the reactive mixture components to frontal curing of epoxy diane oligomers. Khimicheskaya fizika, vol.19, № 9, pp.100–109, 2000.
22. Davtyan, S. P., Shaginyan, A. A., Tonoyan, A. O., Ghazaryan, L. Oscillatory and spin regimes at frontal solidification of epoxy combinations in the flow. Compounds and Material with Specific Properties, Nova Science Publishers, Inc ISBN 978–1-60456–343–6, Editor: Bob, A. Howell et al., p.88, 2008.
23. Davtyan, S. P., Berlin, A. A., Tonoyan, A. O. Advances and problems of frontal polymerization processes. Obzor. J. po khimii, 1(1), 56, 2011.
24. Tonoyan, A. O., Ketyan, A. G., Zakaryan, H. H., Sukiasyan Zh.K., Davtyan, S. P. Polyacrylamide/bentonite, polyacrylamide/diatomite nanocomposites obtained by frontal polymerrization. Khimicheski, J. Armenii, 63(2), 193, 2010.
25. Ivleva, T. P., Merzhanov, A. G. Structure and variability of spinning reaction waves in three-dimensional excitable media. Physical Review, E., 64(3), 036218, 2001.
26. Volpert, A. I., Volpert Vit.A., Volpert Vl.A. Traveling Wave Solution of Parabolic Systems. AMS Books Online, p. 455, 2003.
27. Ivleva, T. P., Merzhanov, A. G. Three dimensional modes of unsteady- solid flame combustion. Chaos, 13(1), 80, 2003.
28. Avetisyan, A. S., Tonoyan, A. O., Ghazanchyan, Davtyan, S. P. The influence of high-temperature polymerization modes on monomer-polymer equilibrium. Vestnik GIUA, Issue 15(2), 32–39, 2012.
29. Chipara Mircea, Cruz Jessica, Vega Edgar, R., Alarcon Jorge, Mion Thomas, Chipara Dorina Magdalena, Ibrahim, A., Tidrow Steven, Hui David. Polyvinylchloride-Single Walled Carbon Nanotube Composites: Thermal and Spectroscopic Properties, Special Issue: Synthesis, Properties, and Applications of Polymeric Nanocomposites. Journal of Nanomaterials, Article ID 435412, 6 pages, 2012.
30. Wu, X. L., Liu, P. Poly(vinyl chloride)-grafted multi-walled carbon nanotubes via Friedel-Crafts alkylation. EXPRESS Polymer Letters, 4(11), 723, 2010.

CHAPTER 21

STUDY OF INFLUENCE OF IONIC ADDITIVES ON THE STRUCTURAL CHANGES OF WATER NANOCAGES CONFINED IN THE AOT REVERSE MICELLES

T. G. BUTKHUZI,[1,2] M. K. KURTANIDZE,[1] and M. D. RUKHADZE[1]

[1]*Faculty of Exact and Natural Sciences, Ivane Javakhishvili Tbilisi State University, 3 I. Chavchavadze ave, Tbilisi, 0128, Georgia, Email: marina.rukhadze@tsu.ge*

[2]*Physics Department, New York City College of Technology, CUNY, Brooklyn, New York 11201, USA*

CONTENTS

ABSTRACT

The relative content of free, bound and trapped water in the water core of reverse micelles in the presence of iodide and fluoride ions may be expressed by 1.7/2.2/5.2 and 5.7/1.0/2.4 ratios respectively. Binding constants of o-NA are 2 times higher in the presence of fluoride ions than in the case of iodide ions. The values of proton chemical shifts in nanocages of reversed micelles under the influence of iodide and fluoride anions are 3.8 and 3.9 ppm, respectively.

21.1 INTRODUCTION

In the past two decades the structure of water confined in the core of reverse micelles has been the subject of intensive research [1]. A reverse microemulsions are thermodynamically stable liquid systems, containing surfactant, co-surfactant, oil and water. These systems have attracted significant attention because of their likeness to biomembranes and living cells and their catalytic properties in chemical and enzymatic reactions [2]. Reverse micelles are the aggregates of surfactant formed in nonpolar solvents. Polar head groups of the surfactants are located toward to the core of micelles and the hydrocarbon chains are located in the nonpolar solvent [3–4]. Size of the micelles is nanometer-scale and can be changed by the water content in the core of micelles, which can be expressed by the molar ratio of water to surfactant ($W=[H_2O]/[surf]$) [3]. One of the most commonly used surfactant is sodium 1,4-bis (2-ethylhexyl) sulfosuccinate (AOT), which has a special ability to solubilize a large amount of water without help of any co-surfactant [3, 4]. This provides a good opportunity to use AOT for studying the properties of water aggregates close to the ionic center [5]. Behavior of water that is confined in water pool of reverse micelles can be considered as anomalous. Encapsulated water displays at least two structures: water, located near the interface of surfactant head groups (peripheral water) and water in the center of the water droplets of the reverse micelles (free water). These two types of water differ from free, pure water [6]. Molecular dynamics simulations were performed in order to reveal the Hofmeister series effect in the context of anion-lipid bilayer interactions. The mentioned simulations showed that the large chaotropic anions penetrate more deeply into the interfacial region of the lipid bilayer interior since the larger anions have poor hydrated layer and approach the surfactant head groups tightly [7]. Thermodynamical and spectroscopic properties of water confined in the reverse micelles have been studied by means of a great number of experimental techniques, such as dynamic and static light scattering, differential scanning calorimetric and conductivity methods, as well as nuclear magnetic resonance (NMR), electron spin resonance, infrared (IR) and UV–visible spectroscopic methods [3, 5, 6, 8–15].

The goal of this chapter was investigation of structural changes of water pools of AOT reverse micelles in the presence or absence of structure-making and structure-breaking ions by IR, UV and NMR spectroscopic methods.

21.2 EXPERIMENTAL PART

21.2.1 MATERIALS

21.2.1.1 Preparation of the Reverse Microemulsions

Reverse microemulsions were prepared on the basis of AOT, hexane, water and 0.05 M water solutions of potassium iodide and sodium fluoride. AOT (98%) was obtained from Fluka and was used without further purification.

21.2.1.2 Technics

IR absorption spectra were taken in a IR spectrophotometer Specord 75 equipped with a 1 cm path length sodium chloride window. All the IR spectra at W=1 were obtained between 4000 and 400 cm^{-1} at room temperature. In order to investigate microstructure of the confined water, the O-H stretching vibrational absorption spectra in the region of 3000–3800 cm^{-1} were fitted into three subpeaks. The curve fitting was performed by using Origin 6.5. Gauss function was chosen to fit the overlapped peaks. Gaussian curve fitting was achieved with a Monte Carlo method.

UV-visible absorption spectra were recorded in a UV-visible spectrophotometer Optizen POP using cells with 1 cm path length. O-nitroaniline (o-NA) was used as a molecular probe. Binding constants of o-NA with AOT micelles were calculated by absorption data of o-NA at wavelengths of 376 and 398 nm in hexane (0.0M and 0.1M AOT). Concentrations of free and bound o-NA were determined by equation systems at intermediate concentrations of AOT.

All ^1H NMR spectra were recorded in n-hexane on Agilent Mercury 300 NMR spectrometer operating at 300 MHz. Tetramethylsilane was added as an internal reference standard for chemical shift measurements. Before measurement, the spectrometer was locked at D_2O frequency for filed/frequency stabilization and shimmed using gradient shimming protocol.

21.3 RESULTS AND DISCUSSION

In order to study the influence of kosmotropic and chaotropic ions on the distribution of free, bound and trapped water in the water core of reversed micelles the O-H stretching vibrational absorption spectra were fitted into three subpeaks in the region of 3000–3800 cm^{-1}. The vibrational characteristics of bound, free and trapped water are different. It has been considered that free water (in the centre of micelle) has the lowest wavenumber 3290 ± 20 cm^{-1}. The 3490 ± 20 cm^{-1} peak is assigned to the bound water fraction. The high wavenumber 3610 ± 10 cm^{-1} corresponds to the trapped water molecules, which are located among the alkyl chains of surfactant [13]. As water inside the micelle is in three different states, it can be assumed that total peak area of O-H vibrational stretching band is the sum of peak areas of the different states of water [16]. Results show, that free water fraction in the presence of fluoride ions in the water pockets of the reverse micelles three times exceeds the free fraction of water in the presence of iodide ions. At the same time, the fractions of bound and trapped water are two times higher in the presence of iodide ions, than in the presence of fluoride ions in water droplets (Figures 21.1 and 21.2).

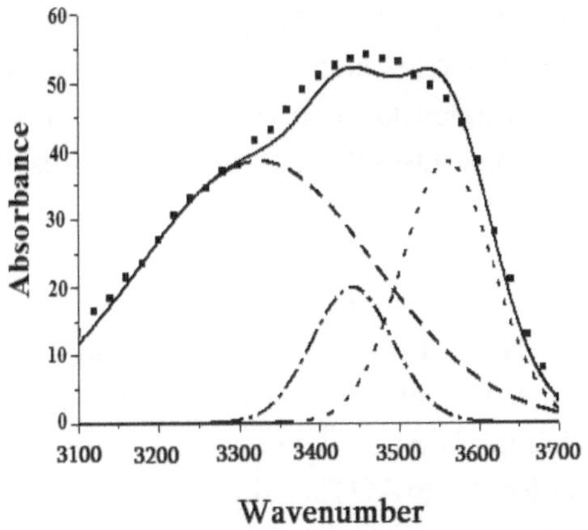

FIGURE 21.1 Deconvoluted infrared spectrum for the 0.05 M sodium fluoride-AOT-hexane system (W=1). Curves with lines - - - - correspond to trapped water, with lines -·-·-·- correspond to bound water and — — — curves represent free bulk water. The dots correspond to experimental data. The solid lines correspond to the sum of the deconvoluted spectrum of the water (sum of surfaces of trapped, bound and free water peaks).

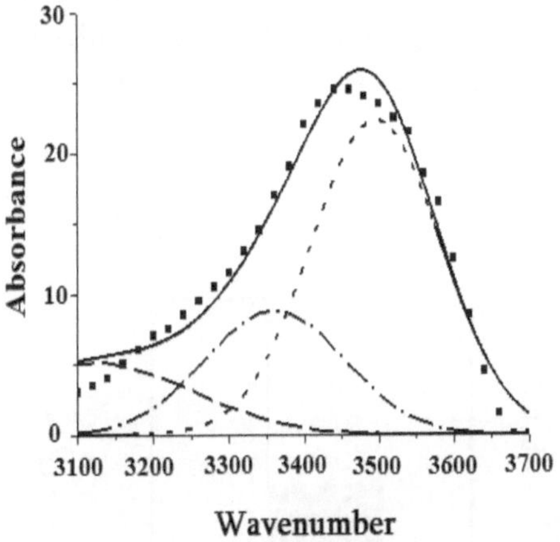

FIGURE 21.2 Deconvoluted infrared spectrum for the 0.05 M potassium iodide-AOT-hexane system (W=1). Curves with lines - - - - correspond to trapped water, with lines -·-·-·- correspond to bound water and — — — curves represent free bulk water. The dots correspond to experimental data. The solid lines correspond to the sum of the deconvoluted spectrum of the water (sum of surfaces of trapped, bound and free water peaks).

The reinforcement of formation of trapped water fraction under the influence of iodide ions can be explained so: as iodide ion is chaotropic, and has big radius and less hydrated layer, it can be located near the interface of surfactant/water [7]. This causes that part of water passes the surfactant head groups and is located among the alkyl chains. This is called trapped water.

This suggestion that iodide ions are located near the surfactant/water interface is somewhat substantiated by UV-visible spectroscopic results, viz. the binding constants of o-NA to the AOT head groups are less in the presence of iodide ions in comparison with fluoride ions in the water core of reverse micelles (Figure 21.3). This indicates that iodide ions interfere the binding of o-NA molecules to AOT head groups and hence the binding constants have the lower values in this case than in the presence of fluoride ions.

The high content of free water (63%) in the presence of fluoride ions is in good correlation with NMR results. As seen from Figure 21.4 the value of chemical shift is higher in the presence of fluoride ions in comparison with the case when water core is modified by the iodide ions. This means that content of free water in water droplet should be higher in the presence of fluoride ions, as free water exhibit proton chemical shifts in downfield direction [9]. The appearance of proton chemical shift to downfield is hindered in case of iodide ions not only by the amount of free water, but also by bound water fraction (bound water exhibit proton chemical shifts in upfield direction), which is more in case of iodide ions than in the presence of fluoride ions.

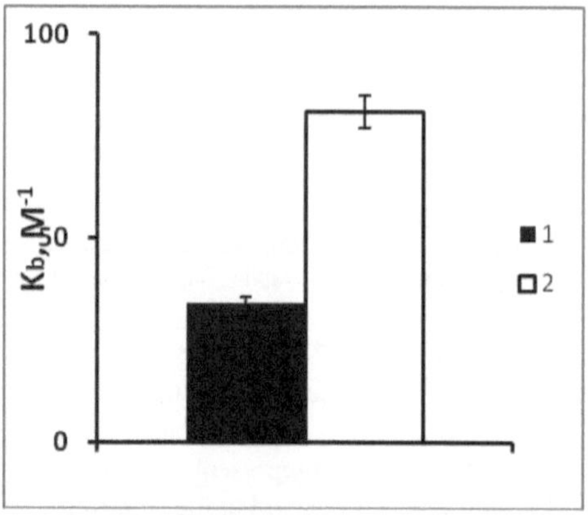

FIGURE 21.3 Diagram of binding constants of o-NA with AOT reverse micelles in the presence of 0.05 M potassium iodide (■) and 0.05M sodium fluoride (□).

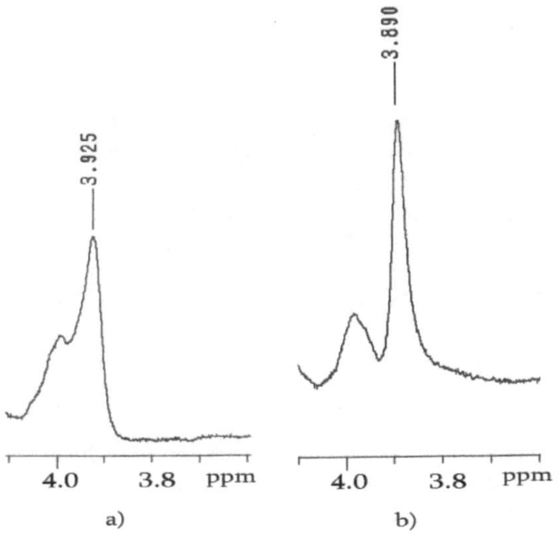

FIGURE 21.4 NMR spectra of sodium fluoride (a) and potassium iodide (b).

21.4 CONCLUSIONS

Microenvironment of AOT reversed micelles has been studied by IR, UV and NMR spectroscopic methods. The influence of kosmotropic and chaotropic ions on the ratio of free, bound and trapped water fractions was investigated. It was found that formation of free water is promoted by fluoride anions in water core, but producing of free water is inhibited in the presence of iodide ions. Binding constants of the optical probe to AOT head groups are higher in the presence of fluoride anions in the water pools of the reverse micelles.

KEYWORDS

- **chaotropic and kosmotropic ions**
- **confined water**
- **nanocages**
- **reverse micelle**

REFERENCES

1. Li, Q., Weng, S. F., Wu, J. G., Zhou, N. F., Comparative study on structure of solubilized water in reversed micelles. 1. FT-IR spectroscopic evidence of water/AOT/n-heptane and water/NaDEHP/n-heptane systems, J. Phys. Chem. B, 102, 3168–3174, 1998.
2. Levinger, N. Water in confinement, Science, 298, 1722–1723, 2002.
3. De, T. K., Maitra, A. Solution behavior of Aerosol OT in non-polar solvents, Adv. Colloid Interface Sci., 59, 95–193, 1995.
4. Moulik, S. P. Paul, B. K. Structure, dynamics and transport properties of microemulsions, Adv. Colloid Inteface Sci. 78, 99–195, 1998.
5. Onori, G., and Santucci, A. IR Investigations of Water Structure in Aerosol OT Reverse Micellar Aggregates, J. Phys. Chem, 97, 5430, 1993.
6. Najjar, R., Microemulsions – A brief Introduction. In: microemulsions – An Introduction to Properties and Applications, InTech, pp. 3–30, 2012
7. Sachs, J. N., Woolf Th.B. Understanding the Hofmeister Effect in Interactions between Chaoropic Anions and Lipid Bilayers: Molecular Dynamics Simulations, J.Am.Chem.Soc., 125(29), 8742–8743, 2003.
8. Falcone, R. D., Baruah, B. Gaidamauskas, E., Rithner, C. D., Correa, N. M., Silber, J. J., Crans, D.C and Levinger, N. E. Layered Structure of Room-Temperature Ionic Liquids in Microemulsions by Multinuclear NMR Spectroscopic Studies, Chem. Eur. J., 17, 6837–6846, 2011.
9. Maitra, A. Determination of Size Parameters of Water-Aerosol OT-Oil Reverse Micelles from Their Nuclear Magnetic Resonance Data, J. Phys. Chem., 88, 5122–5125, 1984
10. Haering, G., Luisi, P. L., Hauser, H. Characterization by Electron Spin Resonance of Reversed Micelles Consisting of the Ternary System AOT-Isooctan-Water, J. Phys. Chem., 92, 3574–358l, 1988.
11. Nazario, L. M.M., Hatton, T. A., Grespo, J. P. S. G. Nonionic cosurfactants in AOT reversed micelles: Effect on percolation, size, and solubilization site, Langmuir, 12, 6326–6335, 1996
12. Zhang, X., Chen, Y., Liu, J., Zhao, J. and Zhang, E. Investigation on the Structure of Water/AOT/IPM/Alcohols Reverse Micelles by Conductivity, Dynamic Light Scattering, and Small Angle X-ray Scattering. J. Phys. Chem. B, 116, 3723- 3734, 2012
13. Jain, T.K, Varshney, M., Maitra, A. Structural studies of aerosol OT reverse micellar aggregates by FT-IR spectroscopy, J. Phys. Chem., 93, 7409–7416, 1989.
14. Correa, N. M., Silber, J. J. Binding of Nitroanilines to Reverce Micelles of AOT n-Hexane, J. Mol. Liq., 72 163, 1997
15. Falcone, R. D., Silber, J. J., Biasutti, M. A., Correa, N. M. Binding of o-Nitroaniline to Nonaqueous AOT Reverse Micelles, Organic Chemistry in Argentina, ARKIVOC, vii, 369, 2011.
16. MacDonald, H., Bedwell, B., Gulari, E. FTIR Spectroscopy of Microemulsion Structure, Langmuir, 2, 704, 1986.

CHAPTER 22

SYNTHESIS AND CHARACTERIZATION OF A NEW NANO COMPOSITE

SHAHRIAR GHAMMAMY,[1] SADJAD SEDAGHAT,[2] MAHSA KHOSBAKHT,[1] and REZA FAYAZI,[3] and AMIR LASHGARI[1]

[1]*Department of Chemistry, Faculty of Science, Imam Khomeini International University, Qazvin, Iran, E-mail: shghamami@yahoo.com*

[2]*Department of Chemistry, Faculty of Science, Islamic Azad University, Malard Branch, Malard, Iran*

[3]*Department of Chemistry, Faculty of Science, Islamic Azad University, Ardebil Branch, Ardebil, Iran*

CONTENTS

ABSTRACT

Nanotechnology has gained a great deal of public interest due to the needs and applications of nanomaterials in many areas of human endeavors. Nano composites are a special group of materials with unique features and extensive applications in diverse fields. Studying these particular features has always been of great interest to many scientists. As a result of this research a new nano composite of titanium dioxide was produced. The nanocomposite, which synthesized is in proportion to the weight of Al_2O_3-TiO_2 was characterized by the FT-IR, XRD and SEM.

22.1 INTRODUCTION

Nanotechnology is the ability to work on a scale of about 1–100 nm in order to understand, create, characterize and use material structures, devices and systems with new properties derived from their nanostructures [1]. All biological and man-made systems have the first level of organization at the nano scale. Nanostructured materials have been extensively explored for the fundamental scientific and technological interests in accessing new classes of functional materials with unprecedented properties and applications [2–6]. In recent years, there has been an increasing interest in the synthesis of nanosized crystalline metal oxides [7]. This is based on positively perceived characteristics of these nanocomposites. Such characteristics include mechanical performance, electric behavior, thermal properties, biodegradability, optical properties, bactericidal effects, magnetic characteristics and transport, permeation and separation properties [8–11]. Two building strategies are currently used in nanotechnology: a "top-down" approach and the "bottom-up" approach. The commercial scale production of nano materials currently involves basically the "top-down" approach, in which nanometric structures are obtained by size reduction of bulk materials, by using milling, nanolithography, or precision engineering. Smaller sizes meaning a bigger surface area, desirable for several purposes. The newer "bottom-up" approach, on the other hand, allows nanostructures to be built from individual atoms or molecules capable of self-assembling [9]. In this research a new titanium nanocomposite was synthesized and characterized. This nano composite has a mixed oxide base and can classify as an inorganic base nano composite.

22.2 MATERIALS AND METHODS

22.2.1 MATERIALS

All material was prepared by Merck Company and used as received without further treatment. Solvents that were used for reactions were purified and dried by standard procedures. In this study, the alumina powder (with purity higher than 90%) and titanium dioxide nano powder with an average grain size 21 nm with a purity of 99.9% of which approximately 80% and 20% rutile/anatazitis with the weight ratio were used. XRD diffractions were studied with X-ray diffraction device Siemens D500 Diffractometer model. In all phases of Cu-Kα radiation with a wavelength of 1.5404 Å was used. The morphology and size of the nanocomposites were studied with a JEOL 2010 scanning electron microscope (SEM).

22.2.2 PREPARATION OF AL2O3-TIO2 NANOPARTICLE

About 3.4 g TiO_2 was taken to the beaker (50 mL) and in added in 50 mL of water. Then this component of the beaker was added to another beaker that is including 5 g of Al_2O_3. The whole solution was mixed; finally the compound was precipitated and separated. The product, dried at room temperate for 3 h. Then this nano composite was taken to the electric furnace at 185°C for 5 h until to calsinate. FT-IR spectrum is the main reason of producing mentioned nano composite.

22.2.3 CHARACTERIZATION OF NANO COMPOSITE

New nanocomposite was characterized by IR, XRD and SEM techniques. IR spectrum shows the bands of components such as TiO_2. (Figure 22.1) The XRD pattern showed the formation of a nanocomposite. The diffraction pattern was similar to other published nanocomposites. (Figure 22.2)

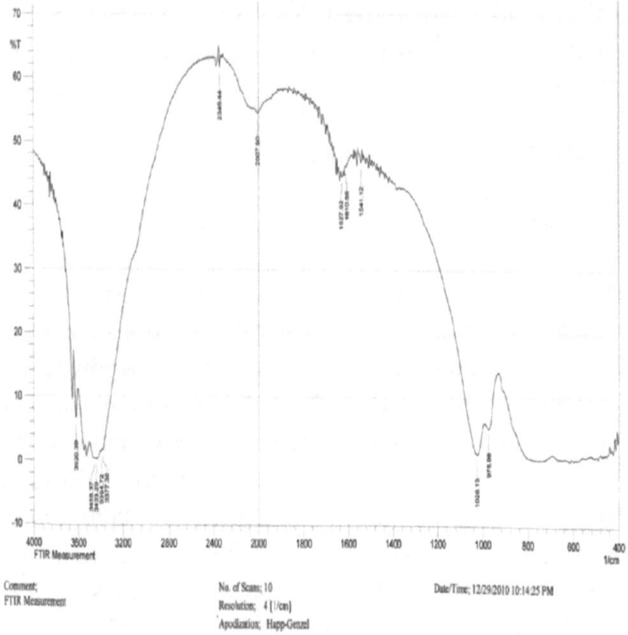

FIGURE 22.1 FT-IR spectrum Al_2O_3-TiO_2nano-composite in KBr disk.

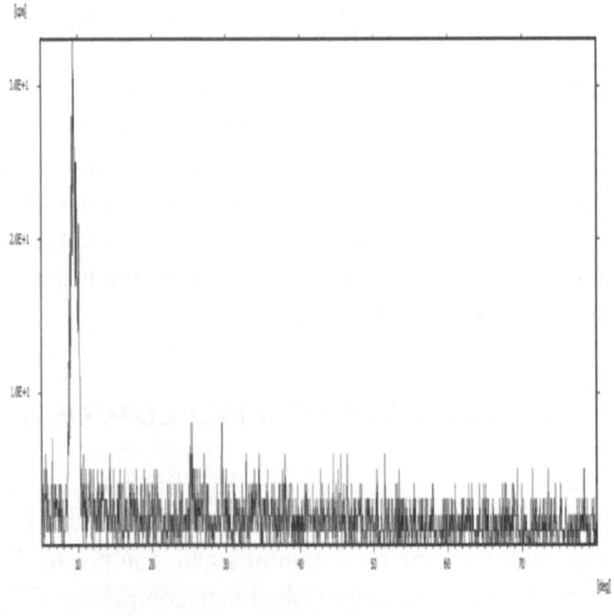

FIGURE 22.2 XRD pattern of Al_2O_3-TiO_2 nanocompositės.

Scanning electron microscope (SEM) pictures shown the composition of nanocomposites in pictures (Figure 22.3). These pictures can show the homogency of a mixture of two mentioned compounds easily shown the morphologies.

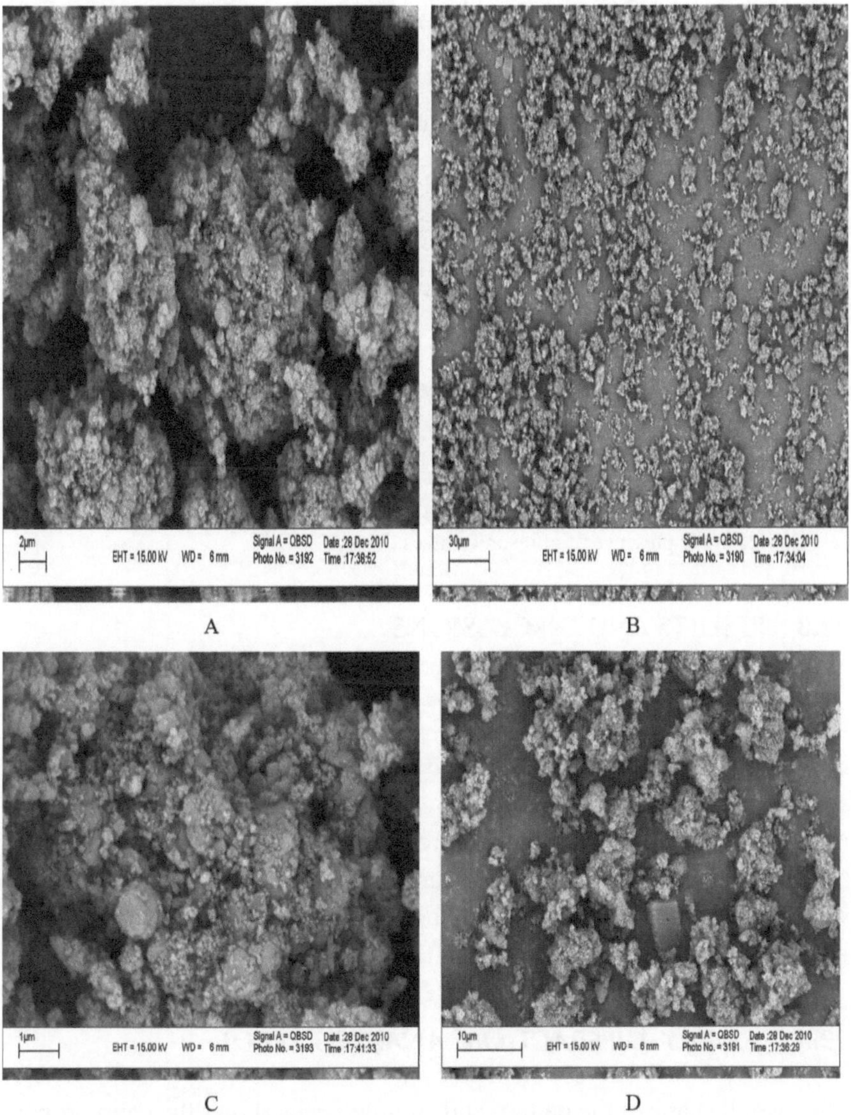

FIGURE 22.3 (Continued)

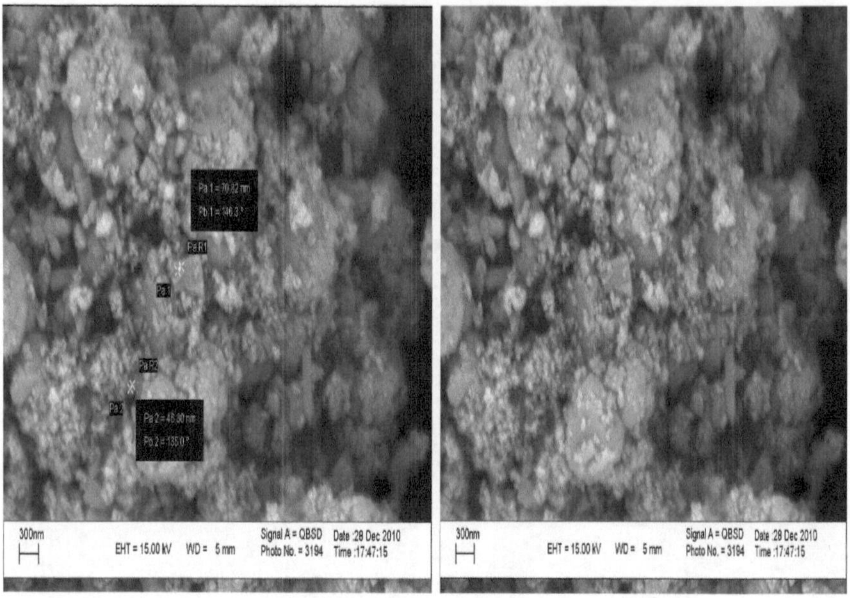

F E

FIGURE 22.3 Al_2O_3-TiO_2nano-composite morphology with Magnification (a) 1000 times, (b) 5000 times, (c) 10000 times, (d) 30000 times, (e) 50,000 times, (f) shows the particle size in the presence of ultrasonic waves.

22.3 RESULTS AND DISCUSSIONS

22.3.1 VIBRATIONAL SPECTRA OF (FT-IR)

FTIR absorption was used in order to check the characteristic bands of the synthesized powder. Similar spectra were observed for other nanocomposites. The bands at 450–900 cm^{-1} were related to Ti-O bonds [10, 11]. The spectrum shows two strong IR absorption bands at 3154 and 1409 cm^{-1} which are characteristic of H–O–H bending of the H_2O molecules revealing the presence hydroxyl groups in the as prepared sample (Figure 22.1).

22.3.2 X-RAY DIFFRACTION ANALYSIS (XRD)

The overall shape of the pattern and angular peaks shows the nano properties of a nano composite. The characteristic pattern of components could be found in the XRD pattern such as absorption 2θ=9.5–10, related to the TiO_2 and peak at 2 θ=29–30, related to Al_2O_3.

XRD spectra using the approximate size of the particles in the nano-composite can be calculated. When particles are smaller than 100 nm XRD peaks are much wider than this factor can be used to estimate the size of the nanoparticles. Shrr equation used for this purpose:

$$D = 0.9\lambda/B\cos\theta$$

In the above equation D in terms of particle diameter Å, B corresponds to the width of the strongest peak at half height in radians, and θ is the angle at which the peak appears. XRD analysis shows the synthesized nano composites. The XRD pattern of nano-composite Al_2O_3-TiO_2 in scattering observed and Nano TiO_2 sharp peak appeared at 2Θ=9.50–10 is clearly seen in all samples.

22.3.3 SEM PICTURES

Scanning Electron Microscope image of Al_2O_3-TiO_2 nanocomposite with different magnification for 5 hours at 185° C has been shown admixture Figure 22.3. Scanning electron microscopic observations of Al_2O_3-TiO_2 nanocomposite shows that the powder particles due to heat released during admixture of the flexible and an increase in average particle size and morphology is irregular. Due to the heat released during admixture irregular particles are particles in different forms (crystalline, includes and snow mass) are observed. With increasing admixture time to 8 hours of powder particles due to higher heat release, the grain size was larger and thus reducing the distance between the atoms and consequently the particles are too large. Figure 22.3 the morphology of this nano composite as seen in SEM pictures is mixed mode and semi spherical. As is know the morphologies sometimes arise spontaneously as an effect of a templating or directing agent present in the synthesis such as miscellar emulsions or anodized alumina pores, or from the innate crystallographic growth patterns of the materials themselves. Amorphous particles usually adopt a spherical shape (due to their microstructural isotropy) whereas the shape of anisotropic microcrystalline whiskers corresponds to their particular crystal habit. At the small end of the size range, nanoparticles are often referred to as clusters. Spheres, rods, fibers, and cups are just a few of the shapes that have been grown.

22.4 CONCLUSION

The nanocomposite results were determined with SEM, FT-IR, and XRD, which confirmed the formation of Al_2O_3-TiO_2nanocomposite. In recent years, reducing the structural weight of the factor considered by manufacturers, especially in aerospace and automobile industries is, therefore, the use of lightweight aluminum and titanium alloys in these industries has grown considerably. As a result of this research a new nano composite of alumina-titanium dioxide was produced. The nanocomposite, which synthesized is in proportion to the weight of Al_2O_3-TiO_2 was characterized by the FT- IR, XRD and SEM. This new nanocomposite can have applications in automobile and medical industries.

ACKNOWLEDGMENTS

We gratefully acknowledge the financial support from the research council of Imam Khomeini International University.

KEYWORDS

- **Al_2O_3-TiO_2**
- **characterization**
- **nano composite**
- **nanoparticle**
- **SEM**
- **synthesis**
- **XRD**

REFERENCES

1. Moraru, C., Panchapakesan, C., Huang, Q., Takhistov, P. Nanotechnology: A New Frontier in Food Science. Food Technology., 12, 24, 2003.
2. Wu, Y., He, Y., Wu, T., Chen, T., Weng, W., Wan, H. Influence of some parameters on the synthesis of nanosized NiO material by modified sol-gel method. J. Mater. Lett., 61(14–15), 3174–3178, 2007.
3. Neuberger, T., Scopf, B., Hofmann, H., Hofmann, M., Rechenberg, B. V. Superparamagnetic Nanoparticle for Biomedical Applications. J. Magn. Magn. Mater., 293, 438, 2005.
4. Schiffrin, D. J. Capped Nanoparticles as Potential Electronic Components with Nanoscale Dimension. MRS Bull., 26, 1015, 2001.
5. Ramesh, T., Vishnu, N., Kamath, P. Synthesis of nickel hydroxide: Effect of precipitation conditions on phase selectivity and structural disorder. J. Power. Sources, 156, 655, 2006.
6. The-Long, L., Youn-Yuen, S., Gim-Lin, H., Chia-Chan, L., Chen-Bin, W. Microwave-assisted and liquid oxidation combination techniques for the preparation of nickel oxide nanoparticles. J. Alloys Compounds, 450, 318, 2008.
7. Ying, W., Yiming, H., Tinghua, W., Weizheng, W., Huilin, W. Effect of synthesis method on the physical and catalytic property of nanosized NiO. J. Mater. Lett., 62, 2679–2682, 2007.
8. Shi, H., Liu, F., Yang, L., Han, E. Characterization of protective performance of epoxy reinforced with nanometer-sized TiO_2 and SiO_2. Progress in Organic Coatings Characterization., 62, 359, 2008
9. Reddy, K.M, Kevin F, Jason, B., Denise, G. W., Cory, H., Alex, P. Selective toxicity of zinc oxide nanoparticles to prokaryotic and eukaryotic systems. J. Appl. Phys. Lett., 90, 1, 2007.
10. Ispir, E., Kurtoglu, M. The d^{10} metal chelates derived from Schiff base ligands having silane: synthesis, characterization, and antimicrobial studies of cadmium(II) and zinc(II) complexes. Synthesis and Reactivity in Inorganic, Metal-Organic and Nano-Metal Chemistry, 36, 627, 2006.
11. Constable, E. C. The coordination chemistry of 2,2′:6′,2″-terpyridine and higher oligopyridines. Academic Press, 30, 69, 1986.

CHAPTER 23

ORGANOMINERAL IONITES

M. B. GURGENISHVILI, I. A. CHITREKASHVILI, G. SH. PAPAVA, SH. R. PAPAVA, V. A. SHEROZIA, N. Z. KHOTENASHVILI, and Z. SH. TABUKASHVILI

P.G. Melikishvili Institute of Physical and Organic Chemistry, Iv. Javakhishvili Tbilisi State University, Georgia, E-mail: marina.gurgenishvili@yahoo.com

CONTENTS

ABSTRACT

Organomineral ionites have been synthesized, in which natural mineral sorbent is chemically bound to organic part of a molecule, containing ionogen groups. Ionogen groups of these ionites contribute to efficient exchange in water solutions. They might be used for cleaning drainage waters and technical solutions, as well as for purification of medicinal preparations from various admixes. Hydrogen forms of natural zeolite – clinoptilolite and bromoacetic acid were used to resolve this problem. Chemical modification of clinoptilolite, by inculcation of ionogen groups into zeolite skeleton, enables us to increase significantly ionite exchange capacity. Static exchange capacity of a cationite with carboxyl ionogen groups increases from 0.1–0.9 (for chemically unmodified zeolite) to 5–6 mg-equiv/g.

23.1 INTRODUCTION

Organic monomers with double links, containing ionogen groups were used to synthesize ionites. Ionites were polymerized, as a result of which organic ionites were obtained with ionogen groups in elementary rings of polymers characterized by spatial structure.

At the application of inorganic natural sorbents, for example zeolites in ionite synthesis, zeolite is presented just as a matrix, playing a role of a skeleton of synthetic organic ionite [1–3].

It was considered interesting to synthesize organomineral ionites, in which natural mineral sorbent would be chemically bound to organic part of a molecule, containing ionogen groups.

In the work [2] zeolite was modified through its treatment in aceto-acetic acid. In that method of ionite synthesis, ionogen groups were not introduced into zeolite skeleton, capable to participate in ion exchange process. As a result of such treatment zeolite exchange capacity fell from 0.9 to 0.3 mg-equiv/g.

The present work pursued to synthesize ionites in which organic molecule, containing ionogen groups, would be chemically bound to zeolite skeleton.

Ionogen group of such ionites is capable to contribute to efficient exchange in water solutions and they might be used for treatment of drainage waters, and technical solutions as well as for purification of medicinal preparations from various admixes.

23.2 EXPERIMENTAL PART

23.2.1 MATERIALS

For the resolution of this problem the authors of the present work used natural zeolite –clinoptilolite of the following composition, expressed in oxides:

$$(Na_2K_2)O \cdot Al_2O_3 \cdot 10\ SiO_2 \cdot 8\ H_2O.$$

The other component was bromoacetic acid ($BrCH_2COOH$), which due to increased activity of bromine atom easily comes into reaction with hydrogen atom of H-form zeolite.

Zeolite easily absorbs bromoacetic acid, but bromoacetic acid molecules are washed out easily from zeolite pores too. Preliminary experiments proved that at the treatment by water, bromoacetic acid is washed out completely.

23.2.2 TECHNICS

Initially zeolite was treated by 5% hydrochloric acid solution. In this way it is transformed into hydrogen form (H-form zeolite). At this moment cations, mainly K and Na, were substituted by hydrogen atoms. After thorough washing it was dried to remove water. At the interaction of bromoacetic acid and H-form zeolite, in organic solvent, at the temperature 60 – 80°C, we obtained ionite with carboxyl ionogen groups. Schematically the reaction can be presented as follows:

$$\text{Z-Na (K)} + \text{HCl} \rightarrow \text{Z} - \text{H} + \text{Na(K)Cl}$$

$$\text{Z} - \text{H} + \text{BrCH}_2\text{COOH} \rightarrow \text{Z} - \text{CH}_2\text{COOH} + \text{HBr}$$

IR-spectral studies of ionites showed absorption bands in the area 1090–1050 cm^{-1}, inherent to Si–O–C chemical bond, referring to the fact that the remainder of acetic acid is chemically bound to zeolite skeleton. Reiterated washing by water showed that nature and intensity of absorption bands doesn't change, which refers to the presence of Si–O–C chemical bond.

Cationite on the basis of modified natural zeolite – clinoptilolite and bromoacetic acid was obtained as follows: natural zeolite–clinoptilolite was crushed into particles of 1–2 mm, than 0.5 N hydrochloric acid solution was added to it and the mix was heated for 4–5 hours. Then it was decanted many times till it showed negative reaction to chlorine atom, and finally, it was dried at 120°C.

H-form zeolite and bromoacetic acid were placed in thick-wall ampoule. Distilled water of the volume equal to that of zeolite was added and was placed in metallic capsule. Ampoule was welded, placed in metallic capsule and heated over silicone bath. Temperature was increased gradually up to 60–65°C. Reaction duration was 6–8 hours. Then the ampoule was opened and its contents was shifted to a beaker and filtered. Remainder on a filter was washed by warm distilled water and then was diluted in hydrochloric acid and again by distilled water till neutral reaction and negative reaction to bromine. The obtained cationite was dried on air and then in drying cabinet at 80°C.

23.3 RESULTS AND DISCUSSION

At the synthesis of ionites we studied impact of various factors on the process of reaction and on static exchange capacity of the obtained cationite, considering conditions of condensation reaction of bromoacetic acid with H-form clinoptilolite. Effect of quantity (mass.%) of bromoacetic acid, temperature and reaction duration on the reaction progress, was studied.

Experiments showed that at the increase of bromoacetic acid quantity from 0.2 to 7 mas./h., static exchange capacity of the zeolite was increased. Results of experiments are given in Figure 23.1. The data given in figure show that optimal ratio of zeolite and bromoacetic acid is 7:1. Further increase of bromoacetic acid composition doesn't lead to any significant increase in static exchange capacity of a cationite, which is limited by composition of cations in clinoptilolite.

Study of temperature effect showed that optimal is temperature 60–65°C. Decrease of increase of reaction temperature leads to reduction of exchange capacity.

Reaction time also affects a process of interaction of bromoacetic acid and H-form zeolite. Optimal is reaction duration 8 hours. Further increase of reaction time doesn't exert any significant influence on cation exchange capacity.

Thus, study of the influence of various factors on the process of interaction, showed that to reach maximum index of static exchange capacity of a cationite, optimal conditions for the reaction process are: zeolite and bromoacetic acid ratio (mas.h) – 7:1, correspondingly; reaction temperature 60–65°C, and reaction duration 8 hours.

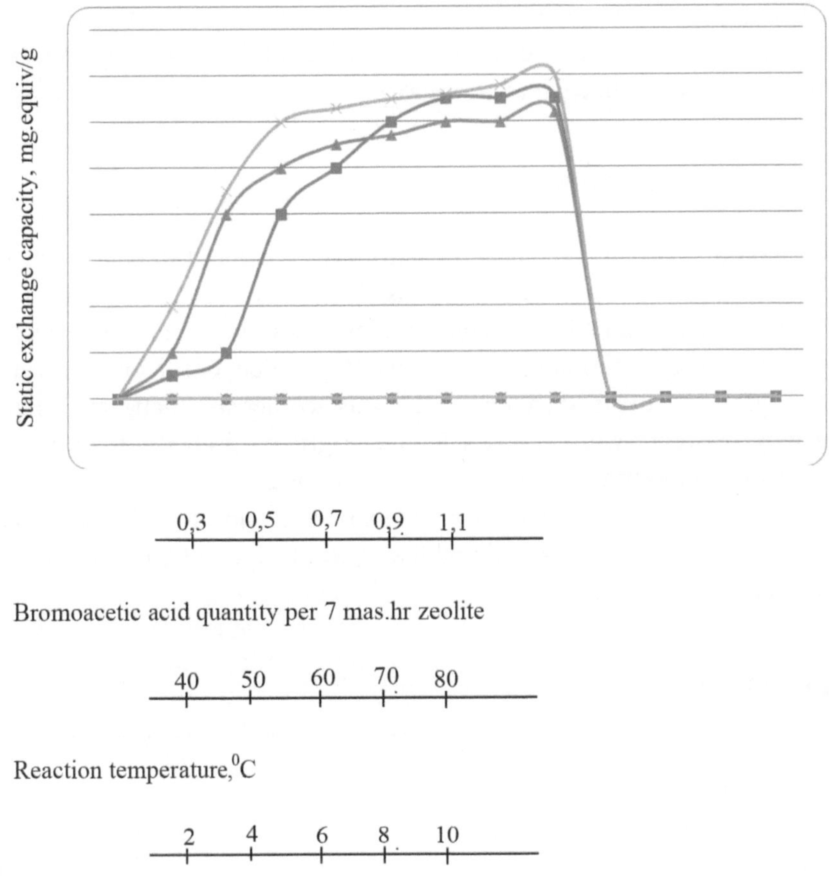

FIGURE 23.1 Dependence of static exchange capacity (SEC) of cationite based on chemically modified clinoptilolite on the conditions of synthesis (■ - Bromoacetic acid quantity, ▲ - Reaction temperature, × - Reaction time).

Static exchange capacity of low-acidity cationite with carboxylic ionogen groups at the treatment in 0.1 N NaOH water solution increase from 0.1–0.9 (for chemically unmodified zeolite) to 5–6 mg-equiv/g.

As is seen from the obtained results, chemical modification of clinoptilolite by means of introduction of ionogen carboxyl groups into the skeleton, enables us to increase significantly their exchange capacity. Carboxyl groups are easily dissociated, which contributes to improvement of kinetics of exchange with various ions.

23.4 CONCLUSIONS

Synthesis of organomineral zeolite on the basis of natural sorbent – clinoptilolite type zeolite, in which sorbent is chemically bound to organic part of a molecule, containing ionogen groups – was performed. Chemical modification of clinoptilolite, by means of introduction of ionogen carboxyl groups into its skeleton, enables us to increase static exchange capacity of ionite with carboxyl ionogene groups from 0.1 to 0.9 (for chemically unmodified zeolite) to 5–6 mg-equiv/g.

Low-acidity cationite with carboxyl ionogen groups in zeolite skeleton was synthesized on the basis of modified H-form clinoptilolite and bromoacetic acid.

KEYWORDS

- **cationite**
- **clinoptillolite**
- **ionite**
- **ionogen group**
- **polymerization**
- **zeolite**

REFERENCES

1. Freeman, D. C., Stamires, D. N. J. Chem. Phys., 35(3), 799, 1961.
2. Papava, G. Sh., Mgeladze, B. M., Papava, K. P., Dokhturishvili, N. S., Maisuradze, N. A., Gavashelidze, E. Sh. Cationites on the basis of zeolites. Bulletin of Acad. Sci. GSSR, 114(2), 303, 1984.
3. Mdivnishvili, O. M., Makharadze, L. V. Bulletin of GSSR, Acad. of Sci., 2, 65, 1972.

CHAPTER 24

ZEOLITE BASED HYBRID CATIONITES

I. A. CHITREKASHVILI, M. B. GURGENISHVILI, G. SH. PAPAVA, V. A. SHEROZIA, K. R. PAPAVA, N. Z. KHOTENASHVILI, and Z. SH. TABUKASHVILI

P. Melikishvili Institute of Physical and Organic Chemistry, Iv. Javakhishvili Tbilisi State University, Georgia, E-mail: marina.gurgenishvili@yahoo.com

CONTENTS

ABSTRACT

Low acidity cationite with carboxyl ionogen groups in zeolite skeleton was synthesized on the basis of modified clinoptilolite (H-form) and chloropropionic acid. At the treatment of the clinoptilolite by water, Na leaves clinoptilolite skeleton, forming water-soluble chloride. At washing of the hydrogen form clinoptilolite that is formed in the process Na is removed and silane groups are formed. At the interaction of these groups with chloropropionic acid the chemical bonds are created between them.

The obtained cationite is stable to acids and alkali; it possesses high mechanical strength and increased exchange capacity.

24.1 INTRODUCTION

It is known that ionites based on natural inorganic sorbents are obtained through their impregnation by monomers containing ionogen groups, by their further polymerization [1]. Skeleton of inorganic sorbents, in this case plays a role of a matrix of organic ionite.

There are ionites, and namely anionites, in which zeolite replaces skeleton of synthesized polymer [2, 3].

Organic ionites with organic carcass (with polymer skeleton) are complexity of technology of their synthesis and high price. Process of obtaining of ionite skeleton, which is an organic polymer, is connected with multi-stage organic synthesis of both original monomers and polymers on their base.

Skeleton of mineral rock (natural) contributes to the increase of thermal resistance of ionites. Such ionites are able to function at increased temperatures, which is most important at their exploitation in extreme conditions.

In special literature we find description of cationite with inorganic skeleton. Widening of assortment of such type ionites is a rather urgent task. Widening of number of possible monomers, able to fulfill the same functions, will increase the possibilities and spheres of their application in the form of ion-exchangers.

It was interesting to synthesize cationite with inorganic skeleton, on the base of natural mineral material. Synthesis of this type ion-exchangers is interesting since a ready skeleton enables us to use all active centers of organic polymer, without their spending for creation of ionite skeleton

24.2 EXPERIMENTAL PART

24.2.1 MATERIALS

In the synthesis of cationites with inorganic skeleton, one of the original components chosen by us was natural zeolite, clinoptilolite, as all other zeolites, is characterized by molecular-screen properties. Size of entry apertures of this zeolite is accessible to molecules with critical diameter 0.35 nm [4]. According to other data diameter of efficient apertures of this zeolite is 0.44 nm [4].

Significant sorption parameter of zeolite is quantity of water, present in intra-crystalline cavities, which to a significant extent depends on the zeolite form. There are many scientific works dealing with the study of physical-chemical characteristics of clinoptilolite-containing tuffs of various deposits. It was revealed that all of them are complex by their structure and the main difference is in their purity, that is, in the presence of incidental rocks, as well as in cation composition.

Idealized composition of elementary cell: $Na_6[(AlO_2)_6(SiO_2)_{30}]$ $24H_2O$;
Typical oxide formula: $(Na_2, K_2)O \cdot Al_2O_3 \cdot 10SiO_2 \cdot 8H_2O$;
Limits of composition alteration: Si/Al = 4.25 – 5.25, Na, K " Ca, Mg;
Density: 2.16g/cm^3; volume of elementary cell: 2100 Å^3, free volume: 0.34c m^3/cm^3;
Channel system, bi-dimensional, 8- and 10 number rings: 0,40x0,55; 0,44 x 0,72 nm.
Skeleton density – 1.71g/cm^3;

Tables 24.1 and 24.2 show the physical-chemical and structural characteristics and the data of elementary composition of clinoptilolite are offered.

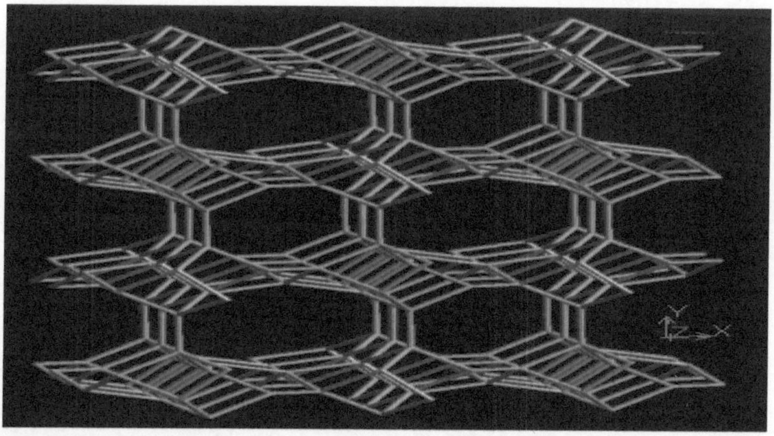

FIGURE 24.1 Clinoptilolite skeleton structure.

TABLE 24.1 Physical-Chemical and Structural Characteristics of Clinoptilolite

Typical composition of elementary cell	Crystalline structure	Volume of elementary cell, nm^3	Free volume cm^3/cm^3	Aperture size, nm	Skeleton density, g/cm^3	KOE, m.equiv/g
$Na_6Al_6Si_{30}O_{72} \cdot 24H_2O$	monocline	2.10	0.34	0.40x0.55 in 8-memb. 0.44x0.72 in 10-memb. rings	1.71	2.16

TABLE 24.2 Chemical Composition of Clinoptilolite-Containing Rocks of Dzegvi Origine, Khekordzula Section

%	Clinoptilolite original sample
50.91	SiO_2
15.42	Al_2O_3
0.78	TiO_2
7.50	Fe_2O_3
0.72	FeO
0.09	MnO
3.32	MgO
1.90	CaO
5.20	Na_2O
2.80	K_2O
0.30	P_2O_5
—	SO_3
6.10	H_2O
4.72	ПППП
99.5	Σ

Si/Al=2.89.

The data of chemical analysis show that a clinopyilolite specimen (sect. Khekordzula) is mainly of Na-form.

Other components are chloropropionic acid ($ClCH_2CH_2COOH$), which reacts with silanol groups of clinoptilolite.

24.2.2 TECHNICS

Cathionites on the basis of modified natural zeolite – clinoptilolite and chloropropionic acid has been obtained as follows: natural zeolite – clinoptilolite, was crushed to the particle size 1–2 mm., 0.5 N hydrochloride solution was added to it and the mix was heated for 4–5 hours. Then it was decanted, washed iteratively till negative reaction to chlorine ion and then it was dried at 120°C.

Hydrogen form clinoptilolite has also been obtained according to other method. Zeolite granules, prepared in advance (0.5–1.0 mm), crushed manually, were treated in 0,1n NH_4Cl solution, then were washed in distilled H_2O and dried in a thermostat at 150–200°C.

IR-spectral studies of modified samples prove the preservation of crystalline skeleton of clinoptilolite after cathion modification. Insertion of ammonium ions (NH_4^+) to the zeolite decreases of "entry apertures" of clinoptilolite. At heating of ammonium form clinoptilolite up to 150°C and higher, it is transformed into hydrogen form and dimensions of entry apertures of zeolites are increased.

The obtained hydrogen form zeolite and chloropropionic acid were placed in thick-walled ampoule. Distilled water was added to the mix, in quantity equal to that of zeolite. Ampoule was welded and placed in metallic capsule and was heated over silicone bath. Then temperature was increased gradually up to 150°C. Reaction duration was 8–10 hours. Then the ampoule was opened and its content was shifted to a beaker and filtered. The remainder on the filter was washed initially by warm distilled water, and then by diluted hydrochloric acid solution, and again by distilled water, up to neutral reaction to rinsing waters and negative reaction to chlorine. The obtained cationite was dried initially on air, and then in drying cabinet, at 80°C.

Schematically the reaction can be presented as follows:

$$Z\text{-}Na(K) + HCl \rightarrow Z - H + Na(K)Cl$$

$$Z - H + ClCH_2CH_2COOH \rightarrow Z\text{-}CH_2CH_2COOH + HCl$$

Ionite IR-spectral studies showed absorption bands in the area 1090–1050 cm^{-1}, inherent to Si–O–C chemical bond, referring to the fact that the remainder of propionic acid molecule is chemically linked with zeolite skeleton. Reiterated washing by water showed that nature and intensity of absorption bands doesn't change, which refers to presence of Si–O–C chemical bond.

24.3 RESULTS AND DISCUSSION

At the treatment of modified clinopyilolite by acid, we are able to remove Na and K from its skeleton, which form water soluble chlorides and at washing of the created hydrogen form clinoptilolite, Na and K are removed, as a result of which silane groups are formed. At the interaction of these groups with chloropropionic acid chemical bonds are formed between them, according to the following scheme:

$$- Si - OH + ClCH_2CH_2COOH \rightarrow - Si - O - CH_2\, CH_2\text{-}\, COOH + HCl$$

Cationite obtained at the interaction with chloropropionic acid and H-form zeolite, is characterized by the following properties: grain size 1–2 mm, apparent density 0.6–0.8 g/mL, static exchange capacity to 0.1 N NaOH solution 5–6 mg-equiv/g.

If we compare exchange capacity of the obtained cationite with that of the natural zeolite/original, we'll see that its exchange capacity 0.2–0.3 increases 4–5 times. Alongside with it, introduction of easily dispersible ionogen carboxyl groups into zeolite skeleton, contributes to improvement of kinetics of exchange of a series of ions. Cationite is stable to acids and alkali, and is characterized by high mechanical strength.

Thermal stability of a cation was evaluated at its heating for 24 hours, at various temperatures. Investigations showed that at the increase of temperature from 60 to 200°C exchange capacity of a cationite decreases from 5 to 2 mg-equiv/g.

Thermal stability of an ionite was determined in dynamic conditions of heating. It appeared that thermal stability remains at the level of results obtained at the terms of their testing in static conditions.

Thus cationite obtained on the basis of modified natural zeolite – clinoptilolite (H-form) is characterized by high thermal stability and mechanical strength, high static exchange capacity and thus can be used at high temperatures.

24.4 CONCLUSIONS

Low acidity cationite with carboxyl ionogen groups was obtained on the basis of modified clinoptilolite (H-form) and chloropropionic acid.

Spectroscopic studies proved that in IR-spectra of a cathionite there is an absorption band in the region of 1090–1050 cm^{-1}, that is characteristic to Si–O–C bond, that refers to the creation of ether bond.

Static exchange capacity with respect to 0.1 N NaOH, compared with that of natural clinoptilolite increase from 0.2–0.3 to 5–6 mg-equiv/g. Simultaneously, introduction of easily dissociable ionogen groups into zeolite composition, contributes to improvement of kinetics of exchange capacity of a series of other ions.

The obtained cationite is stable to the impact of acids and alkali, possesses high mechanical strength and increased thermal stability.

KEYWORDS

- **carboxyl group**
- **cationite**
- **clinoptilolite**
- **hybrid**
- **ionogen group**
- **zeolite**

REFERENCES

1. Cationites on the basis of vinyl acetate. Patent of Great Britain. 1456974, 1976.

2. Zeolite based anionite. Auth. Certificate. 710960, publ. in Bull. Inf., 3, 90, 1981.
3. Papava, G. Sh., Khotenashvili, N. Z., Dokhturishvili, N. S., Gelashvili, N. S., Gavashelidze, E. Sh. Zeolite based anionites. Plast. Masses, Moscow, 7, 51, 1988.
4. Mumpton, F. A. La roca magica: Uses of Natural Zeolites in Agriculture and Industry. Proc. Nat. Acad. Sci. USA, 96, 3463–3470, 1999.

CHAPTER 25

COMPOSITE MATERIALS BASED ON COAL TAR PITCH

I. KRUTKO, V. KAULIN, and K. SATSYUK

Donetsk National Technical University, Donetsk, Ukraine,
E-mail: techlab@ukr.net

CONTENTS

ABSTRACT

Coal tar pitch is a unique product with a rich set of properties, among which are polymeric ones. Modification of coal tar pitches provides a wide opportunity to adjust and change their properties. This opens a new direction – the use of coal tar pitch chemical potential for creating the disperse-filled compositional polymeric materials.

The obtained pitch composite has relatively high mechanical properties and thermal stability. The obtained material for all parameters can be attributed to the hard-burning polymer composite what gives it a cogent advantage comparably to thermoplastic polymers.

Tests of modified coal tar pitch as a polymer matrix showed that it can successfully compete with classical polymers in order to create compositional materials.

25.1 INTRODUCTION

Creating a composite polymeric material in recent years is considered as the main reserve of production of new materials with improved properties. Polymer composites consist of a plastic base (matrix), which is reinforced with fillers with high strength, hardness, etc. By varying the composition of matrix and filler, their ratio, the orientation of filler, a wide range of materials with desired set of properties can be received.

Properties of polymeric composite materials are determined by the properties of the polymer matrix, the nature of filler and the interfacial interaction between them.

One of the promising application of coal tar pitch chemical potential is creation of new pitch composite materials.

Coal tar pitch is a multicomponent mixture of condensed aromatics hydrocarbons and heterocycles, which form a permolecular structure as a result of intermolecular interactions (hydrogen bond, dipole interaction, dispersion interaction).

Coal tar pitch can be used as polymeric matrix. This statement is substantiated by the fact that coal tar pitch has some properties inherent to polymers [1]. Despite the fact that coal tar pitch is not the classic polymer it is characterized by three states of amorphous polymers and plastic range. A characteristic feature of both classical polymers and coal tar pitch is the ability to form permolecular structures. Morphological forms of structural formations of polymers (fibrils, spherulites, globular) are observed in coal tar pitch.

The low mechanical strength and heat resistance of coal tar pitch does not allow to produce appropriate composite materials.

Coal tar pitch due to the nature of its composition and structure is a highly reactive material.

The whole system is in nonequilibrium state and any physical or chemical effects lead to irreversible changes in ratio of monomeric, oligomeric and highly condensed components.

Modification of coal tar pitch by active polymer additives has shown ability to adjust and modify the properties of coal tar pitch.

Active polymer additives have a great influence on thermo-chemical and structural transformations of coal tar pitch. Their effect is the change in structural-group composition, thermal, rheological and mechanical properties of coal tar pitch [2–4].

Thermochemical conversion at low temperatures (up to 200°C), taking place in coal tar pitch under the influence of active polymer additives, lead to viscosity increase and softening point of coal tar pitch on 15–20°C. Due to structural changes mechanical properties of coal tar pitch are improved: compressive strength is increased in 7 times, bending strength – 2 times, modulus of elasticity in compression – 7 times, flexural modulus in 3.5 times [5].

Introduction the filler to pitch polymeric matrix improves technological and operational properties of pitch composite [5]. The important factor in the process of composite materials creating is the nature of filler. Most interesting is the fibrous filler – chrysotile. Chrysotile is used in polymer compositions as a reinforcing filler, giving increased strength and flexural modulus to products, improving heat resistance.

One of the main tasks when creating composite pitch polymers is to optimize the filler content in order to keep a balance between the mechanical and thermal properties of material.

The results of studies concerning the effect of filler amount and the time-temperature conditions for obtaining the pitch composite on its mechanical properties and heat resistance are presented in article.

25.2 EXPERIMENTAL PART

For laboratory research coal tar pitch with the following fractional composition was used (%): α-fraction 34.9; α1-fraction 8.0; β-fraction 34.1; γ-fraction 31.0. Softening temperature: by method stem-ring 80°C, by Vicka 53°C. Viscosity at 135°C is 10 Pa·s, the density is 1300 kg/m^3, volatile matter content is 55%.

For coal tar pitch modification polyvinyl chloride (PVC) and polymer of polyolefin row with grafted maleic functional groups (PM – EV) were selected.

Polyvinyl chloride – amorphous polar polymer with density 1380–1400 kg/m^3, white amorphous powder with a glass transition temperature of 70–80°C, $K_F = 63$, the bulk density of 500 kg/m^3.

As the filler of pitch polymeric matrix chrysotile was used.

Chrysotile belongs to a class of disperse fillers in origin – mineral, in the form of fibers. On the chemical composition of chrysotile is highly hydrated magnesium silicate with general formula $Mg_6[(OH)_4Si_2O_5]_2$.

In accordance with practice of new materials development in laboratory plant experimental batches of pitch composite materials were received. The laboratory plant was a reactor equipped with auger mixers, and a circulation thermostat to heat the reactor.

The experimental procedure was as follows.

Coal tar pitch and modifiers (PVC 3% of coal tar pitch, PM- EV 5% of coal tar pitch) were mixed with a filler in an amount up to 40% of coal tar pitch. All operations were carried out in a high speed blending mixer.

Next, the resulting homogeneous mixture was charged into the reactor, which was preheated to desired temperature of experiment (t=140–180°C) and covered with a lid. After a certain period of time (depending on the temperature of experiment), when the mixture was sufficiently melted, the stirrer was engaged and experiment took place during required timed (τ=20–60 min). At the end of appropriate time the stirres was turned off and resulting pitch composite was unloaded. Next, by hot pressing using molds and hydraulic laboratory press to turn out batch of samples for testing them on heat resistance and mechanical strength.

Heat resistance refers to ability of sample to maintain its shape when a certain temperature and/or not exceed the predetermined threshold distortion while temperature test.

Determination of thermal stability of polymers by Vicka softening temperature (standard ISO 306) is that the sample is exposed to a predetermined load (A50 load 10N) and heated at a certain rate (500°C/h). During the test the temperature is determined, expressed in C, at which the indenter head is immersed in the sample to a depth of 1mm.

Multisamples for mechanical testing were prepared in accordance with standard ISO 3167. This approach provides a sample with a similar internal structure (orientation, residual stresses) and allows to compare results from different types of tests.

Compression testing and static bending were carried out on a universal testing machine M500 -CT (Testometric, UK). Mechanical tests were carried out in full compliance with current standards.

Flexural deformation – the most common types of polymer composites loading. Bending strength – the maximum effort that a sample can withstand when tested on a bend, determined according to ISO 178 (three-point method). Test procedure for compression sets by standard ISO 604.

25.3 RESULTS AND DISCUSSION

Behavior of pitch composite materials (PCM) at high temperatures is of particular interest in terms of their practical application.

Dependencies the softening point by Vicka (SPV) on conditions of pitch composite obtaining (mixing time and temperature) and the amount of filler are shown in Figures 25.1–25.2. As it seen from figures heat resistance of PCM increases with increasing the amount of filler (Figure 25.1), temperature and time of mixing (Figure 25.2).

As shown in Figure 25.1, when filling with chrysotile the significant increase of PCM heat resistance is achieved. While filler introduction in amount of up to 40% the SPV of pitch composite increases by 20–50°C

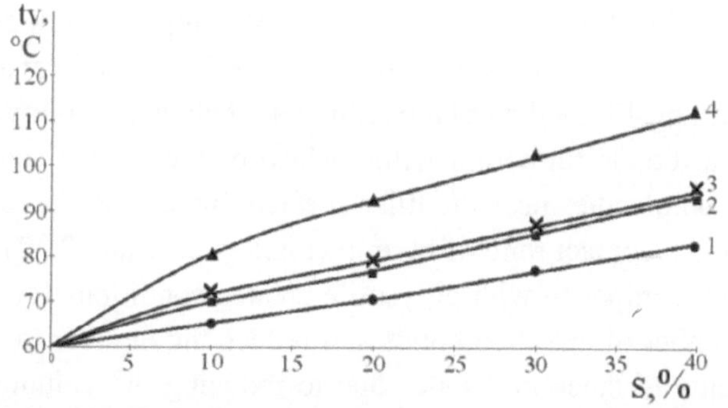

FIGURE 25.1 Influence of the amount of filler (s) on softening point by Vicka (tv) at: t=140°C, τ=20 min (1); t=140°C, τ=60 min (2); t=180°C, τ=20 min (3); t=180°C, τ=60 min (4).

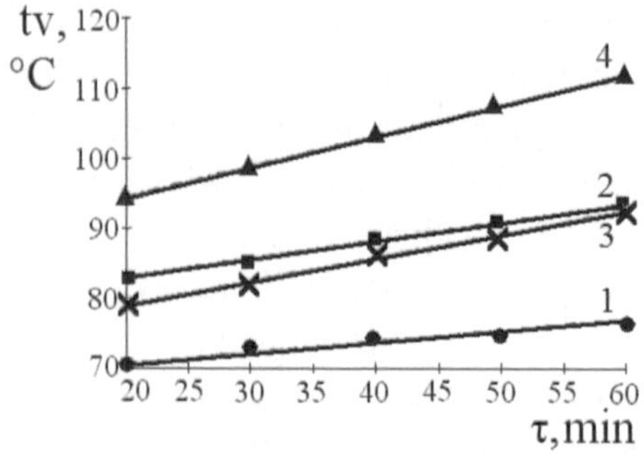

FIGURE 25.2 Influence of the time of experiment (τ) on softening point by Vicka (tv) at: t=140°C, s=20% (1); t=140°C, s=40% (2); t=180°C, s=20% (3); t=180°C, s=40% (4).

depending on the temperature-time conditions of its production. This explained by the fact that surface of filler adsorb segments of pitch polymer macromolecules. This forms a bond pitch polymer-filler and significantly reduces the segmental mobility, which increases the SPV.

With an increase of filler content the viscosity of pitch composites increases. Adhesive interaction between filler and matrix inhibits the fluidity of matrix, and therefore reduces the creep flow or deformation for time application.

According to modern concepts, the nature of heat resistance change is caused due to complexity of relaxation processes in the surface layers of polymers, adjacent to the solid surface.

Resistance deformability is directly caused by molecular motion while high temperatures. Rapid Vicka temperature rise on 50°C at high processing temperatures (180°C, Figure 25.2, line 4) compared with an increase of 20°C (processing temperature 140°C, Figure 25.2, line 2) at a degree of chrysotile filling 40% indicates the reduction of chain flexibility as a result of additional bonds formation with surface or the reduce conformation set of macromolecules near the filler surface. Increasing the temperature to 180°C and a contact time of 60 min (Figures 25.1 and 25.2) when filling the pitch composite with chrysotile creates conditions for increasing the total number of effective bonds and reduce the number of conformation segments of macromolecules due to reducing the volume in which segmental motion took place. This leads to a significant increase in heat resistance of PCM.

The maximum temperature of pitch composite by Vicka of 112°C is achieved at the mixing temperature of 180°C, contact time of 60 minutes and content of chrysotile 40%.

During pitch composite obtaining chemical and physical transformations in pitch polymer matrix take place. Besides, in formation of pitch composite properties intensity of interfacial interaction plays an important role. Volume of pitch polymer matrix immediately adjacent to the interface pitch polymer – filler has a structure and properties different from those in volume. The interface defines the intensity of interaction between pitch polymer and filler by two structural parameters: direct adhesive interaction of contacting phases and modification of structure of matrix pitch polymer in the contact area.

The first step of pitch composite production is pitch melting. Due to the low viscosity coal tar pitch wets chrysotile, which has a rough surface covered with hydroxyl groups, which improve adhesion interaction process (wetting).

In second step, the active polymer additives are melted, interdiffusion occurs and interaction with coal tar pitch, which is distributed on the surface of chrysotile fibers. The total contact area is increased in several times, which leads to intensification of coal tar pitch modification process. This process is accompanied by adsorption and a change in permolecular structure of adjacent layer.

Chemical reactions between pitch composite components having reactive functional groups strengthen interfacial interaction.

The dependence of pitch composite flexural strength on temperature experiment, the contact time and filler content is shown in Figures 25.3–25.5. Analysis of results showed that the temperature does not affect on flexural strength of pitch composite at 20 minutes of stirring (lines 1 and 2 in Figure 25.3). However, increasing the contact time to 60 minutes as the temperature rises to 180°C impairs the flexural strength of material (lines 3 and 4 in Figure 25.3).

The greatest influence on maximum bending stress has time and amount of chrysotile. With mixing time increasing the flexural strength decreases (Figure 25.4), and the greater with amount of filler increasing (lines 2 and 4).

Flexural strength of pitch composite increases with increasing the content of chrysotile (40%) at a contact time of 20 minutes, regardless of the temperature (curves 1 and 3 in Figure 25.5).

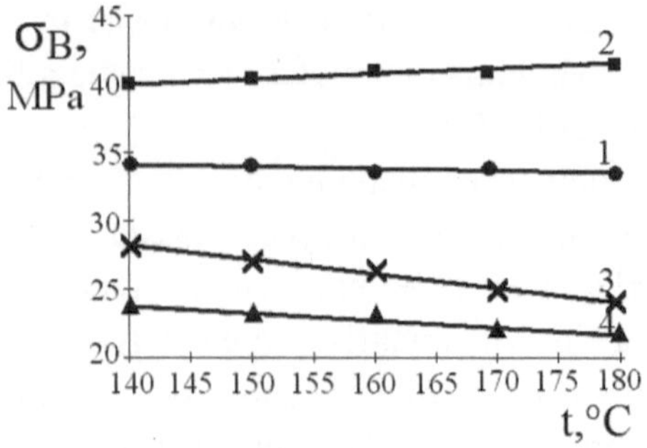

FIGURE 25.3 Influence of temperature (t) of experiment on maximum bending stress (σ_B) at: τ=20 min, s=20% (1); τ=20 min, s=40% (2); τ=60 min, s=20% (3); τ=60 min, s=40% (4).

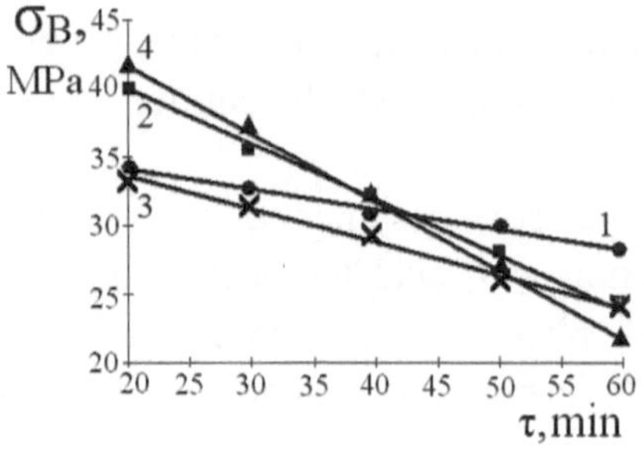

FIGURE 25.4 Influence of the time of experiment (τ) on maximum bending stress (σ_B) at: t=140°C, s=20% (1); t=140°C, s=40% (2); t=180°C, s=20% (3); t=180°C, s=40% (4).

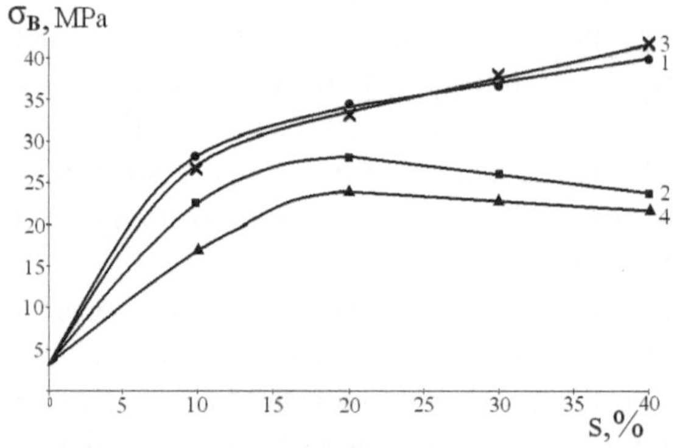

FIGURE 25.5 Influence of the amount of filler (s) on maximum bending stress (σ_B) at: t=140°C, τ=20 min (1); t=140°C, τ=60 min (2); t=180°C, τ=20 min (3); t=180°C, τ=60 min (4).

If the contact time increases to 60 minutes, the bending strength at first increases with increasing filler content to 20%, and then decreases during the filling with chrysotile to 40% (lines 2 and 4 in Figure 25.5). Flexural strength of pitch composite filled with chrysotile at 180°C (curve 4 Figure 25.5), is less than filled at 140 C (curve 2 in Figure 25.5).

Increase the contact time to 60 minutes leads to unwanted degradation processes and polycondensation of pitch polymer matrix, which causes strong structuring of system. This leads to disappearance of ability of pitch polymeric matrix to highly elastic deformation. Pitch polymeric matrix becomes brittle. Changes in the structure and properties of pitch polymer in volume of IFS changes adhesive interaction at the interface. Deterioration of adhesive interaction is particularly noticeable when chrysotile content increases over 20%, when there is a decrease in flexural strength.

Maximum flexural strength of pitch composite is 42 MPa achieved at 180°C, 40% filler content, and a contact time of 20 minutes.

The dependence of maximum compressive stress on the content of filler and the time-temperature conditions of pitch composite obtaining shown in Figures 25.6 and 25.7.

The influence of temperature is observed at a high filling degree (40%) increase in temperature to 180°C reduces the compressive strength (lines 2 and 4 in Figure 25.6). With a low content of filler (20%), the temperature has little effect on compressive strength of pitch composite (lines 1 and 3 of Figure 25.6).

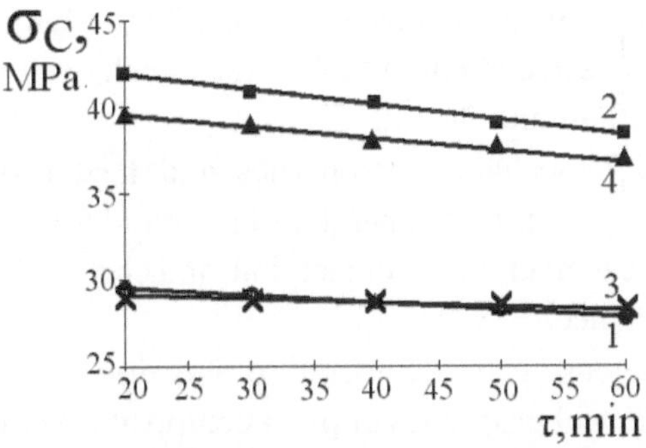

FIGURE 25.6 Influence of the time of experiment (τ) on maximum compressive stress (σ_C) at: t=140°C, s=20% (1); t=140°C, s=40% (2); t=180°C, s=20% (3); t=180°C, s=40% (4).

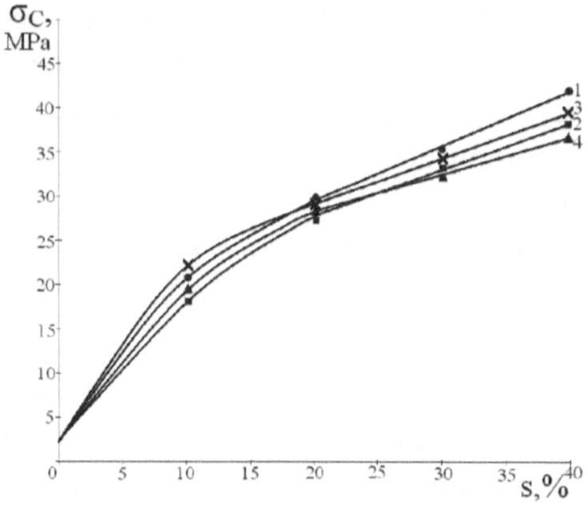

FIGURE 25.7 Influence of the amount of filler (s) on maximum compressive stress (σ_C) at: t=140°C, τ=20 min (1); t=140°C, τ=60 min (2); t=180°C, τ=20 min (3); t=180°C, τ=60 min (4).

Increase the contact time reduces the compressive strength independently of degree of filling and temperature (Figure 25.6).

Compressive strength is determined by the adhesion between filler and matrix and the cohesive strength of the weakest part of composition. Increase of contact time intensifies processes leading to a change of properties and deterioration of pitch polymeric matrix and adhesive interactions with filler. Moreover, the rigid links which formed while matrix structuring lead to the fact that pitch polymeric matrix loses its ability to redistribute stresses and starts to lose strength. All this leads to a deterioration of mechanical strength of pitch composite.

The greatest influence on the maximum compressive stress has the content of filler, the greater the amount of chrysotile, the higher the compressive strength of pitch composite – Figure 25.7. Moreover, the greatest increase in compressive strength to 43 MPa is achieved with a minimum contact time of 20 minutes at 140°C.

Therefore, the mechanical properties and heat resistance of pitch composite depends on time-temperature factor and filler content.

Increasing the temperature, contact time and chrysotile content leads to higher heat resistance.

However, increasing the temperature and time of mixing adversely affects on mechanical properties of pitch composite. Moreover, this effect is most pronounced with a high degree of filling. This suggests that in the presence of chrysotile thermo-chemical and structural transformations of coal tar pitch under the influence of an active polymer additives occur intensively. The direction of thermochemical transformations of coal tar pitch shifted toward the formation of the most condensed α-faction due to a more valuable component of coal tar pitch, which is responsible for its polymer properties, β-fraction. Pitch polymeric matrix loses fluidity and ability of stresses redistribution, which adversely affects on mechanical properties of pitch composite.

25.4 CONCLUSIONS

The influence of the temperature- time conditions of a composite material production on its properties are very important because of thermochemical transformations and structural changes in system: coal tar pitch – active additives – filler. This factor should be taken into account when creating of pitch polymeric structural materials. Executed studies have shown that entirely new pitch composite materials based on coal tar pitch by its modification and incorporation of the filler can be created.

Pitch composite unlike many polymeric materials characterized by high heat endurance. Selection of the optimal time-temperature conditions of mixing of pitch composite components will receive a material with high mechanical properties.

Pitch composite filled with chrysotile belongs to the hard-burning materials [5], what gives it a cogent advantage comparably to thermoplastic polymers.

The results of studies presented in this article served as the basis for the construction of a pilot plant for production of new pitch-polymeric composite material.

KEYWORDS

- **coal tar pitch**
- **filler**
- **modification**
- **pitch composite**
- **polymer**

REFERENCES

1. Krutko, I., Kaulin, V. Teoretychni peredumovy vykorystannya kam'yanovugilnogo peku yak polimernogo material. Naukovi pratsi DonNTU. Khimia i khimichna tehnologiya, 15(163), 103–107, 2010.
2. Krutko, I., Kaulin, V. Vplyv chlorvmisnyh polimeriv na sklad ta strukturu kam'yanovugilnogo peku. Vestnik natsionalnogo tehnicheskogo universiteta KhPI., 10, 148–151, 2010.
3. Krutko, I., Kaulin, V., Satsyuk, K. Rheologichni doslidjennya modyfikovanyh kam'yanovugilnyh pekiv. Naukovi pratsi DonNTU. Khimia i khimichna tehnologiya, 16(184), 150–158, 2011.
4. Krutko, I., Kaulin, V., Satsyuk, K. Termichnyy analiz modyfikovanyh kam'yanovugilnyh pekiv. Naukovi pratsi DonNTU. Khimia i khimichna tehnologiya, 19(199), 133–138, 2012.
5. Krutko, I., Kaulin, V., Satsyuk, K. Testing of modified coal tar pitch as polymer matrix in composite materials. Naukovi pratsi DonNTU. Khimia i khimichna tehnologiya, 2(21), 161–167, 2013.

CHAPTER 26

TOOLS FOR MODELING ADVANCED MATERIALS

KAKHA TSERETELI[1] and KHATUNA KAKHIANI[2]

[1]*Independent Expert, P. Melikishvili Institute of Physical and Organic Chemistry, Iv. Javakhishvili Tbilisi State University, Georgia*

[2]*Ivane Javakhishvili Tbilisi State University, I. Chavchavadze Ave., 1, Georgia, E-mail: ktsereteli@gmail.com*

CONTENTS

ABSTRACT

Cost-efficient and environmental friendly energy harvest and storage materials can change the way energy is generated, stored and distributed for decades to come. Despite the billions of dollars companies invest, the discovery and optimization of new promising materials is too slow to answer modern needs. First principle physics modeling of materials on supercomputers is a new trend in material science to accelerate the discovery and commercialization of new materials. Equations of motion (EOM) that carry out simultaneous dynamics of nuclei and electrons are derived and implemented in a computer program. This approximation to the time-dependent Schrödinger equation is obtained using the Dirac-Frenkel time-dependent variational principle with Kohn-Sham (KS) electronic Hamiltonian. The total molecular wave function is defined similar to a Born-Huang series, but with Gaussian wave-packets for the nuclei and a single complex spin unrestricted determinant for the electrons in Thouless parameterization, build from time-dependent KS orbitals. Another major difference is the use of derivatives of one, the reference electronic wave function, instead of all electronic Eigen states, as is the case of an original Born-Huang series. This approach has been applied to atom-molecule reactive collisions of a hydrogen atom with the hydrogen molecule at laboratory energy of 30 eV and the simulation of fundamental vibrations of the water molecule.

26.1 INTRODUCTION

Novel, cost-efficient and environmental friendly energy harvest and storage materials and devices have the potential to enhance the quality of human life over the next decades.

Materials development is a long-lasting and resource-costly process. Despite the large number of companies investing in designing novel materials, the success stories are few and infrequent. On average, it takes 15 to 20 years to commercialize a promising new material, and as a rule the technical progress only slows down this process because of tighter requirements.

Today, the process of engineering materials into novel and predefined forms is on the verge of a revolution. Enormous, century-long progress in first-principles physics, combined with a supercomputer's power, has made it possible. The idea is to computationally design *in silico* new materials which have given properties such as electronic and ionic conductivity, hardness, density, shininess, etc., that are determined by quantum mechanical behavior of the consisting atoms.

Usually, one computationally screens a set of compounds – containing tens of thousands of existing and hypothetical, mostly inorganic, materials – to find the one with the best given property or function, be it a new photovoltaic compound, metal alloy or battery electrode. This seems ambitious, but it is now well understood that almost all properties of materials can be predicted by solving the corresponding Schrödinger equation (SE). The most successful and practical approach to solving the SE is the Density Functional Theory (DFT), for which its authors were awarded the 1998 Nobel Prize in Chemistry [1]. So DFT is the main tool for such high-throughput calculations of materials.

The promise is big and based on the assumption of a direct correlation between the material's properties or functions and its fundamental descriptors. The only real obstacles seem to be the computational time and available funding.

To justify the feasibility of the previous statement, we do not need to go too far: chemists have already made considerable progress in developing multi-scale modeling of chemical reactions recognized by 2013 Nobel Prize in chemistry [2], such as the photosynthesis in green leaves. The main leitmotif of their success can be understood and explained by two statements made by two famous physicists: Paul Dirac and Richard Feynman. Dirac stated that "the underlying physical laws necessary for the mathematical theory of a large part of physics and the whole of chemistry are thus completely known, and the difficulty is only that the exact application of these laws leads to equations that are much too complicated to be soluble. It therefore becomes desirable that approximate practical methods of applying quantum mechanics should be developed, which can lead to explanation of the main features of complex atomic systems without too much computation [3]." Feynman later added that "…everything that living things do can be understood in terms of the jiggling's and wiggling's of atoms [4]."

However, unveiling nature's secrets to energy efficiency is not an easy, nor an intuitive task. It needs a tool that discovers nature's "designs" *in situ* at an atomic level and preferably in action.

Quantum effects are very important to our understanding: for example, it was recently discovered that the solar light-harvesting macromolecules, chromophores, capture solar energy very efficiently only because of quantum effects [5]. The efficient coherent exchange of a single quantum energy only happens when chromophores' electronic and vibrational degrees of freedom are collectively, although temporarily, in a superposition of quantum states, a feature that can never be predicted with classical mechanics.

In this account, we present a tool which has the features to unlock nature's secrets to "design" efficient materials and devices, visualize underlying processes and report the results in a manner that is intuitive to the broad audience of material scientists.

The tool by design incorporates quite sophisticated methods, including non-adiabatic electronic terms and quantum effects, for light nuclei (hydrogen for instance) to be truly predictive. In this way, new materials search will be accelerated through its use, as scanning multiple promising materials at the same time is enabled.

26.2 THEORETICAL PART: MULTI-SCALE HYBRID METHOD

According to new trends described in the introduction, we developed a multi-scale hybrid method of simultaneous quantum dynamics of electrons and nuclei based on time dependent DFT for electrons and quasi-classical coherent state (CS) theory for nuclei (Figure 26.1). The method is founded on mathematically rigorous separation of length scales and physically meaningful interface between quantum (finite as well as periodic) and classical regions. Rather than being relayed on the predefined potential energy surface (PES) from electrons, our method calculates the

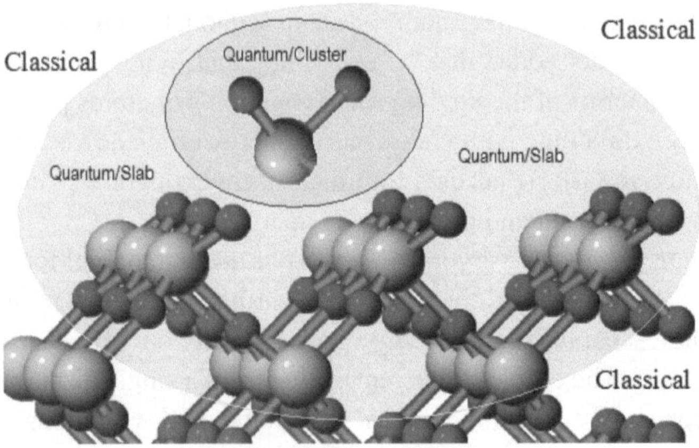

FIGURE 26.1 Cartoon of multi-scale hybrid method.

DFT energy of simultaneously moving electrons "on the fly" in atomic, or plane-wave, basis sets. These two basis sets enable the realistic treatment of molecular systems attached to a surface.

On the other hand, nuclei are treated classically using frozen Gaussian wave packet approximation [6], or quantum-mechanically using Gaussian wave packet approximation with evolving width. The latter is capable of accounting for quantum effects for light atoms, such as hydrogen.

The following accomplishments made feasible multi-scale modeling of advanced materials based on the simultaneous quantum dynamics electrons and nuclei:

- Development of an adaptive radial quadrature grid for efficient and accurate integration of time evaluation electron density in the outer regions of the molecular systems that has been implemented into the code [7].
- The program tool has been implemented on hybrid parallel architectures with Central Processing Unit (CPU) and Graphical Processing Unit (GPU). Due to the block matrix forms of derived equation of motion (EOM) formulas and NVIDIA CUDA technology [8], we have achieved an excellent convergence and overall speedup of a factor of 5 over conventional codes, which will allow us to scan promising materials for energy harvest and storage on the NVIDIA CUDA enabled personal supercomputers in a regular manner.

26.2.1 EQUATION OF MOTION (EOM)

EOM of system of interest can be obtained from time-dependent SE by means of time-dependent variational principle as well as least quantum action and in a block matrix notation has the following form [9]:

$$\begin{bmatrix} i\mathbf{C} & \mathbf{0} & \mathbf{0} & \mathbf{0} & i\mathbf{C}_{\mathbf{R}} & i\mathbf{C}_{\mathbf{P}} \\ \mathbf{0} & -i\mathbf{C}^{*} & \mathbf{0} & \mathbf{0} & -i\mathbf{C}_{\mathbf{R}}^{*} & -i\mathbf{C}_{\mathbf{P}}^{*} \\ \mathbf{0} & \mathbf{0} & \mathbf{0} & \mathbf{0} & \mathbf{0} & \mathbf{I} \\ \mathbf{0} & \mathbf{0} & \mathbf{0} & \mathbf{0} & -\mathbf{I} & \mathbf{0} \\ i\mathbf{C}_{\mathbf{R}}^{\dagger} & -i\mathbf{C}_{\mathbf{R}}^{T} & \mathbf{0} & \mathbf{I} & \mathbf{C}_{\mathbf{RR}} & -\mathbf{I}+\mathbf{C}_{\mathbf{RP}} \\ i\mathbf{C}_{\mathbf{P}}^{\dagger} & -i\mathbf{C}_{\mathbf{P}}^{T} & -\mathbf{I} & \mathbf{0} & \mathbf{I}+\mathbf{C}_{\mathbf{RP}} & \mathbf{C}_{\mathbf{PP}} \end{bmatrix} \begin{bmatrix} \dot{\mathbf{z}} \\ \dot{\mathbf{z}}^{*} \\ \dot{\mathbf{u}} \\ \dot{\mathbf{v}} \\ \dot{\mathbf{R}} \\ \dot{\mathbf{P}} \end{bmatrix} = \begin{bmatrix} \partial E / \partial \mathbf{z}^{*} \\ \partial E / \partial \mathbf{z} \\ \partial E / \partial \mathbf{u} \\ \partial E / \partial \mathbf{v} \\ \partial E / \partial \mathbf{R} \\ \partial E / \partial \mathbf{P} \end{bmatrix} \tag{1}$$

In Eq.(1) $\mathbf{R}=\{\mathbf{R}_k(t)\}$, $\mathbf{P}=\{\mathbf{P}_k(t)\}$, $\mathbf{z}=\{z_{ph}(t)\}$, $\mathbf{z}^*=\{z^*_{ph}(t)\}$, $\mathbf{u}=\{\mathbf{u}_k(t)\}$, $\mathbf{v}=\{\mathbf{v}_k(t)\}$ all are dynamical variables and **I** is an identity matrix. **R** stands for nuclear degrees of freedom and **P** stands for nuclear momentum.

Besides, **u** and **v** are nuclear wave packet parameters according to the following formula:

$$|\Psi_n\rangle \equiv |\mathbf{R},\mathbf{P},\mathbf{u},\mathbf{v}\rangle = \prod_k \exp\left[-\left(\frac{1-2i\mathbf{u}_k\mathbf{v}_k}{4\mathbf{v}_k^2}\right)(\mathbf{X}_k-\mathbf{R}_k)^2 - i\mathbf{P}_k(\mathbf{X}_k-\mathbf{R}_k)\right] \quad (2)$$

whereas **z** and **z*** stand for Thouless [10] parameters and its complex conjugate for electronic wave function given in terms of basis field (fermion creation and annihilation) operators $\{b^\dagger, b\}$, a set of complex parameters $\{\mathbf{z}\}$ and a reference state $|0\rangle = \prod_{i=1}^{N} b_i^\dagger |vac\rangle$ as

$$|\Psi_e\rangle \equiv |\mathbf{z};\mathbf{R},\mathbf{P}\rangle = \exp\left[\sum_{h=1}^{N}\sum_{p=N+1}^{K} z_{ph} b_p^{\circ\dagger} b_h^{\bullet}\right]|0\rangle = \det\left[\chi_h(\mathbf{x}_p)\right] \quad (3)$$

where $\chi_h = \varphi_h + \sum_{p=N+1}^{K} \varphi_p z_{ph}$ are K spin-orbitals in traveling Gaussian basis set divided into N *{φh}* occupied, i.e., the ones present in $|0\rangle$, and K-N $\{\varphi_p\}$ unoccupied parts.

And finally, **C**, $\mathbf{C_R}$, $\mathbf{C_P}$, $\mathbf{C_{RR}}$, $\mathbf{C_{PP}}$ are all kinds of non-adiabatic, electron-nuclear, and nuclear-nuclear coupling matrices respectively, which arise because of total wave function approximation by the Born-Huang [11] like formula as:

$$|\Psi\rangle \approx |\mathbf{R},\mathbf{P}\rangle|\Psi_e\rangle + \sum_k |\mathbf{R}_k,\mathbf{P}_k,\mathbf{u}_k,\mathbf{v}_k\rangle \frac{\partial}{\partial \mathbf{R}_k}|\Psi_e\rangle \quad (4)$$

The total energy is given by the following formula:

$$E = \frac{\langle\Psi|\hat{H}|\Psi\rangle}{\langle\Psi|\Psi\rangle} = \underbrace{\sum_{i=1}^{N_n}\frac{\mathbf{P}_i^2}{2M_i} + \sum_{i=1}^{N_n}\frac{\mathbf{u}_i^2}{2M_i} + \sum_i^{N_n}\hat{V}^{nn}(\mathbf{R}_i,\mathbf{v}_i)}_{E_n} + \underbrace{\frac{\langle\mathbf{z}|\hat{H}_e|\mathbf{z}\rangle}{\langle\mathbf{z}|\mathbf{z}\rangle}}_{E_e} \quad (5)$$

And the central quantity, electron charge density following classical/quantum trajectories is given by the following formula:

$$\rho(\mathbf{r}) = \sum_{i,j}^{K}\varphi_i(\mathbf{r})\,\Gamma_{ij}\varphi_j(\mathbf{r}) = \sum_{i,j}^{K}\varphi_i(\mathbf{r})\begin{pmatrix} I^\bullet \\ \mathbf{z}\end{pmatrix}\left(I^\bullet + \mathbf{z}^\dagger\mathbf{z}\right)^{-1}\begin{pmatrix} I^\bullet & \mathbf{z}^\dagger\end{pmatrix}\varphi_j(\mathbf{r}) \quad (6)$$

where $\Gamma_{ij} = \frac{\langle \mathbf{z} | b_j^{\dagger} b_i | \mathbf{z} \rangle}{\langle \mathbf{z} | \mathbf{z} \rangle}$ stands for one-density matrix and has the following form:

$$\Gamma = \begin{pmatrix} \Gamma^{\bullet} & \Gamma^{>} \\ \Gamma^{\vee} & \Gamma^{\circ} \end{pmatrix} = \begin{pmatrix} I^{\bullet} \\ z \end{pmatrix} \left(I^{\bullet} + \mathbf{z}^{\dagger}\mathbf{z} \right)^{-1} \left(I^{\bullet} \quad \mathbf{z}^{\dagger} \right) \tag{7}$$

Detailed derivation of Karn-Sham energy terms and its derivatives w.r.t. dynamical variables for EOM is given elsewhere [12].

26.3 RESULTS AND DISCUSSION

By using our tool, it is possible to perform several phases of materials design *in silico*. Suppose scientists are looking for a better thermoelectric

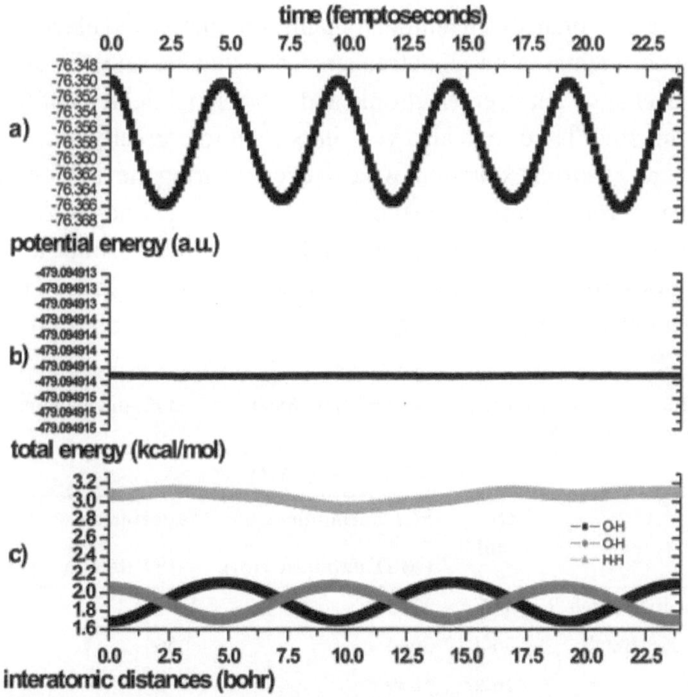

FIGURE 26.2 Simulation of Water vibration: Asymmetric stretch B_2 symmetry time evolution of potential (a) and total (b) energies and interatomic distances (c).

material than lead telluride, which is toxic and expensive. The performance of a thermo electric material depends on its Seebeck coefficient, electronic and thermal conductivity, which are controlled by the dynamics of phonons and charge carriers. From quantum dynamics simulations, one can derive and study these very properties, gaining critical insights into the relationship between the geometric and electronic structures of lead telluride and carrier transport mechanisms. This information will provide guidance to material scientists for controlling carrier/phonon interactions and their transport in other, heaper and benign classes of bulk or nanostructured materials by playing with changes in dimensionality and optimizing the thermo electric performance.

In this regard, it is very important to ensure correct concurrent dynamics of all particles. It can be validated by comparison of experimental and fundamental vibrational frequencies extracted from the simple simulations of water vibrations by fitting periodic exponentials to the position and momentum coordinates (Prony's method) (see Table 26.1). As Table 26.1 shows, our simultaneous quantum dynamics of nuclei and electrons reproduces experimental vibrational frequencies so that the same is expected for energy exchange between electronic and vibrational degrees of freedom.

On the other hand, it is also very important to correctly reproduce and visualize an electron excitation, when it occurs during the time evolution of the system. From chemistry textbooks, we know that bond breaking occurs when an electron moves from the bonding molecular orbital to the antibonding or non-bonding orbitals, and these orbitals are less compact and occupy much more space around the nuclei than bonding ones. Now we can

TABLE 26.1 Vibrational Frequencies (cm^{-1}) of Water Derived From Quantum Dynamics Simulations of Water Vibrations*

	DFT-BLYP-6–31**		
	Our tool	**SCF harmonic (NIST database [13])**	**Experiment (NIST database [13])**
Asymmetric stretch B_2	3750	3755	3756
Symmetric stretch A_1	3640	3644	3656
Bend A_1	1620	1639	1595

*Initial geometries are the equilibrium ones with 10% distorted normal modes corresponding to each vibration.

check if the time evolution of electron charge density following the classical nuclei, which moves according to EOM of Eq. (1), can capture electron excitation when it occurs. For this, we simulated Hydrogen atom collision to Hydrogen molecule at E_{lab}=30eV. This energy is more than enough to break covalent bond between the two Hydrogen atoms, when an atom strikes the middle point of the bond. And indeed, we see that after collision, all the interatomic distances are steadily increasing (Figure 26.3c). Collision starts at 7.5 fs with overleaping electron charge densities, and all the sudden, at 11.25 fs, we see a big red cloud of electrons – electron charge density covering the entire space around the nuclei. At the same time, as we can imagine, electron excitation happens and bond breaking occurs (see Figure 26.4). Such kind of visualization of electron excitations in light-harvesting macromolecules can help to understand the detailed

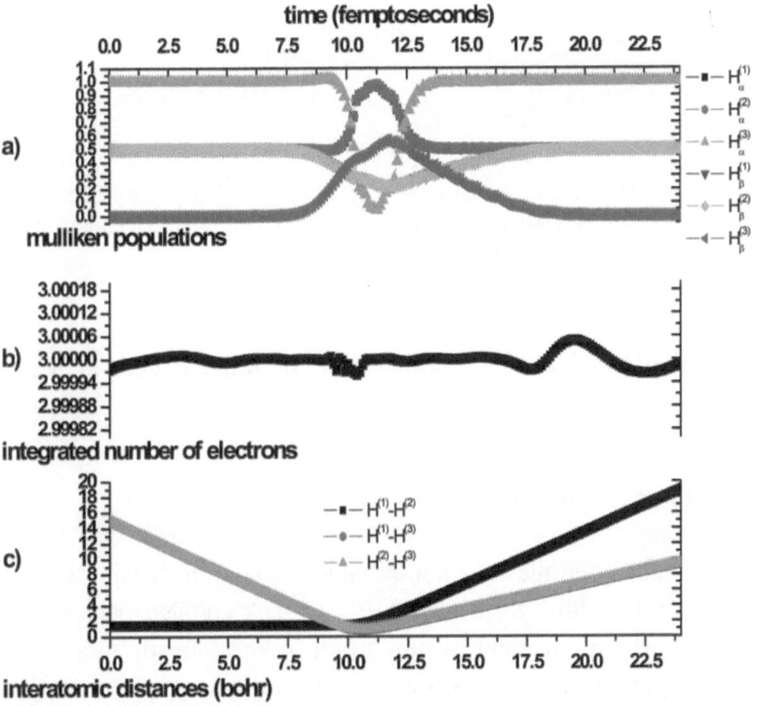

FIGURE 26.3 H + H_2 collision at E_{lab}=30 eV time step=0.12 fs: Mulliken populations [14] (a), integrated number of electrons (b) and interatomic distances (c) (DFT functional: BLYP, Basis set: 6–31**).

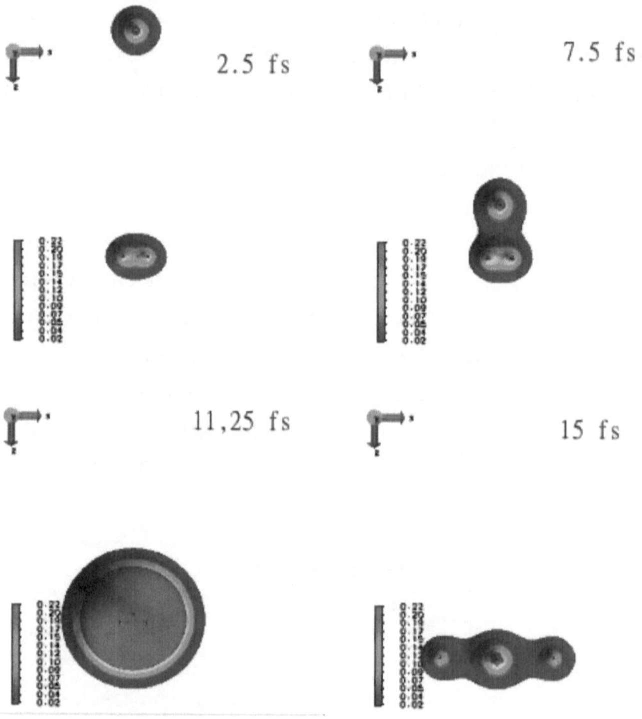

FIGURE 26.4 H + H_2 collision at E_{lab}=30 *eV* time step=0.12 fs: snapshots of electron charge density following classical trajectories at 2.5, 7.5, 11.25 and 15 femtoseconds. (DFT functional: BLYP, Basis set: 6–31**).

mechanism of efficient energy transfer and optimize structures of promising materials.

26.4 CONCLUSIONS

The multi-scale, "on the fly," non-adiabatic quantum dynamics approach in combination with hybrid CPU/GPU CUDA enabled personal supercomputers, which represents a new trend in an atomistic modeling, can play an important role in developing and optimizing new materials for renewable energy harvesting and storage.

KEYWORDS

- **density functional theory**
- **molecular dynamics**
- **non-adiabatic**
- **quantum dynamics**

REFERENCES

1. The Nobel Prize in Chemistry 1998.
2. (a) The Nobel Prize in Chemistry 2013; (b) Karplus, M., Development of Multiscale Models for Complex Chemical Systems From H+H_2 to Biomolecules. 2013; (c) Levitt, M., Nobel Lecture: Birth & Future of Multi-Scale Modeling of Biological Macromolecules.
3. Dirac, P. A. M., Quantum Mechanics of Many-Electron Systems. *Proceedings of the Royal Society of London. Series A, Containing Papers of a Mathematical and Physical Character (1905–1934)*, *123* (792), 714–733, 1929.
4. Feynman, R. P., Leighton, R. B., Sands, M., *The Feynman Lectures on Physics, Desktop Edition* Science: 2013; Vol. I.
5. O'Reilly, E. J., Olaya-Castro, A., Non-classicality of the molecular vibrations assisting exciton energy transfer at room temperature. *Nat Commun, 5*, 2014.
6. (a) Heller, E. J., Frozen Gaussians: A very simple semiclassical approximation. *J. Chem. Phys.*, *75* (6), 2923–2931, 1981; (b) Heller, E. J., Time-dependent approach to semiclassical dynamics. *J. Chem. Phys.*, *62* (4), 1544–1555, 1975; (c) Heller, E. J., Wavepacket path integral formulation of semiclassical dynamics. *Chemical Physics Letters*, *34* (2), 321–325, 1975.
7. Kakhiani, K., Tsereteli, K., Tsereteli, P., A program to generate basis set adaptive radial quadrature grid for density functional theory. *Computer Physics Communications 180* (2), 256–268, 2009.
8. CUDA Parallel Computing Platform. http://www.nvidia.com/object/cuda_home.html.
9. Deumens, E., Diz, A., Longo, R., Öhrn, Y., Time-dependent theoretical treatments of the dynamics of electrons and nuclei in molecular systems. *Rev. Mod. Phys.*, *66* (3), 917 – 983, 1994.
10. Thouless, D. J., Stability Conditions and Nuclear Rotations in the Hartree-Fock Theory. *Nucl. Phys.*, *21*, 225–232, 1960.
11. (a) Born, M., Huang, K., *Dynamical theory of crystal lattices*. Clarendon Press: Oxford, 1954; (b) Deumens, E., Öhrn, Y., Complete Electron Nuclear Dynamics. *J. Phys. Chem. A*, *105* (12), 2660–2667, 2001.

12. Tsereteli, K., *unpublished work.*
13. NIST Standard Reference Database 101 Computational Chemistry Comparison and Benchmark Database. In *Release 16a*, 2013.
14. Mulliken, R. S., Electronic Population Analysis on LCAO–MO Molecular Wave Functions. I. *J. Chem. Phys. 23* (10), 1833–1840, 1955.

CHAPTER 27

MODELING OF THE PHYSICAL MECHANISM OF ACTIVATION (OPENING) OF ION CHANNELS IN NERVE IMPULSE TRANSMISSION

N. S. VASSILIEVA-VASHAKMADZE, R. A. GAKHOKIDZE, and I. M. KHACHATRYAN

Iv. Javakhishvili Tbilisi State University, Department of Bioorganic Chemistry, Georgia, E-mail: nonavas@rambler.ru

CONTENTS

ABSTRACT

The mechanism of activation (opening) of the ion channels conjugated with the neuroreceptors in nerve impulse transmission was studied in this work in macroscopic approach.

It is assumed that the physical mechanism of functioning of the axonal, synaptic and sensory neuroreceptors is based on their intramolecular interaction between the mechanical elasticity and electrical polarizability. It is shown that the activation (opening) of the ion channels at the transmission of nerve impulse is the result of the action of the mutual repulsive force between α-helical subunits deforming the protein component of ion channel conjugated with neuroreceptor. This force has an electromagnetic nature and arises when the polarization process take place. It acts as a mechanical force, which causes the opening of ion channel.

27.1 INTRODUCTION

Many scientists had interested in the device and functioning of the nervous system since ancient times and this interest continues to this day. Modern scientists have linked the beginning of the nervous system study with names that have become common nouns, like Hippocrates (460–356 BC), Aristotle (384–322 BC), Galen (Claudius Galenus, 130–210 BC), Luigi Galvani (Luigi Aloisio Galvani, 1737–1798), René Descartes (Renatus Cartesius, 1596–1650) and many others. But the present study of neural processes started after the discovery (and especially after improvement) of microscope: Rudolf Virchow (Rudolph Carl Virchow, 1821–1902), Ranvier (Louis-Antoine Ranvier, 1835–1922), Camillo Golgi (Bartolomeo Camillo Emilio Golgi, 1843–1926), Santiago Ramón y Cajal (1852–1930), Walther Hermann Nernst (1864–1941), William Waldeyer (1836–1921), Vladimir Bekhterev (1857–1927), Natalia Bekhtereva (1924–2008), Alan Lloyd Hodgkin (1914–1998), Andrew Fielding Huxly (1917–2012), Bernard Katz (1911–2003), Robert J. Lefkowitz (1943), Clay M. Armstrong (1934), Francisco Bezanilla.

But despite the vast number of studies, both experimental and theoretical, nevertheless there are still many unsolved problems, and especially the question of what mechanisms underlie of the functioning of CNS [1, 2].

Resolving these issues can be useful both in developing methods for the treatment of pathologies of the nervous system, as well as in the field of controlled analogues of the nervous system, in organization of which is possible to use such mechanisms.

The structural elements of neurons – neuroreceptors, via discrete sequence of which the nerve impulses are transmitted, provide the necessary communication between the space-separated parts of the brain that determines the concerted functioning of CNS – is a necessary condition for the process of thinking, the most important function of consciousness [3, 4].

The CNS neuron net is located in the glia – the system of specialized cells (Rudolph Carl Virchow), namely oligodendral cells, astrocytes and Schwann cells. Wrapping the axon they form a myelination shell, which among other functions plays a role of an electrical insulator layer with interruptions in the nodes of Ranvier (Louis-Antoine Ranvier).

27.2 RESULTS AND DISCUSSION

Let consider the functioning of the acetylcholine nicotinic neuroreceptor, for which many of the parameters obtained on the basis of various studies are known, due to *X*-ray analysis, electron microscopy data and results of biochemical experiments [5, 6].

The acetylcholine nicotinic neuroreceptor is a glycoprotein complex, whose protein component with its helical structure penetrates the cytoplazmic membrane. Therefore, the length of α-helical protein subunits of neuroreceptors approximately is equal to the thickness of cytoplasmic membrane. At rest the cytoplasm and exoplasm of neuron are characterized by the difference of ionic composition that causes a transmembrane potential difference described by the known Nernst–Goldman equation [7]:

$$\varphi = \frac{RT}{F}\frac{\ln \sum P_k^{(out-in)} \; c_k^{(out)} Z_k}{\ln \sum P_k^{(in-out)} \; c_k^{(in)} Z_k} \tag{1}$$

where R is the universal gas constant; T – absolute temperature; F – the Faraday's number; $c_k^{(in)}$ and $c_k^{(out)}$ are the concentrations of *k*-type ions inside and outside the cell, correspondingly; P_k is the cell membrane permeability for *k*-type ions from outside to inside and vice versa, correspondingly.

The localization of neuroreceptor is such, that the end groups are immersed correspondingly in cytoplasm and exoplasm, which as known are characterized by the different ion composition, and according to Eq. (1), by the potential difference. In the stationary state the transmembrane potential (the resting potential) of neurons of human CNS φ_0 ~ 70–90 mV. In the depth of the cytoplasmic membrane, this potential difference creates an electric field $E_0 \sim 10^6$ V/m. Thus the receptor macromolecule possesses the initial dipole moment:

$$\vec{D} = k\vec{E} \quad \vec{E} = -\vec{\nabla}\varphi \quad E \approx \frac{\varphi}{l} \tag{2}$$

where k – polarizability, $\vec{E}$ – electric field intensity, $\vec{D}$ – electric dipole moment, φ – transmembrane potential, l – length of α-helical subunit of neuroreceptor.

According to Eq. (1) any variation of the ionic composition of the medium near the receptor's mouth changes the potential difference between its ends. This, in turn, is associated with the dipole moment of the receptor (2) and the short-term polarization process – gating current. From Eqs. (1) and (2):

$$\vec{J} = \dot{\vec{D}} \quad \vec{J}(t) = -k\vec{\nabla}\dot{\varphi}(t) \tag{3}$$

where $\vec{D}$ – electric dipole moment of neuroreceptor, which changes with transmembrane potential change. Since the mediator output the ionic composition in the synapse varies in a time interval (τ) and the polarization current is changed, as is seen from relation (3).

The appearance of gating current, which occurs before spike and causing the phase transition of the protein component of the receptor was predicted by Hodgkin, Huxley, and Katz in 1952 [8, 9]. The main difficulty in the registration of gating currents is the screening effect of ion currents. To eliminate this interference the blockers of ion currents were used. First the gating currents were experimentally recorded in the giant axon of the squid by Armstrong and Bezanilla [10].

Important works in this area were carried out by Skulachev [11], Rubin [12], Shaitan [13]. In recent years the view that gating currents occur due to changes of the dipole moments was reinforced. When the output of the mediator (acetylcholine) with its charged ammonium groups – $N{+}(CH_3)_3$ from the presynaptic zone in synapse take place, the transmembrane potential changes (first increases and then decreases) in the local area around the mouth of the channel and short polarization process in the receptor's macromolecule which accompanies the gating current $\vec{J}(t)$ occurs [10, 11].

According to works [10, 11], for α-subunits of neuroreceptors along with such known electrical characteristics (C, R, J…) let introduce [14, 15] one more parameter depending on the spatial α-helical configuration of subunits of neuroreceptor: L – the self-inductance [15]:

$$L = \frac{\mu_0 \mu}{4\pi} \frac{N^2 S}{l} \tag{4}$$

where N – the number of coils; S – the cross section; l – the length of the helical segment of the receptor; μ_0 – the magnetic constant; μ – magnetic permeability of the environment ($\mu \approx 1$) (in SI system).

The particular features of the receptor's structure, namely – the existence of helical sections, that creates favorable conditions for the emergence of an axial compressing force [14, 15]. This force is applied along the major axis, and it can be defined according to [16–18]:

$$F(t) = \frac{\mu \mu_0 N^2 S}{l^2(t)} J^2(t) \tag{5}$$

where μ_0 is the magnetic constant; μ – magnetic permeability, N – the number of coils; S – the cross section; $l(t)$ – the length of the helical segment; $J(t)$ – the gating current.

The force $F(t)$ is equivalent to an external mechanical force capable to deform the helical sections [14, 15].

The ratio of Eq. (5) shows that force $F(t)$ is actuated only if there are moving charges $\vec{J}(t)$ in the system, and its motion ceases with the attenuation of currents. Taking into account the relatively large sizes of the neuroreceptor subunit (l~10^4A, N~10^4) and its deformability [5,6], it can be considered in the classical approximation, i.e., as an elastic filament for which elasticity equation is valid. Since the force $F(t)$ (5) depends on $J^2(t)$, it arises only if $\vec{J}(t) \neq 0$, i.e., when the potential in the local area is changing and in this case it means the release of the acetylcholine with its charged ammonium groups into the synaptic zone.

The ratio (5) describes the internal interrelation between electric polarization and mechanical deformation processes, and the energy is transferred alternately between the polarization and deformation. It is caused by the features of neuroreceptors, because the force $F(t)$ (5), which causes the mechanical deformation of an electromagnetic nature, and is valid only when there is the polarization process – the gating current. As soon as the current ceases ($J=0$), the force $F(t)$ immediately disappears. This force is created by an electromagnetic process, but it acts as a mechanical compressive force causing the deformation of the α-helical subunits of protein component of neuroreceptor [14, 19].

The transmembrane potential causes the polarization current only when it is being changed in short-time. And if the transmembrane potential is not changing the polarization current arises. Therefore, we can conclude that the source of energy that makes neuroreceptors function is not just transmembrane potential $\varphi(t)$ but its change in time. This condition can be by changing the ion composition in the exoplasm or cytoplasm. In the case of the chemical synapse – when the mediator quantum release (e.g., of acetylcholine) with its charged groups NH_3^+ in synaptic zone near the receptor's mouth localized in postsynaptic membrane.

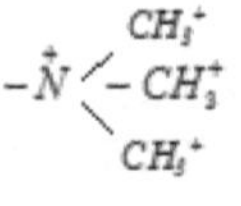

The polarization process is short and corresponds to an appearance of the gating current, which lasts for ~10^{-12} sec and precedes the opening of an ion channel [20, 21]. The time of open state of ion channel is about ~10^{-4} sec. The release of the mediator in synaptic zone changes the transmembrane potential difference between internal and external surfaces of cytoplazmic membrane that in turn is accompanied by polarization of α-helical subunits of protein component of neuroreceptor, accompanied by deformation of α-helical subunit – a structural component of the ion channel, and promotes the opening the ion channel permeable ion current. When the gating current occurs between the α-helical subunits of neuroreceptors, forming the ion channel, arises the mutual repulsive force of an electromagnetic nature, the action of which is causes the opening of the channel [22]:

$$\left|\vec{F}_{ij}\right| \approx \frac{\mu_0 \mu}{2\pi}(\vec{J}_i \vec{J}_j)\frac{\lambda}{R_{ij}} \tag{6}$$

where $\left|\vec{F}_{ij}\right|$ – the mutual repulsive force between the subunits of neuroreceptor, μ_0, μ – accordingly, magnetic constant and magnetic permeability, R_{ij} – the distance between the α-helical subunits of ion channel, $\vec{J}_i, \vec{J}_j$ – the gating current in α-helical subunits, λ – full length of the interacting fragments.

If we accept $\frac{\mu_0 \mu}{2\pi} = 2 \times 10^{-7}$ Hn/m, the mass of the α-helical subunit $m \sim 10^{-27}$ kg, the time of the open state of the ion channel $\tau \sim 10^{-4}$ sec, the quality estimate to increase of the diameter of the ion channel $\Delta R \sim 2.5 \times 10^{-10}$ m can be obtained, that indicates a sufficient value for the transmission of hydrated ions Na^+ [23].

Applying the program *HyperChem* to simulate molecular processes [24], the model scheme of influence of acetylcholine molecule on the triad of active center of acetylcholine nicotinic neuroreceptor was considered. A clear picture, that shows the influence of acetylcholine on the triad of active center (tripeptide His, Ser, Glu) was received [25]. The analysis shows that the acetylcholine causes the redistribution of the electron density on the atoms of triad, that is, polarization, and the spatial configuration change of the triad, which corresponds to deformation processes of active center. Based on this analysis, it can be concluded that this system has an internal interrelation between polarization and deformation processes, which confirms the correctness of the considered model (Figure 27.1).

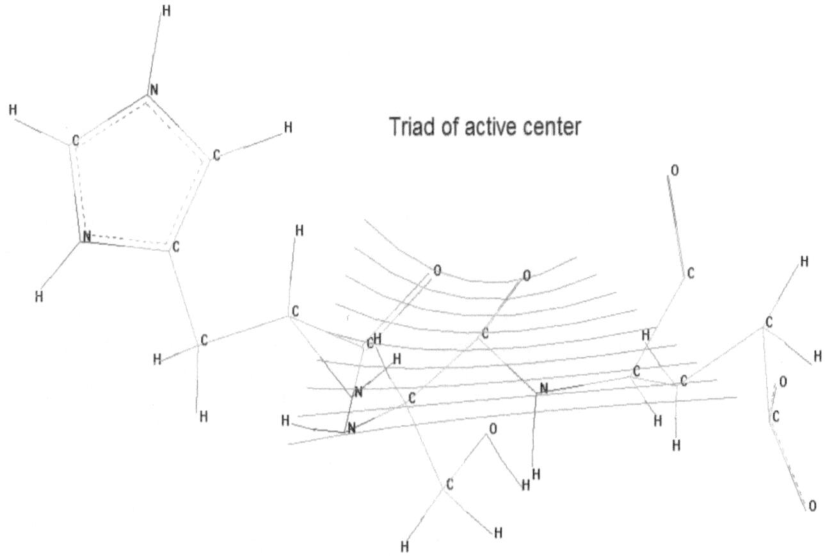

(a)

FIGURE 27.1 Continued

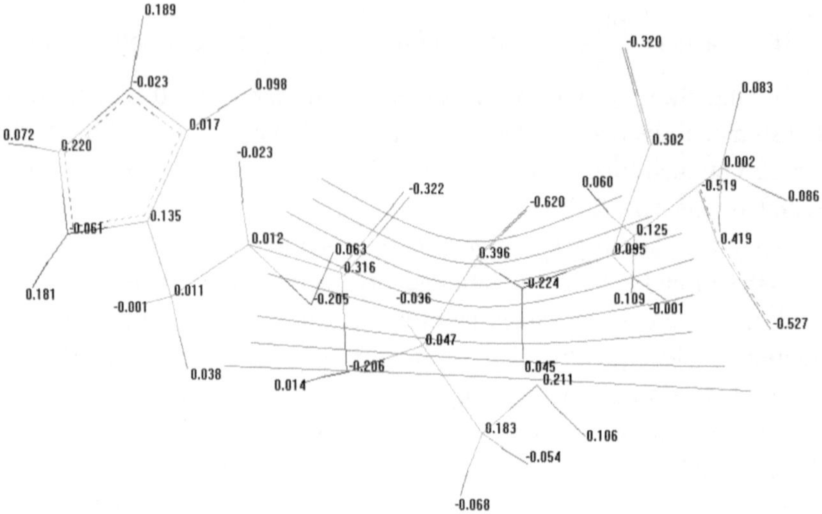

(b)

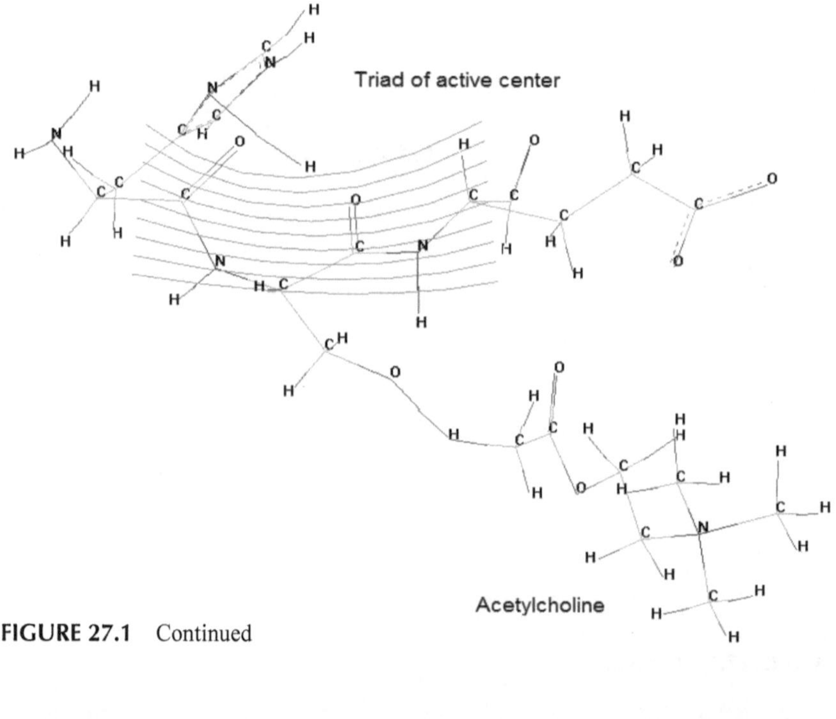

FIGURE 27.1 Continued

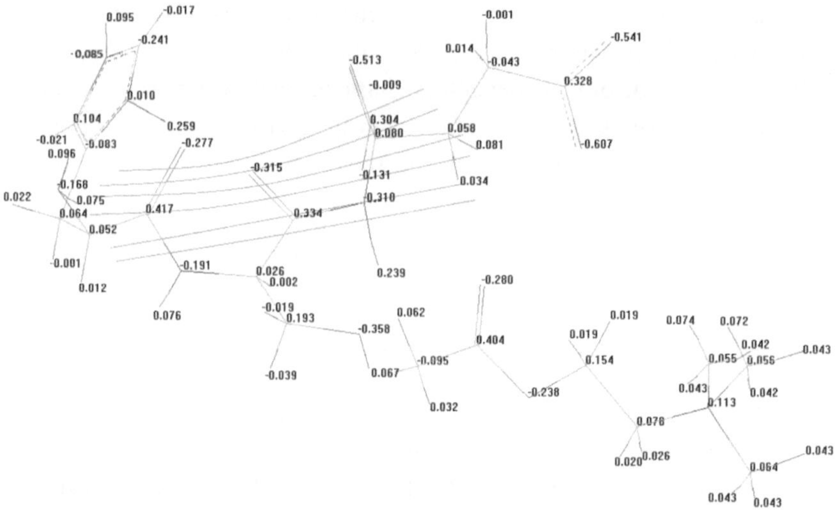

FIGURE 27.1 The comparison of (a) and (b) shows the effect of the polarization of the α-helical fragment of triad of active center of acetylcholine nicotinic neuroreceptor (*Hys-Ser-Glu*) under influence of acetylcholine: (a) shows a model (by using the *HyperChem* program) of the spatial configuration of the triad of active center in free state; (b) shows the effect of acetylcholine on triad of active center of acetylcholine nicotinic neuroreceptor. The redistribution of electron density on the atoms (polarization) and spatial configuration change of the active center (deformation) is noticeable.

27.3 CONCLUSION

On the basis of application of the macroscopic method can be concluded that due to the intrinsic connection between polarization and deformation processes in neuroreceptors at short transmembrane potential change occurs "pack" of the short rapidly decaying interconnected modulated oscillations (l – deformations and J – gating current) prior to the opening of the channel, the specifics of which depends on the structural parameters of neuroreceptors. Based on the analysis, it can be concluded that the system has an internal interrelation between polarization and deformation processes, which confirms the correctness of the considered model

The opening of the ion channel may be a result of the mutual repulsion appearing between he α-helical subunits of the ion channel in the case of the gating currents. It is the electromagnetic force, an important feature of which is that it is generated by local currents and acts as a mechanical force, which disappears with the attenuation of the currents.

Since each type of neuroreceptors characterized by inherent unique set of parameters, we can conclude that each type of receptor correspond its inherent "marker" of nerve impulse in the form of a well-defined modulated signal – the "key" that opens the ion channel.

KEYWORDS

- **electro-magneto-elasticity of neuroreceptors**
- **electromechanical properties of neuroreceptor**
- **mechanism of opening (activation) of ion channels of neuroreceptors**
- **neuroreceptor**

REFERENCES

1. Khukho, F. Neurochemistry: fundamentals and principles. Peace Mir, Moscow, 1990.
2. Shepherd, G. M. Neurobiology. Oxford University Press, UK, 1987.
3. Luders, E., Narr, K. L., Tompson, P. M., Toga, A. W. Neuroanatomical Correlates of Intelligence. Intelligence, 37(2), 156–163, 2009.
4. Kandel, E. R. In search of memory: The Emergence of a New Science of Mind. W. W. Norton & Company. New York, 2006.
5. Karlin, A. On the Application of a "plausible model" of allosteric proteins to the receptors for acetylcholine. J. Theor. Biol., 16(2), 306–320, 1967.
6. Changeux, J.-P. The acetylcholine receptor: "allosteric" membrane protein. Harvey Lect., 75, 85–254, 1981.
7. Goldman, D. E. Potential, Impedance and Rectification in Membranes. J. Gen. Physiol., 27, 37–60, 1943.
8. Hodgkin, A. L., Katz, B. The effect of sodium ions on the electronical activity of the giant axon of the squid. J. Physiol. 108(1), 37–77, 1949.
9. Hodgkin, A. L., Huxley, A. F. A quantitative description of membrane current and its application to conduction and excitation in nerve. J. Physiol., 117, 500–544, 1952.
10. Armstrong, C. M., Bezanilla, F. Charge Movement Associated with the Opening and Closing of the Activation Gates of the Na Channels. J. Gen. Physiol., 63(5), 533–552, 1974.

11. Skulachev, V. P. Evolution of biological mechanisms of energy conservation. Moscow, Soros Educational Journal, Biology, 5, 11–19, 1997.
12. Rubin, A. B. Biophysics. University Press, Moscow, 1999, 2000.
13. Shaitan, K. V. How electron moves in the protein. Moscow, Soros Educational Journal, Biology, 3, 55–62, 1999.
14. Vasilieva-Vashakmadze, N. S. Dynamic model of functioning of neuroreceptors and mechanism of hopping transfer of a nerve impulse. The II Congress of Biophysics of Russia, Moscow, 399, 1999.
15. Vasilieva-Vashakmadze, N. S. Tbil. Bull. Georgian Acad. Sci., 153(1), 102–104, 1996.
16. Kalashnikov, S. G. Electricity. Nauka, Moscow, 1964.
17. Landau, L. D., Lifshitz, E. M. The Classical Theory of Fields (Volume 2 of Course of theoretical physics). Pergamon Press, Oxford, 1971.
18. Sivukhin, D. V. The course of general physics (Volume 3 of Electricity). Nauka, Physmatlit, Moscow, 1996.
19. Vasil'eva-Vashakmadze, N. S. Dynamics of Functioning of Neuroreceptors in Chemical Synapses. Biophysics, 45(4), 662–668, 2000.
20. Hille, B. Gating in sodium channels of nerve. Annu. Rev. Physiol., 38, 139–152, 1976.
21. Levinson, S. R., Sather, W. A. Structure and mechanism of voltage-gated ion channels. In: Sperelakis N (ed.). Cell Physiology Sourcebook, 27, Academic Press, San Diego, 2001.
22. Vasilieva-Vashakmadze, N. S. A physical model of the mechanism of opening ion channels and reverberation nerve impulses in the neural circuits. In: Rubin A (ed) Physical basis of physiological processes. IV Congress of Russian Biophysicists, II Symposium, Nizhny Novgorod, 30, 2012.
23. Vassilieva-Vashakmadze, N. S., Gakhokidze, R. A., Vashakmadze-Veronese, D. T. A Physical Model of Neuroreceptor Functioning. The Journal of Biological Physics and Chemistry, 11(1), 18–25, 2011.
24. Vassilieva-Vashakmadze, N. S., Gakhokidze, R. A., Gakhokidze, A. R., Khachatryan, I. M. The Role of Strained State in *Cis-Trans* Isomerisation Process of Retinal. In: Use of Secondary Raw Materials and Natural Resources in Service of Human and Technological Progress, Abstracts, 49–51, 2011.
25. Netrebko, A. V., Kroo, S. V., Romanovsky, Y. M. Model of Acetylcholinesterase: Diffusion Limitations. In: Abstracts of II Congress of Russian Biophysicists, 2, 432–433, 1999.

CHAPTER 28

INFLUENCE OF THE PHASE STRUCTURE OF DOUBLE AND TRIPLE COPOLYMERS OF ETHYLENE MODIFIED BY GLYCIDOXYALKOXYSILANE ON THE PROPERTIES OF THE COMPOSITIONS

N. E. TEMNIKOVA,[1] A. E. CHALYKH,[2] V. K. GERASIMOV,[2] S. N. RUSANOVA,[1] O. V. STOYANOV,[1] and S. YU. SOFINA[1]

[1]*Kazan National Research Technological University, K.Marx str., 68, Kazan, 420015, Tatarstan, Russia, E-mail: ov_stoyanov@mail.ru*

[2]*Frumkin Institute of Physical Chemistry and Electrochemistry, Russian Academy of Sciences, Leninskii pr. 31, Moscow, 119991, Russia, E-mail: vladger@mail.ru*

CONTENTS

ABSTRACT

The mutual solubility of the components was investigated and the phase diagrams in a wide range of temperatures and compositions in the systems EVA (EVAMA) – glycidoxyalkoxysilane were constructed. The effect of structural heterogeneity of silanol modified EVA (EVAMA), associated with the chemical interaction of the components, on the properties of the compositions was identified.

28.1 INTRODUCTION

Introduction into the polymer of the reactive additives, chemically interacting with macromolecules, not only changes the chemical nature of the material, but also naturally affects the complex of its properties [1–4]. Graft structures formed in the matrix, increase molecular weight of the polymer, thereby affecting the process of the melt flow and solutions of the compositions.

Previously, it has been found [5] that the introduction into the copolymers of ethylene with vinyl acetate (EVA) of small amounts (up to 3%) of ethyl silicate (ETS) leads to increase in intrinsic viscosity of EVA. Herewith further increase in the concentration of ETS did not affect the process of solutions flow. Dependence of melt flow index of modified EVA on the concentration of the modifier has an extreme character with a minimum. On the amount of the minimum significantly affects the proportion of vinyl acetate in the copolymer.

The aim of this work was to study the effect of the phase structure of the modified EVA (EVAMA) on their properties. To achieve the aim the solubility of the components was studied and the phase diagrams in a wide range of temperatures and compositions for the systems EVA (EVAMA) – glycidoxyalkoxysilane were constructed.

28.2 SUBJECTS AND METHODS

Copolymers of ethylene with vinyl acetate Evatane2020 (EVA20) and Evatane2805 (EVA27) with a vinyl acetate content of 20 and 27 wt%, respectively; copolymers of ethylene with vinyl acetate and maleic anhydride brand Orevac9307 (EVAMA13) and Orevac9305 (EVAMA26) with a vinyl acetate content of 13 and 26% by weight were used as the objects of the study. Main characteristics of the copolymers are given in Table 28.1.

As the modifier was used silane, containing glycidoxy group – (3-glycidoxypropyl)trimethoxysilane (GS). It is clear, colorless liquid with a molecular weight of 236. Density is 1070 kg/m^3, refractive index n^D_{20} = 1.4367, content of glycidoxy groups 31%. Melting point is –70°C, flashpoint 135°C, boiling point 264°C. Production of Dow Corning Corporation, USA.

Modification of the copolymers was carried out in the melt on laboratory micro-rollers during 10 minutes in the temperature range from 100 to 120°C (the rotational speed of the rolls is 12.5 m/min, friction is 1:1.2).

The melt flow rate (MFR) was measured in accordance with GOST 11645–73 at 190°C and under a load of 2.16 kg.

Viscosity was measured by viscosimetric method by dissolving the compositions in carbon tetrachloride.

Determination of the composition of coexisting phases and the interdiffusion coefficients was carried out by a processing of series of interferograms obtained by microinterference method. Measurements were performed at a range of temperatures from 50 to 150°C. To construct profiles of concentrations by interference patterns the temperature dependencies of the refractive index of the components are required [6]. Refractive index measurements were carried out by an Abbe refractometer IRF-454 BM at a range of temperatures from 20 to 150°C.

TABLE 28.1 Characteristics of the Copolymers of Ethylene

Polymer	Symbol	VA content, %	MA content, %	Melting temperature, °C	M_v	MFR, g/10min 125°C
Evatane 2020	EVA20	20	-	80	44,000	2.23
Evatane 2805	EVA27	27	-	72	57,000	0.74
Orevac 9305	EVAMA26	26	1.5	47	20,000	11.13
Orevac 9307	EVAMA13	13	1.5	92	73,000	1.1

The structure of the modified copolymers was investigated by transmission electron microscopy. Identification of the phase structure of the samples was carried out by etching of the surface in high-oxygen plasma discharge with the subsequent preparation of single-stage carbon-platinum replicas. View of samples was performed on PEM EM-301 ("Philips," Holland).

28.3 RESULTS AND DISCUSSION

The temperature dependencies of the solubility of (3-glycidoxypropyl)tri-methoxysilane (GS) in the initial double and triple copolymers of ethylene were identified in diffusion experiments by direct bringing into a contact of EVA (EVAMA) and GS.

Studies of mutual solubility of the components are shown in Figures 28.1 and 28.2.

There are two binodal curves on all phase diagrams: the right branch of the binodal corresponds to the solubility of the copolymer in the modifier and is located in the area of infinitely dilute solutions. The second binodal curve represents the solubility of the modifier in the copolymers and is located in a fairly wide concentration area. The solubility of the modifier in the copolymer

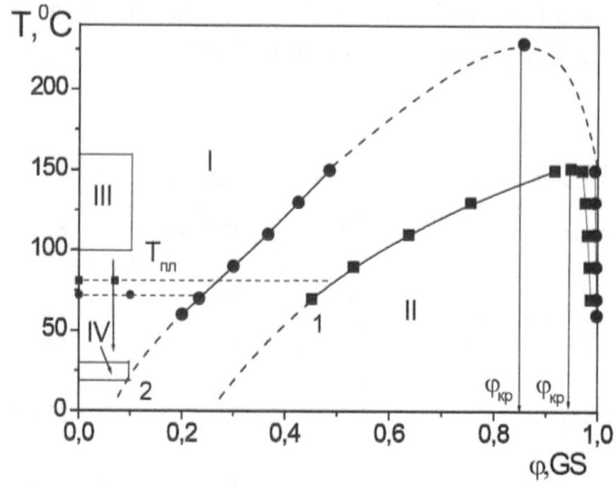

FIGURE 28.1 Phase diagrams of the systems copolymer of ethylene with vinyl acetate – GS: 1 – EVA20; 2 – EVA27: I, II – the areas of true solutions, heterogeneous condition; III – the area of preparation of the compositions; IV – the area of study of the structure and physical properties

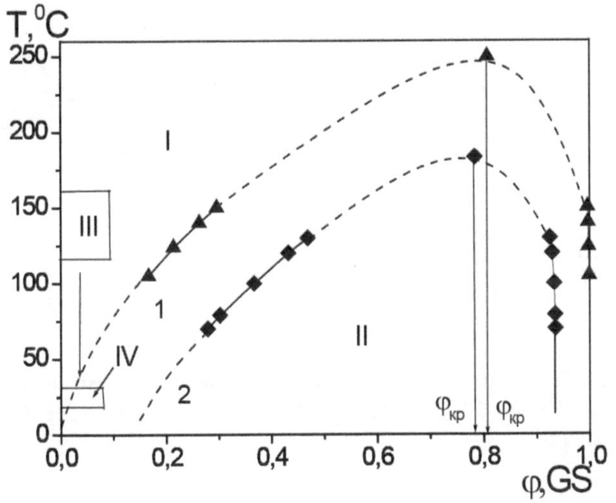

FIGURE 28.2 Phase diagrams of the systems copolymer of ethylene with vinyl acetate – GS: 1 – EVAMA13; 2 – EVAMA26: I, II – the areas of true solutions, heterogeneous condition; III – the area of preparation of the compositions; IV – the area of study of the structure and physical properties.

is increased as the temperature rises. In the systems EVA – GS the solubility of modifier in the system reduces with the increase of VA content. However, for the systems EVAMA – GS this is not observed. In these systems the solubility of GS in the copolymer increases with the increase of VA content.

It was found that the rate of a chemical reaction is greater than or comparable to the diffusion rate, and the movement of the modifier occurs in a chemically modified matrix, so the appearance of the "hourglass" on the diagram, we do not see [5]. The resulting matrix is soluble in the copolymer, and therefore we do not observe the phase decomposition during reheating [7].

Thus, it can be argued that these mixtures are characterized by diagrams with upper critical point of solubility (UCPS). This is also evidenced by the temperature dependencies of the pair interaction parameters (Figure 28.3), calculated according to the equation, assuming that the right branch of the binodal is located on the axis $\varphi_2 = 1$:

$$\varphi_{2,\text{крИТ}} = \frac{\sqrt{r_1}}{\sqrt{r_1} + \sqrt{r_2}},$$

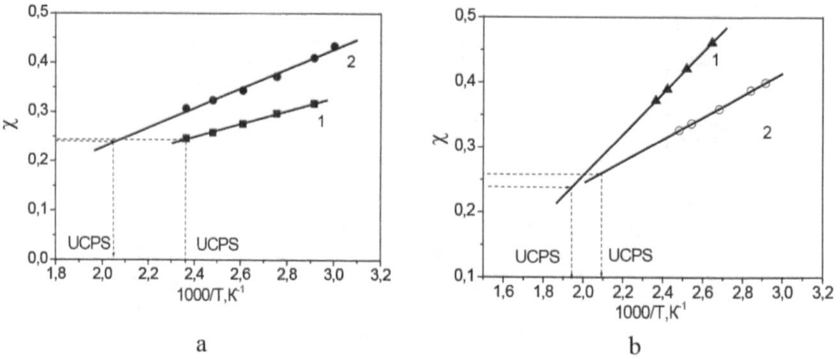

FIGURE 28.3 Temperature dependence of the pair interaction parameter of the systems: a – 1 – EVA20 – GS; 2 – EVA27 – GS; b – 1 – EVAMA13 – GS; 2 – EVAMA26 – GS. Arrows show UCPS.

$$\chi = \frac{1}{2}\left(\frac{1}{\sqrt{r_1}} + \frac{1}{\sqrt{r_2}}\right)^2$$

where χ – pair interaction parameter; φ_1 and φ_2 – concentration of EVA and GS respectively; r_1, r_2 – their degrees of polymerization.

In all phase diagrams (see Figures 28.1 and 28.2) there are the areas corresponding to temperature-concentration areas of the components mixing (area III), and the areas corresponding to the structure of the compounds and their physical properties (area IV). The diagrams show that the preparation of the mixtures takes place in the single-phase area (area I in the diagrams) and the study of the part of the system takes place in a heterogeneous area (area II in the diagram).

After etching of the cooled samples in plasma of high-oxygen discharge appearing some particles, protruding above the surface, which have a lower etching rate compared with the dispersion medium. Earlier it has been shown that the lowest etching rate among carbo- and heterochain polymers have polysiloxanes. Thus, it can be concluded that the dispersed phase is enriched by siloxanes (Figure 28.4).

Since the reaction mixing was accompanied by intense shear impacts in the presence of oxygen and air moisture, the obtained composite material is the result of the processes of diffusion mixing and chemical

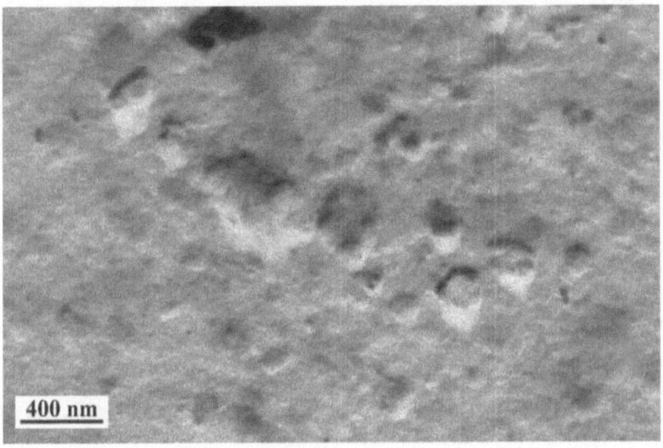

FIGURE 28.4 Microphotograph of EVAMA26 – GS (10%).

interactions between the components. This may explain the presence of the dispersed phase, although based on the phase diagrams it should not be here [8].

Formation of grafted siloxane units affects viscosity and rheological characteristics of the compositions. With the introduction of glycidoxysilanes, there is a decrease of the melt flow (see Figure 28.5) of the

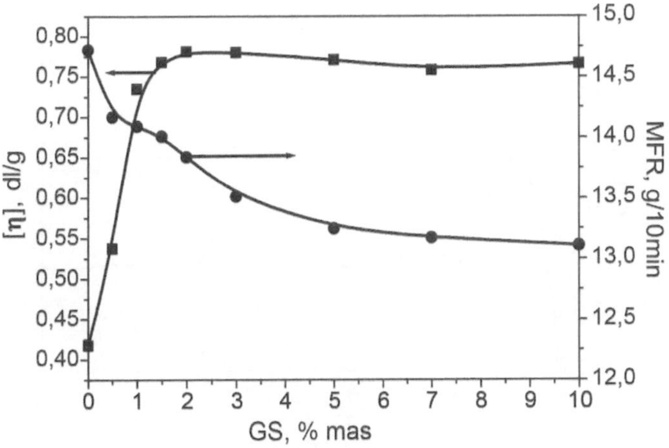

FIGURE 28.5 The dependence of the intrinsic viscosity and MFR from the modifier content: EVAMA26 – GS (viscosity at 40°C, MFR 125°C – 2.16kg).

modified polymers (by 30%) and an increase in the intrinsic viscosity (in 1.3–1.5 times), indicating that the preferential formation of branching and intermolecular bridges, increasing the length of the macromolecule, takes place.

Isolation of siloxane into a separate phase leads to an enrichment by them of the surface of the composition, and, consequently, to an increase of the adhesion characteristics [9].

Glycidoxypropyltrimethoxysilane is used as an additive for polyesters, polyacrylates, polysulfides, urethanes, epoxy and acrylic resins to improve their adhesion to glass, aluminum, steel and other substrates. Aqueous and alcoholic solutions of glycidoxysilane are used to improve an adhesion of epoxy resin to aluminum plates.

Thus, the system EVAMA26 − GS has a good adhesion to PET. For this system at the moment of the substrate rupture an adhesion strength at the content of GS of 1.5 wt.% increased in 3.4 times (see Figure 28.6). In this case the rapture has a cohesion nature and accompanied by the rapture of the substrate [10].

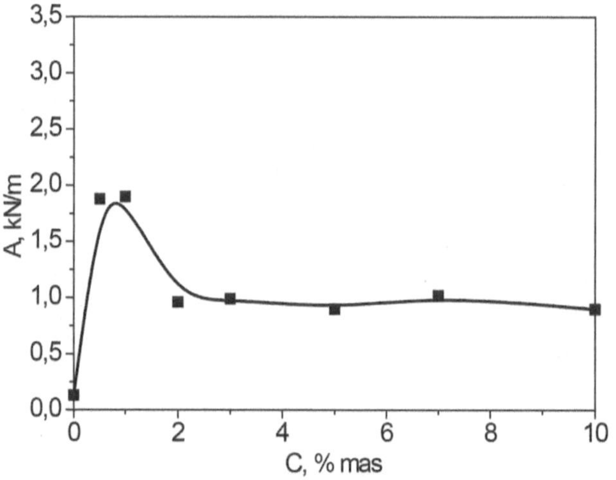

FIGURE 28.6 Adhesion strength of the polymer – PET: EVAMA26 – GS. Condition of the formation at 160°C for 10 minutes.

28.4 CONCLUSION

Thus, the data of these phase diagrams can be used in the study of the compositions, as well as selecting regimes for their preparation. From the diagram we can define the aggregate state, the amount and chemical composition of the phases, as well as structural-phase state of the compositions according to the temperature and concentration of its constituent components.

The triple copolymer, modified by glycidoxysilane, can be used in the production of laminated multilayer films as it is simultaneously has a good flowability and a high value of the adhesion strength.

ACKNOWLEDGEMENT

This work was financially supported by the Ministry of Education and Science of Russia in the framework of the theme №693 "Structured composite materials based on polar polymer matrices and reactive nanostructured components."

KEYWORDS

- **adhesion strength**
- **aminoalkoxysilane**
- **copolymers of ethylene**
- **phase diagrams**
- **rheology**
- **viscosity**

REFERENCES

1. Temnikova, N. E. The Effect of an Amino-Containing Modifier on Properties of Ethylene Copolymers. N. E. Temnikova, S. N. Rusanova, Yu.S. Tafeeva, S.Yu. Sof'ina, and O. V. Stoyanov. Polymer Science Series D. 2012. Vol.5 №4. 259–265.

2. Temnikova, N. E. Influence of aminoalkoxy- and glycidoxyalkoxysilanes on adhesion characteristics of ethylene copolymers. N. E. Temnikova, S. N. Rusanova, S.Yu. Sofina, O. V. Stoyanov, R. M. Garipov, A. E. Chalykh, V. K. Gerasimov, G. E. Zaikov.. Polymers Research Journal. 2014. Vol.8 №4. 305–310.
3. Temnikova, N. E. The Effect of Aminoalkoxy and Glycidoxyalkoxy Silanes on Adhesion Characteristics of Double and Triple Copolymers of Ethylene. N. E. Temnikova, S. N. Rusanova, S.Yu. Sof'ina, O. V. Stoyanov, R. M. Garipov, A. E. Chalykh, and V. K. Gerasimov. Polymer Science Series D. 2014. Vol.7 №3. 84–187.
4. Temnikova, N. E. Study of modification of ethylene copolymers by aminosilanes by IR spectroscopy FTIR. N. E. Temnikova, S. N. Rusanova, Yu.S. Tafeeva, O. V. Stoyanov. Bulletin of Kazan Technological University. 2011. № 19. 112–124.
5. Rusanova, S. N. Influence of the phase structure of copolymers of ethylene with vinyl acetate, modified by ethyl silicate, on their rheological properties. S. N. Rusanova, O. V. Stoyanov, V. K. Gerasimov, A. E. Chalykh. Bulletin of Kazan Technological University. 2006. № 1. 156–163.
6. Phase structure of silanol-modified ethylene-vinyl acetate copolymers. Chalykh A. E., Gerasimov V. K., Petukhova O. G., Kulagina G. S., Pisarev S. A., Rusanova S. N. Polymer Science Series A. 2006. Vol.48 №10. 1058 1066.
7. Temnikova, N. E. Effect of amino and glycidoxyalkoxysilanes on the formation of the phase structure and properties of ethylene copolymers: Thesis PhD. N. E. Temnikova. Kazan. 2013. 154 p.
8. Rusanova, S. N. IR spectroscopic study of the interaction of glycidoxy silane and copolymers of ethylene. S. N. Rusanova, N. E. Temnikova, O. V. Stoyanov, V. K. Gerasimov, A. E. Chalykh. Bulletin of Kazan Technological University. 2012. № 22. 95–96.
9. Chalykh, A.E. Effect of structural heterogeneity of ethylene-vinylacetate copolymers modified by ethyl silicate on their stress-strain characteristics. Chalykh A. E., Gerasimov V. K., Rusanova S. N., Stoyanov O. V. Polymer Science Series D. 2011. Vol.4, №2. 85–89.
10. Temnikova, N. E. The Effect of an Amino-Containing Modifier on Properties of Ethylene Copolymers. N. E. Temnikova, S. N. Rusanova, Yu.S. Tafeeva, S.Yu. Sof'ina, and O. V. Stoyanov. Polymer Science Series D. 2012. Vol.5, №4. 259–265.

CHAPTER 29

PHASE EQUILIBRIUM AND DIFFUSION IN THE SYSTEMS OF ETHYLENE COPOLYMERS – AMINOPROPYLTRIETHOXYSILANE[1]

N. E. TEMNIKOVA, O. V. STOYANOV, A. E. CHALYKH, V. K. GERASIMOV, S. N. RUSANOVA, and S. YU. SOFINA

P. Melikishvili Institute of Physical and Organic Chemistry, Iv. Javakhishvili Tbilisi State University, Georgia

CONTENTS

[1]This work was financially supported by the Ministry of Education and Science of Russia in the framework of the theme №693 "Structured composite materials based on polar polymer matrices and reactive nanostructured components."

ABSTRACT

The solubility of components has been studied in a wide range of temperatures and compositions in the systems copolymers of ethylene – aminoalkoxysilane. Phase diagrams have been constructed. Temperature and concentration areas of changes in solubility have been identified and structure of the modified copolymers has been studied.

29.1 INTRODUCTION

Copolymers of ethylene are widely used for obtaining materials and products for various purposes including coatings and adhesives. In this connection there is a need for continuous improvement in the properties of existing materials, because the synthesis of new polymers is difficult. Thus, to extend the scope of industrial-produced copolymers of ethylene is possible by their modification. One of the effective ways of modification is the introduction of organosilicon compounds. Introduction of such additives allows to achieve various changes in polymer properties [1–3], including adhesive characteristics [4, 5].

So γ-aminopropyltriethoxysilane (AGM-9) is used in fiberglass and paint industries to improve the adhesion of different polymers and coatings (acrylates, alkyds, polyesters, polyurethanes) to inorganic substrates (glass, aluminum, steel, and others) and to increase water resistance and corrosion stability of paint materials. AGM-9 is also used as pigmenting additives (enhance the interaction of the pigment with the polymeric matrix of composite material or paint material).

In order to optimize the composition of the polymer compounds and conditions of the structure formation of their mixtures the information about the phase organization of these systems is of considerable interest [5].

Despite the fact that there are papers devoted to improving adhesion to various substrates when modifying copolymers of ethylene by aminosilanes [6, 7], the information about the influence of monoaminofunctional silane on phase balance and phase structure of the polyolefin compositions in the scientific literature is not available.

The aim of this work was to study the formation of the phase structure of silanol-modified ethylene copolymers with vinyl acetate and vinyl acetate and maleic anhydride in a wide range of temperatures and compositions.

29.2 SUBJECTS AND METHODS

As the objects of study were used copolymers of ethylene with vinyl acetate Evatane 2020 (EVA20) and Evatane 2805 (EVA27) with vinyl acetate content of 20 and 27 wt%, respectively; copolymers of ethylene with vinyl acetate and maleic anhydride brand Orevac 9307 (EVAMA13) and Orevac 9305 (EVAMA26) with vinyl acetate content of 13 and 26 wt%. Main characteristics of the copolymers are given in Table 29.1.

As the modifier was used silane, containing an amino group – γ-aminopropyltriethoxysilane (AGM-9). Transparent, colorless liquid with a molecular weight 221. Density is 962 kg/m^3. Refractive index n^d_{20} = 1.4178, the content of amine groups is 7–7.5%. The melting temperature is -70°C.

Determination of the composition of coexisting phases and the interdiffusion coefficients was carried out by a processing of series of interferograms obtained by microinterference method. Interferometer ODA-3 was used for the measurements. Measurements were performed at a range of temperatures from 50 to 150°C. To construct profiles of concentrations by interference patterns the temperature dependencies of the refractive index of the components are required [6–8]. Refractive index measurements were carried out by an Abbe refractometer IRF-454 BM at a range of temperatures from 20 to 150°C.

The structure of the modified copolymers was investigated by transmission electron microscopy. Identification of the phase structure of the samples was carried out by etching of the surface in high-oxygen plasma discharge with the subsequent preparation of single-stage carbon-platinum replicas. View of samples was performed on PEM EM-301 ("Philips," Holland).

29.3 RESULTS AND DISCUSSION

29.3.1 KINETICS OF MIXING THE COMPONENTS

Typical interferograms of interdiffusion zones of the systems EVA (EVAMA) – modifier are shown in Figure 29.1.

Preliminary studies have shown that at high temperatures, the interdiffusion process is completely reversible, that is, the phase structures,

TABLE 29.1 Characteristics of the Copolymers of Ethylene

Polymer	Symbol	VA content, %	MA content, %	Melting temperature, °C	M_v	MFR, g/10min 125°C	Density, g/cm³
Evatane 2020	EVA20	20	-	80	44,000	2.23	0.936
Evatane 2805	EVA27	27	-	72	57,000	0.74	0.945
Orevac 9305	EVAMA26	26	1,5	47	20,000	11.13	0.951
Orevac 9307	EVAMA13	13	1,5	92	73,000	1.1	0.939

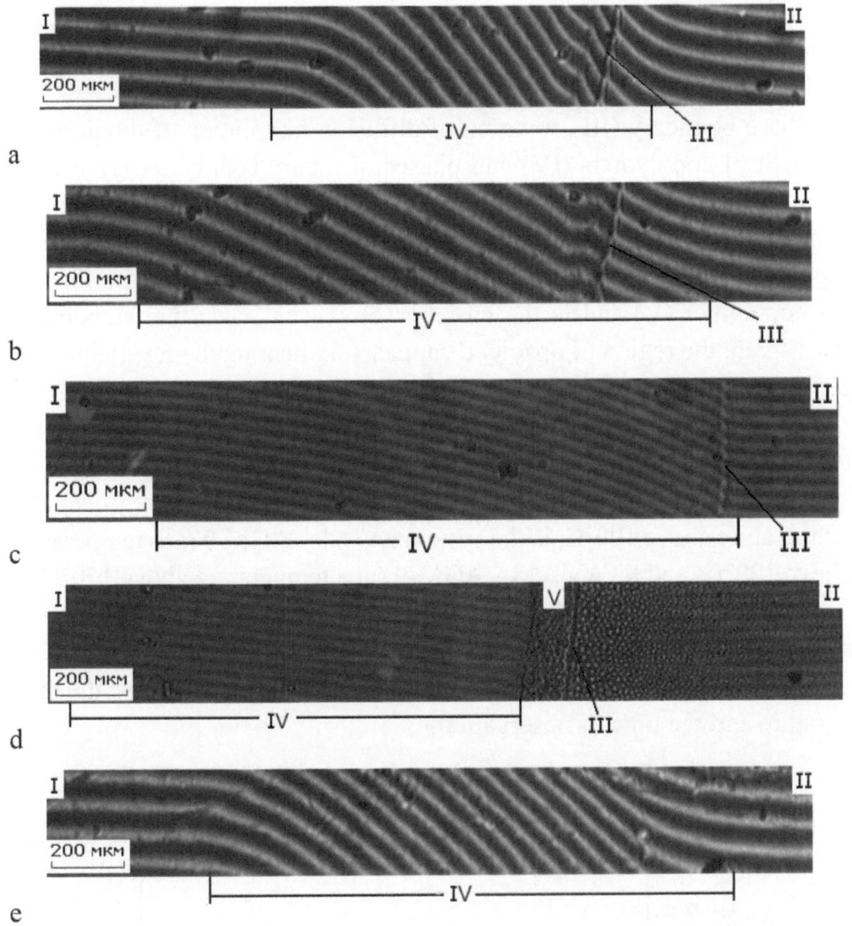

FIGURE 29.1 The interferograms of interdiffusion zones of the systems: a – EVAMA26–AGM-9 (100°C); b – EVAMA26–AGM-9 (60°C – cooling); c – EVA20–AGM-9 (120°C); d – EVA20–AGM-9 (90°C – cooling), e – EVAMA26–AGM-9 (140°C).

occurred when the temperature is lowered, dissolved again when the temperature rises. This means that the net of the diffusion experiment is not formed or it is formed but broken during the diffusion of the modifier, which is unlikely.

It is known that during the interaction of polymers with a modifier transitional zones appear, within which the structure, composition and properties vary continuously at the transition from one phase to another.

For all systems the general picture is characteristic for partially compatible systems with a primary dissolution of alkoxysilanes in the melt of copolymers.

Phase boundary (III), a region of diffusion dissolution of modifiers in the melt of copolymers (IV) and phases of a pure copolymer (I) and the modifier (II) are clearly expressed on interferograms. Situation in the systems changes with temperature decreasing: near the interface there is a region of opacity with separation of the dispersed phase in the melt of the copolymer (V) and in the modifier. However, when the temperature rises again, the region of opacity disappears, indicating the reversibility of phase transformations occurring in the systems.

However, there were registered also the compatible systems and the temperatures corresponding to this dissolution. AGM-9 is compatible in the systems EVA27 – AGM-9, EVAMA26 – AGM-9 (see Figure 29.1e) at a temperature above 100°C, in the system EVA20 – AGM-9 (at a temperature above 120°C) and EVAMA13 – AGM-9 (at a temperature above 150°C).

Typical profiles of the concentration distribution in these systems are shown in Figure 29.2.

The size of the diffusion zone is influenced by several factors: the temperature and the time of observation.

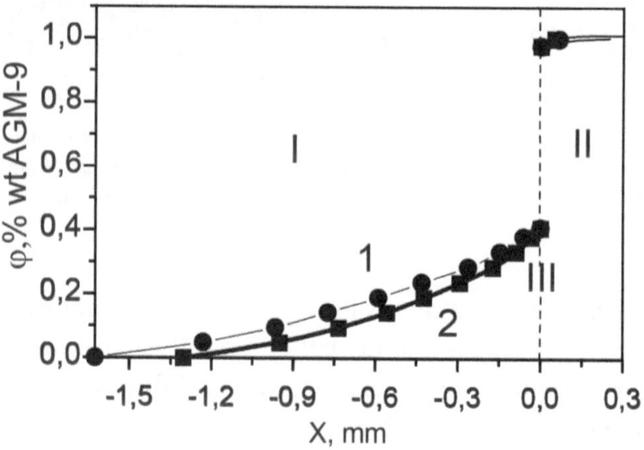

FIGURE 29.2 Profiles of the concentration distribution in the system EVAMA13 – AGM-9 at 135°C. Diffusion time: 1 – 11min, 2 – 7 min. I – diffusion zone of AGM-9 in EVAMA13, II – diffusion zone of EVAMA13 in AGM-9, III – phase boundary.

It can be seen that the sizes of the diffusion zones on both sides of the phase boundary increases in time, whereas the values of concentrations near the interphase boundary in isothermal process conditions do not change their values.

The influence of temperature has a number of characteristic features. The higher the temperature, the greater the distance the molecules of the modifier diffuse for equal periods of time and the larger the diffusion zone. This distribution of the concentration profiles when the temperature changes indicate that the system belongs to a class of systems with upper critical point of mixing [8].

Figure 29.3 shows the kinetic curves of isoconcentration planes moving at different temperatures for the systems studied.

Despite the fact that the kinetics of movement of the modifier front in the matrix of the copolymer has a linear dependence, IR-spectroscopy showed that all of the systems chemically react [9]. We can assume that the method of interferometry was unable to fix a chemical reaction under the given observation time. The chemical reaction rate is comparable or slightly higher than the rate of diffusion, and the movement of the modifier already occurs into a chemically modified matrix, which is the reason for the lack of bending motion of the modifier front into the copolymer.

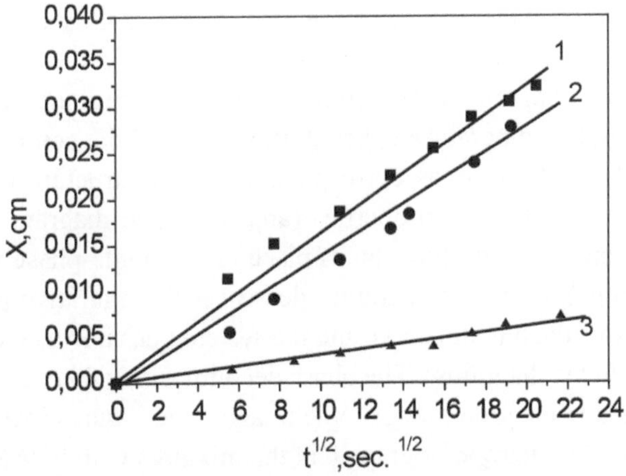

FIGURE 29.3 The time dependence of the size of the diffusion zone EVAMA26 – AGM-9 at various temperatures: 1 – 140°C, 2 – 120°C, 3 – 100°C.

The resulting matrix is soluble in the copolymer, so we do not observe the phase decomposition by reheating.

As the temperature increases, the nature of the concentration distribution in the zone of diffusion mixing of the components is maintained. Only the velocity of the isoconcentration planes movement changes. The angle of inclination of these relationships varies with temperature: the higher the temperature, the greater the angle of inclination of the line in the coordinates $X - t^{1/2}$. The slope of the kinetic lines is proportional to the coefficient of the modifier diffusion into the matrix. Therefore, the greater the angle of inclination, the higher the numerical value of the diffusion coefficient.

29.3.2 PHASE EQUILIBRIA

Consideration of diffusion zones of interacting copolymers and modifiers allows us to obtain not only the concentration profiles, but also the phase diagrams of the studied systems by quantitative analysis of interferograms obtained at different temperatures.

There are two binodal curves on all phase diagrams: the right branch of the binodal corresponds to the solubility of the copolymer in the modifier and is located in the area of infinitely dilute solutions. The second binodal curve represents the solubility of the modifier in the copolymers and is located in a fairly wide concentration area. The solubility of the modifier in the copolymer is increased as the temperature rises.

In all phase diagrams (see Figures 29.4 and 29.5) there are the areas corresponding to temperature-concentration areas of the components mixing (area III), and the areas corresponding to the structure of the compounds and their physical properties (area IV). The diagrams show that the preparation of the mixtures takes place in the single-phase area (area I in the diagrams). As the temperature decreases the figurative point of the systems crosses the binodal curve and the system goes into a heterogeneous area (area II in the diagrams). The phase decomposition takes place and it is uniquely fixed in electron microscopic images (see Figures 29.6 and 29.7). Judging from the microphotographs, in the mixtures with EVAMA13, precipitated phases have a size of from 0.1 to 1 micron, whereas for EVAMA26 the dispersed particles have a size of from 50 to 100 micron.

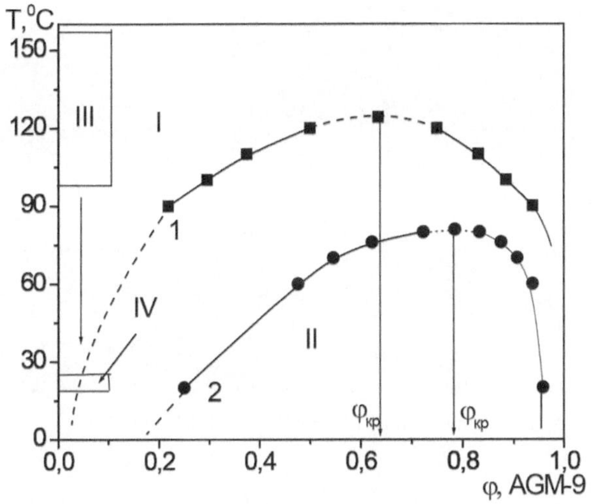

FIGURE 29.4 Phase diagrams of the systems a copolymer of ethylene with vinyl acetate – AGM-9: 1 – EVA20; 2 – EVA27: I, II – the areas of true solutions, heterogeneous condition; III – the area of preparation of the compositions; IV – the area of study of the structure and physical properties.

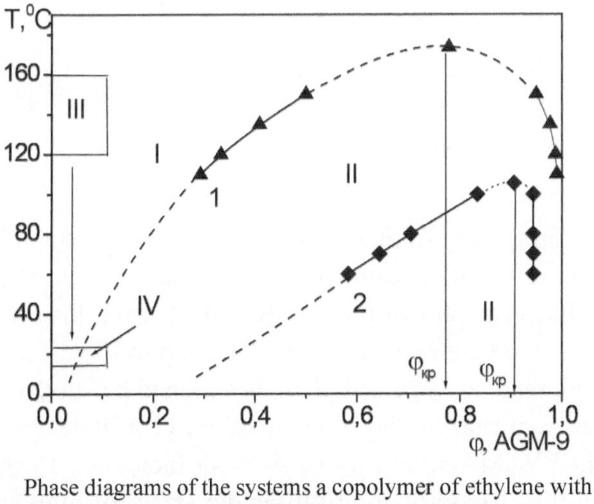

FIGURE 29.5 Phase diagrams of the systems a copolymer of ethylene with vinyl acetate and maleic anhydride – AGM-9: 1 – EVAMA13; 2 – EVAMA26: I, II – the areas of true solutions, heterogeneous condition; III – the area of preparation of the compositions; IV – the area of study of the structure and physical properties.

FIGURE 29.6 Microphotograph of EVAMA13 – AGM-9 (10%).

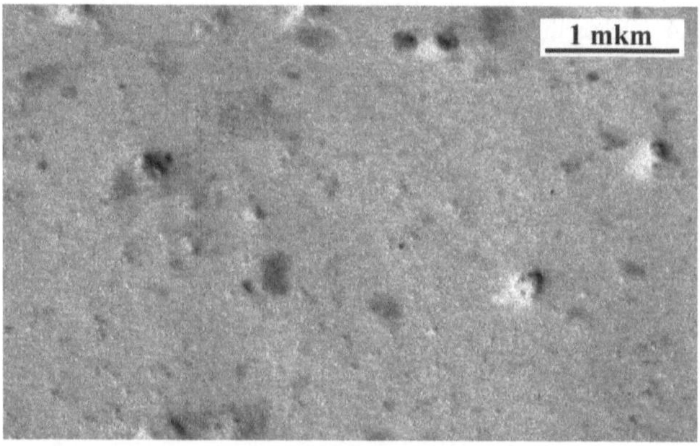

FIGURE 29.7 Microphotograph of EVAMA26 – AGM-9 (10%).

Particles, protruding from the surface, etched in plasma of high-oxygen discharge, have smaller etching rate compared with the dispersion medium. It has been shown previously [10] that the lowest etching rate among carbo- and heterochain polymers have polysiloxanes. Thus, it can be concluded, that the dispersed phase is enriched by siloxanes.

With the increase of the content of vinyl acetate groups both in EVA and in EVAMA solubility of AGM-9 increases. In this case, the tendency in the change of solubility is the same for high and low temperature areas.

29.4 CONCLUSION

Thus, comprehensive studies of diffusion, phase and structural-morphological characteristics of the compositions allow us to identify the contribution of chemical reactions into the change of phase equilibrium and into the phase structure formation.

KEYWORDS

- **copolymers of ethylene**
- **aminoalkoxysilane**
- **phase diagrams**

REFERENCES

1. Stoyanov, O. V. Modificatsya promyshlennykh etilenvinilatsetatnykh sopolimerov predelnymi alkoksisilanami. O. V. Stoyanov, S. N. Rusanova, R. M. Khyzakhanov, O. G. Petykhova, A. E. Chalykh, V. K. Gerasimov. Vestnik Kazanskogo tekhnologicheskogo universiteta. 2002. № 1–2. 143–147.
2. Chalykh, A. E. Formirovanie fazovoi structury silanolno-modifitsirovannykh polimerov etilena s vinilatsetatom. A. E. Chalykh, V. K. Gerasimov, S. N. Rusanova, O. V. Stoyanov, O. G. Petykhova, G. S. Kulagina, S. A. Pisarev. VMS seriya A. 2006. V. 48 № 10 S. 1801–1810.
3. O. V. Stoyanov, S. N. Rusanova, R. M. Khuzakhanov, O. G. Petuhova, T. R. Deberdeev. Russian Polymer News, 7 (4), 7 (2002).
4. Temnikova, N. E. The Effect of Aminoalkoxy and Glycidoxyalkoxy Silanes on Adhesion Characteristics of Double and Triple Copolymers of Ethylene. N. E. Temnikova, S. N. Rusanova, S.Yu. Sof'ina, O. V. Stoyanov, R. M. Garipov, A. E. Chalykh, and V. K. Gerasimov. POLYMER SCIENCE SERIES D. 2014. Vol. 7 № 3. P. 84–187.
5. Temnikova, N. E. Influence of aminoalkoxy- and glycidoxyalkoxysilanes on adhesion characteristics of ethylene copolymers. N. E. Temnikova, S. N. Rusanova, S.Yu. Sofina, O. V. Stoyanov, R. M. Garipov, A. E. Chalykh, V. K. Gerasimov, G. E. Zaikov. Polymers Research Journal. 2014. Vol. 8 № 4. P. 305–310.
6. Osipchik, V. S. Razrabotka i issledovanie svoistv silanolnosshitogo polietilena. V. S. Osipchik, E. D. Lebedeva, L.G Vasilets. Plasticheskie massy. 2000. № 9. S. 27–31.

7. Kikel, V. A. Sravnitelnyi analiz structury i svoistv sshitykh razlichnymi metodami polietilenov. V. A. Kikel, V. S. Osipchik, E. D. Lebedeva. Plasticheskie massy. 2005. № 8. S. 3–6.
8. Polimernye smesi. V 2-kh tomakh. Pod red. D. R. Pola i K. B. Baknella. Per. s angl. pod red. Kulezneva V. N. Spb.: Naychnye osnovy i tekhnologii, 2009. 1224 s.
9. Rusanova, S. N. IK-spektroskopicheskoe issledovanie vzaimodeistviya glitsidoksisilana i sopolimerov etilena. S. N. Rusanova, N. E. Temnikova, O. V. Stoyanov, V. K. Gerasimov, A. E. Chalykh. Vestnik Kazanskogo tekhnologicheskogo universiteta. 2012. № 22. S. 95–96.
10. Temnikova, N. E. Vliyanie amino- i glitsidoksialkoksisilanov na formirovanie fazovoi structury i svoistva etilenovykh sopolimerov: dis. kand. tekh. nauk. N. E. Temnikova. Kazan. 2013. 154 s.

CHAPTER 30

CHARACTERIZATION OF CS-BASED NANOFIBROUS WEB FOR ANTIBACTERIAL FILTER APPLICATIONS

MOTAHAREH KANAFCHIAN and MOHAMMAD KANAFCHIAN

University of Guilan, Rasht, Iran

CONTENTS

ABSTRACT

The Chitosan (CS) based nanofibers web is a biocompatible, biodegradable, antimicrobial and non-toxic structure, which has both physical and chemical properties to effectively capture and neutralize toxic pollutants from air and liquid media. The purpose of this study is to characterize CS-based nanofibers web for filtration. Antibacterial experiments were performed to examine the amount of bacteria reduction. Nanofibers analyzed with FTIR and DSC instruments. In antibacterial test, Turbid metric method was more suitable than agar diffusion method and the web with 60/40 weight ratio was demonstrated most bacteria reduction. FTIR analysis demonstrated that there were strong intermolecular hydrogen bonds between CS and PVA molecular. In DSC analysis, it was known that filters made of CS/PVA nanofibers are not suitable and applicable for high temperatures.

30.1 INTRODUCTION

Recently, membrane filtration in water treatment and air cleaning has been used worldwide [1]. Filters have been widely used in both households and industry for removing substances from air or liquid. Filters for environment protection are used to remove pollutants from air or water. In military, they are used in uniform garments and isolating bags to decontaminate aerosol dusts, bacteria and even virus, while maintaining permeability to moisture vapor for comfort. Respirator is another example that requires an efficient filtration function. Similar function is also needed for some fabrics used in the medical area [2]. Central to this application is also the ability of the various membrane filtration processes to remove pathogenic microorganisms such as protozoa, bacteria and viruses [3]. Among the membrane processes, nanofiltration is the most recent technology, having many applications, especially for drinking water and wastewater treatment and air filtration [4]. Fibrous media in the form of non-wovens have been widely used for filtration applications. Non-woven filters are made of randomly laid micron-sized fibers, which provide a physical, sized-based separation mechanism for the filtration of air and water borne contaminants [5]. Non-woven nanofibrous filter media (nanofiber is defined as having diameter < 0.5 μm by non-wovens industry [6]) would offer a unique advantage as they have high specific surface area, good interconnectivity of pores, and ease of incorporation of specific functionality on the surface effectively filtering out contaminants by both physical and chemical mechanisms [6]. Nonwoven nanofibrous media have low basis weight, high permeability, and small pore size that make them appropriate for a wide range of filtration applications. In addition, nanofibers web offers unique properties like high specific surface area (ranging from 1 to 35 m^2/g depending on the diameter of fibers) and the potential to incorporate active chemistry or functionality on nanoscale [7].

30.1.1 CHITOSAN

Over the recent years, interest in the application of naturally occurring polymers such as polysaccharides and proteins, owing to their abundance in the environment, has grown considerably [8, 9]. CS is a modified natural amino polysaccharide derived from chitin, known as one of the most abundant organic materials in nature. Chitin is the major structural component in the exoskeleton of arthropod sand cell walls of fungi and yeast [10]. Commercial chitin is mainly prepared from crab, lobster and shrimp shells, which are the massive waste products of seafood industries [11]. Applications for chitin are very limited because of its poor solubility in common solvents resulting mainly from its highly extended hydrogen-bonded semi-crystalline structure [12]. Chitin, the second most polysaccharide found on earth next to cellulose, is a major component of the shells of crustaceans such as crab, shrimp and crawfish. The structural characteristics of chitin are similar to those of glycosaminoglycans. When chitin is deacetylated over about 60% it becomes soluble in dilute acidic solutions and is referred to CS or poly(N-acetyl-D-glucosamine). CS and its derivatives have attracted much research because of their unique biological properties such as antibacterial activity, low toxicity, and biodegradability [8, 9]. Thus, chitin is often converted to its more deacetylated derivative called CS. Chitin is very similar to cellulose, except for the hydroxyl group at C_2 position that is replaced by the acetylamino group. Depending on the chitin source and the methods of hydrolysis, CS varies greatly in its molecular weight (MW) and degree of deacetylation (DDA). The MW of CS can vary from 30 kDa to well above 1000 kDa and its typical DDA is over 70%, making it soluble in acidic aqueous solutions. At a pH of about 6–7, the biopolymer is a polycation and at a pH of 4.5 and below, it is completely protonated. The fraction of repeat units, which are positively charged is a function of the degree of deacetylation and solution pH. A higher degree of deacetylation would lead to a larger number of positively charged groups on the CS backbone [13]. As mentioned above, CS has several unique properties such as the ability to chelate ions from solution and to inhibit the growth of a wide variety of fungi, yeasts and bacteria. Although the exact mechanism with which CS exerts these properties is currently unknown, it has been suggested that the polycationic nature of this biopolymer that forms from acidic solutions below pH 6.5 is a crucial factor. Thus, it has been proposed that the positively charged amino groups of the glucosamine units interact with negatively charged components in microbial cell membranes altering their barrier properties, thereby preventing the entry of nutrients or causing the leakage of intracellular contents. Another reported mechanism involves the penetration of low MW CS in the cell, the binding to DNA, and the subsequent inhibition of RNA and protein synthesis. CS has been shown also to activate several defense processes in plant tissues, and it inhibits the production of toxins and microbial growth because of its ability to chelate metal ions [14, 15]. To date, the most successful method of producing nanofibers is electrospinning.

30.1.2 ELECTROSPINNING OF CS

CS is insoluble in water, alkali, and most mineral acidic systems. However, though its solubility in inorganic acids is quite limited, CS is in fact soluble in organic acids, such as dilute aqueous acetic, formic, and lactic acids. CS also has free amino groups, which makes it a positively charged polyelectrolyte. This property makes CS solutions highly viscous and complicates its electrospinning [16]. Furthermore, the formation of strong hydrogen bonds in a 3-D network prevents the movement of polymeric chains exposed to the electrical field [17]. Different strategies were used for bringing CS in nanofiber form. The three top most abundant techniques include blending of favorite polymers for electrospinning process with CS matrix [18, 19] alkali treatment of CS backbone to improve electrospinnability through reducing viscosity [20] and employment of concentrated organic acid solution to produce nanofibers by decreasing of surface tension [21]. Electrospinning of polyethyleneoxide (PEO)/CS [18] and polyvinyl alcohol (PVA)/CS [19] blended nanofiber are two recent studies based on first strategy. In the second protocol, the MW of CS decreases through alkali treatment. Solutions of the treated CS in aqueous 70–90% acetic acid produce nanofibers with appropriate quality and processing stability [20]. Using concentrated organic acids such as acetic acid [21] and triflouroacetic acid (TFA) with and without dichloromethane (DCM) [22] has been reported exclusively for producing neat CS nanofibers. They similarly reported the decreasing of surface tension and at the same time enhancement of charge density of CS solution without significant effect on viscosity. This new method suggests significant influence of the concentrated acid solution on the reducing of the applied field required for electrospinning. The electrospinning process uses high voltage to create an electric field between a droplet of polymer solution at the tip of a needle and a collector plate. When the electrostatic force overcomes the surface tension of the drop, a charged, continuous jet of polymer solution is ejected. As the solution moves away from the needle and toward the collector, the solvent evaporates and jet rapidly thins and dries. On the surface of the collector, a nonwoven web of randomly oriented solid nanofibers is deposited [7].

30.2 EXPERIMENTAL PART

30.2.1 MATERIALS

CS (degree of deacetylation 0.85) and medium molecular weight was supplied by SIGMA-ALDRICH. PVA (degree of hydrolysis, 98%) and acetic acid (AA) purchased from MERK. Nutrient Broth and Nutrient Agar was supplied from LIOFILCHEM COMPANY. *Staphylococcus aureus* bacteria used for this research.

30.2.2 PREPARATION OF CS/PVA SOLUTIONS

CS/PVA solution was prepared by blending of CS and PVA solution with concentration 20 wt% and 3 wt%, respectively. PVA solution was prepared by dissolving PVA polymer in warm water (80°C) with magnetic stirring apparatus until a clear solution be made. Also, CS solutions were prepared by dissolving CS in aqueous 2%v/v acetic acid under magnetic stirring overnight at room temperature to obtain homogeneous solutions. The weight ratios of CS to PVA were selected as ranging from 10/90 to 70/30, respectively. These blends were stirred for 6 hours.

30.2.3 ELECTROSPINNING

After the preparation of spinning solution, it was imported in a 2 mL syringe with a stainless steel needle (Inner diameter 0.4 mm) and then the syringe was placed in a metering pump from WORLDPRECISION INSTRUMENTS (Florida, USA). The electrospinning instrument is shown in Figure 30.1. The needle was connected with a high voltage power supply, which could generate positive DC voltages. A piece of aluminum foil was selected as a collector. The electrospun webs were obtained at 10 kV voltage, 10 cm tip to collector distance and 0.1 mL/hour feed rate.

30.2.4 SCANNING ELECTRON MICROSCOPY (SEM)

The morphology of the electrospun fibers of CS/PVA was observed under a LEO1455VP scanning electron microscope after gold coating.100 fibers randomly selected in SEM micrographs and their diameter measured by IMAGE J software. Finally, the average fiber diameter and diameter distribution were reported.

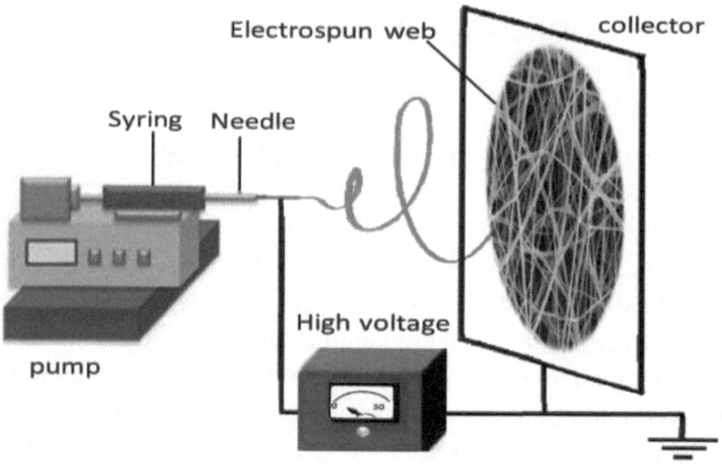

FIGURE 30.1 Electrospinning setup.

30.2.5 FOURIER-TRANSFORM INFRARED SPECTROSCOPY (FTIR)

A sample of electrospun fibers was prepared by electrospinning of CS/PVA solutions at 10 kV, 10cm collection distance. FT-IR measurements were performed in a FT-IR Spectrometer IR 560 (Nickolet Magan) to obtain functional groups and the formed chemical bonds between PVA and CS in the fiber.

30.2.6 DIFFERENTIAL SCANNING CALORIMETRY (DSC)

To investigate the melting point and the shift of endothermic peaks of electrospun web, it was placed in a *BAHR Thermo Analyze* DSC 302 at 10°C/min heating rate from room temperature to 250°C. The sample was stored in a desiccators prior to analysis.

30.2.6.7 Antibacterial Assessment

30.2.6.7.1 Nutrient Broth (NB) Solution Preparation

NB solution was prepared by adding NB powder to distilled water and agitating this mixture to reach a clear solution. Then, the solution was sterilized in a steam autoclave under these conditions, steam at a pressure of about 15 pounds per square inch (121°C) in about 15 minutes.

30.2.6.7.2 Nutrient Agar (NA) Preparation

Preparation of NA solution was performed in the same way as described for NB. To make homogeneous solution, NA added to heated distilled water and then sterilized. Then, it was pour into a plate to cool the solution and to achieve a solid Agar medium.

30.2.6.8 Antimicrobial Activity

To evaluate antibacterial activity, CS/PVA webs with different weight ratio were tested. Before bacterial testing, the NA and NB solutions were sterilized under UV laminar flow and autoclave apparatus. At this research, *Staphylococcus aureus* bacteria isused. Two method used for antimicrobial test: Agar plate and Turbidimetric method. In Agar plate, First, CS/PVA webs were cut in 1×1 cm. Then, the bacteria suspension was filled in Agar plate and the samples placed on the agar medium. Finally, the Agar plate was tacked into incubator with 37°C for 4 hours.

For Turbidimetric method, bacteria culture was performed into liquid medium. In this method, 5ml solution which it had included NB and *Staphylococcus aureus* of bacteria and salt solution was prepared for each electrospun web. Then, the webs imported into this solution and then absorbance of solution was read by spectrophotometric UV-visible at wavelength of 600 nm after 3 hours.

30.3 RESULTS AND DISCUSSIONS

30.3.1 EFFECT OF BLEND WEIGHT RATIO

Table 30.1 shows SEM images of CS/PVA webs with different weight ratio of CS to PVA and its fiber diameter distribution. In this study, we prepared the electrospun web of CS/PVA using acetic acid-water solution as a spinning solvent. As we know, CS is a cationic polysaccharide with amino groups at the C_2 position, which are ionizable under acidic or neutral pH conditions. Therefore, the morphology and diameter of electrospun fibers will be seriously influenced by the weight ratio of CS/PVA. As seen at Table 30.1, the fiber diameter gradually decreased with increasing CS content in the blend. When CS content was more than 60%, an electrospun web was created with a lot of beads. These behaviors can be explained as the following. CS is ionic polyelectrolytes, therefore a higher charge density on the surface of ejected jet forms during electrospinning. This aggregation of charge causes a higher elongation force imposed to the jet under the electrical field and the diameter of final fibers becomes smaller.

For filtration application, we needed a nanofiber web that it have smaller diameter, uniform and beadles fibers. Thus, the nanofiber web of CS/PVA with weight ratio 40/60,50/50 and 60/40 selected, because, these weight ratios have smaller diameters in comparison with others. The average fiber diameter of these webs was 97.70, 129.53 and 176.79 nm, respectively.

TABLE 30.1 SEM Photographs of Nanofiber Web in Different Weight Ratios of CS to PVA

Weight ratio (CS /PVA)	SEM image	Average Diameter(nm)	Distribution of nanofibers diameter
70/30		54.13± 14	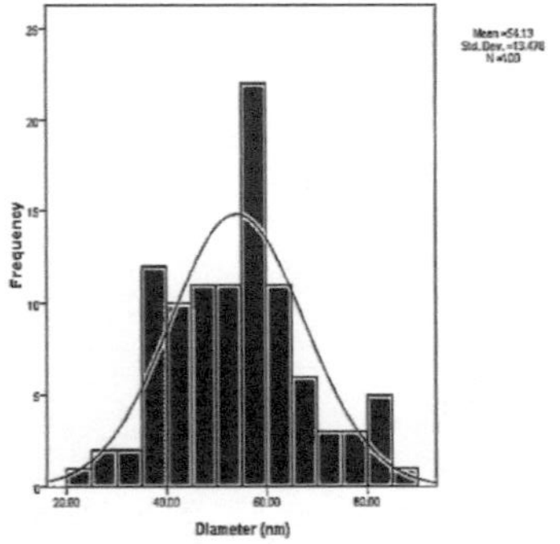
60/40		97.70 ±35	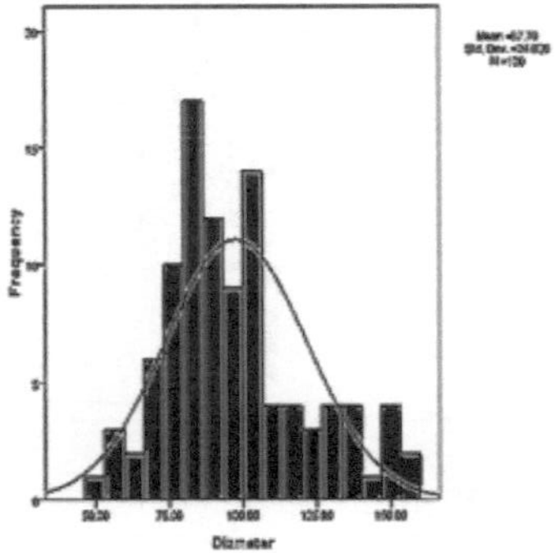

TABLE 30.1 Continued

Weight ratio (CS /PVA)	SEM image	Average Diameter(nm)	Distribution of nanofibers diameter
50/50	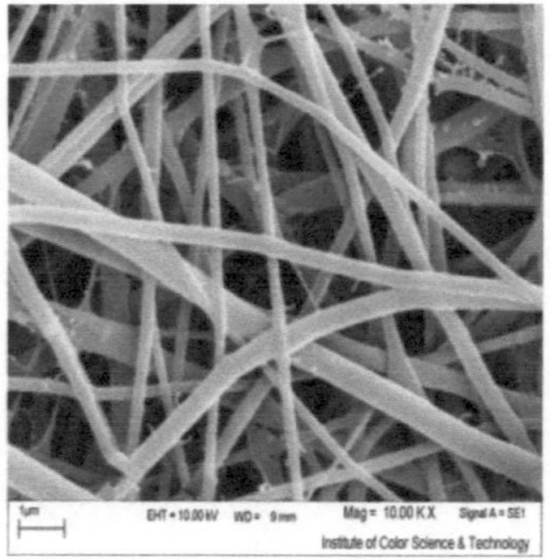	129.53 ± 24	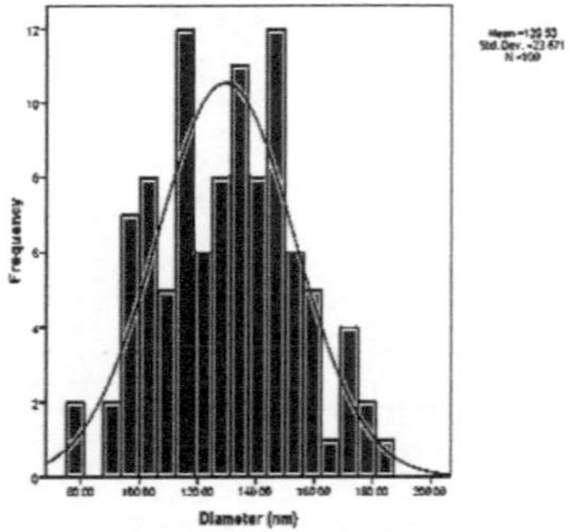
40/60		176.79 ± 28	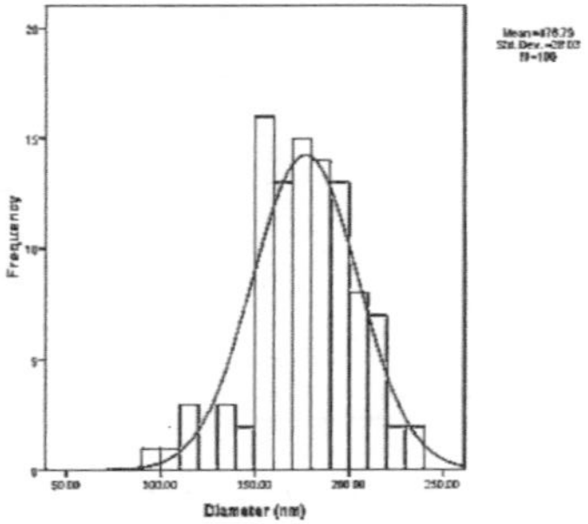

TABLE 30.1 Continued

Weight ratio (CS /PVA)	SEM image	Average Diameter(nm)	Distribution of nanofibers diameter
30/70		181.01± 35	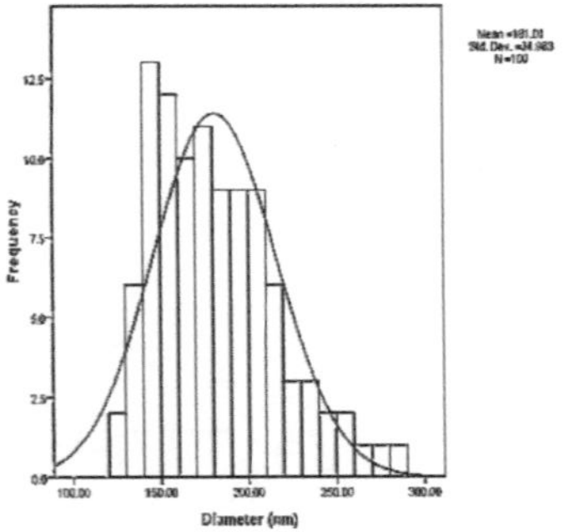
20/80	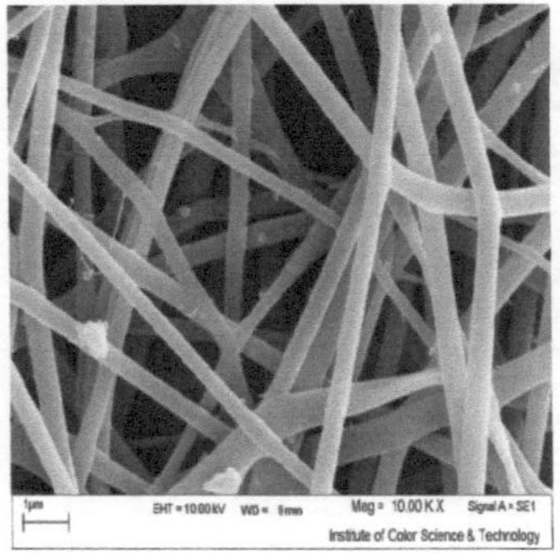	299.37±100	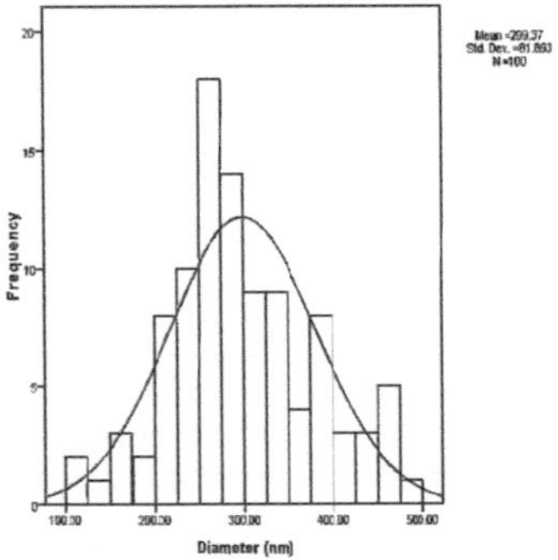

TABLE 30.1 Continued

Weight ratio (CS /PVA)	SEM image	Average Diameter(nm)	Distribution of nanofibers diameter
10/90		395.11±75	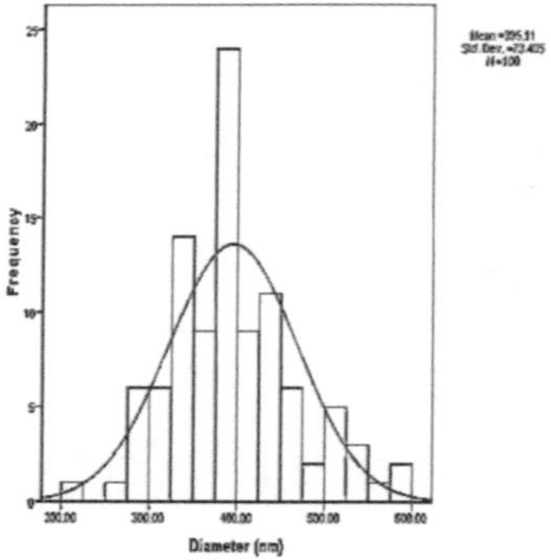

Finally after evaluation of antibacterial activity, the best weight ratio of nanofiber web can determine for filter application.

30.3.2 ANTIBACTERIAL ACTIVITY

In antibacterial assessment, 3 wt% CS/20 wt% PVA electrospun fibers with different weight ratio prepared from electrospinning process at 10 kV applied electrical potential, 10 cm collection distance and 0.1 mL/hr Feed rate. These electrospun fibers exposed to *Staphylococcus aureus* (*S. aureus*) bacteria. Before exposing, all samples had been sterilized with Ultraviolet (UV) because the electrospun fibers can be deformed by alcohol or autoclave. There are many methods to determine the antibacterial activity Such as Agar plate diffusion, Agar tube diffusion assay for Agar medium, and Turbidimetric, pH change and broth dilution assay for broth medium. In this research, the antimicrobial activity of CS in CS/PVA fibers was tested by an Agar plate diffusion method and Turbidimetric method. Antibacterial activities determine with *inhibition zone* around the samples in Agar plate diffusion. Figure 30.2 shows that there is no *inhibition zone* around electrospun fiber with different weight ratio but bacteria is grew on sample 10/90 (when amount of CS was minimum) completely. It is possible that CS in the electrospun web dissolves in bacteria suspension or has no diffusion. In Agar plate diffusion need a chemical active material to be able to diffuse out of the matrix into the plate in order to prohibit microorganism growth. This experiment indicates that the Agar plate diffusion method was not suitable for testing antibacterial assessment for CS/PVA fiber system because, it is able to show antibacterial activity and is not a quantitative method and do not exhibit *inhibition zone* easily.

Therefore, turbidimetric method was used for antibacterial assay. In this method, the absorbency of solution which it is included of bacteria, brine and webs measured by spectrophotometer UV-visible. Figure 30.3 shows the relationship of bacteria absorbance and CS content at 600 nm wavelength. This diagram presents that solution absorbency decreased by increasing of CS content to 60/40 weight ratio. In other words, maximum reduction of bacteria was occurred at 60/40 weight ratio, because, bacteria can prevent of light transmission from the cell. Thus, the

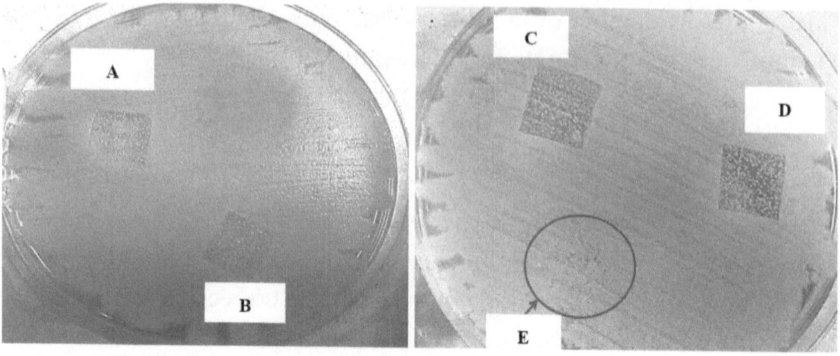

FIGURE 30.2 Agar plate diffusion method for CS /PVA electrospun web at different weight ratio A: 50/50, B: 40/60, C: 30/70, D: 20/80, E: 10/90.

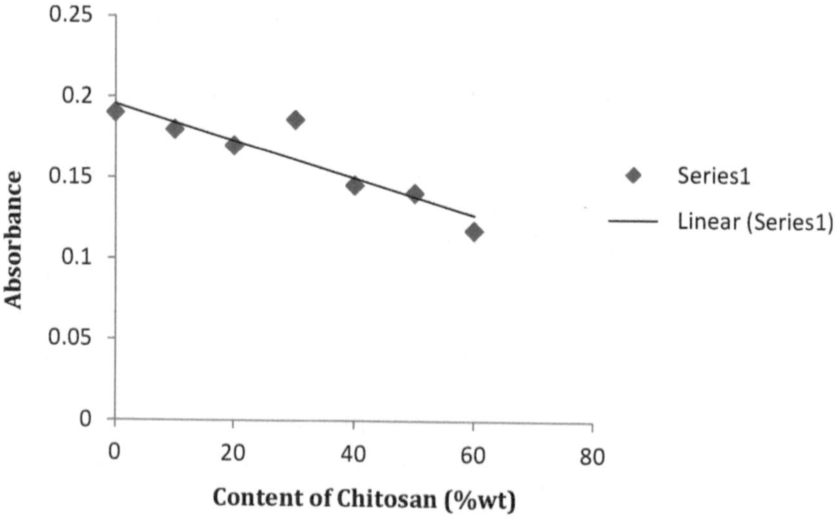

FIGURE 30.3 The relationship of bacteria absorbance and CS content at 600 nm wavelength.

percent of bacteria reduction increases by increasing of CS content and electrospun web of CS/PVA with 60/40 weight ratio has better bacteria reduction.

30.3.3 CHARACTERIZATION OF ELECTROSPUN FIBERS

To confirm the existence of PVA and CS in electospun fibers, an electrospun web was prepared from spinning solution 20wt% PVA/3wt% CS and evaluated by FTIR and DSC experiments.

30.3.4 FTIR ANALYSIS

The FTIR test of composites was carried out in order to characterize the participated functional groups in formation of CS/PVA nanofibers web. Figure 30.4 presents the FTIR spectra of electospun web CS/PVA with weight ratio 60/40 in comparison with pure PVA and CS.

The FTIR spectra of PVA reveals peak sat 1430 cm^{-1} (CH_2 bending) and 3354cm^{-1} (-OH stretching). It represents the characteristic broad band at 2900–3000 cm^{-1} for CH_2 group and CH_3 group, respectively. The CS exhibited characteristic broadband of OH group at 3400–3500 cm^{-1}[23]. The bands of NH_2 group (1560–1640) and O-C-NH_2 (1600–1640) group can be observed at 1634 cm^{-1}. The broad bands of CH_3 group and CH_3-O group can be observed at 1000–1200 cm^{-1} [24]. It can be found the peaks over the wave

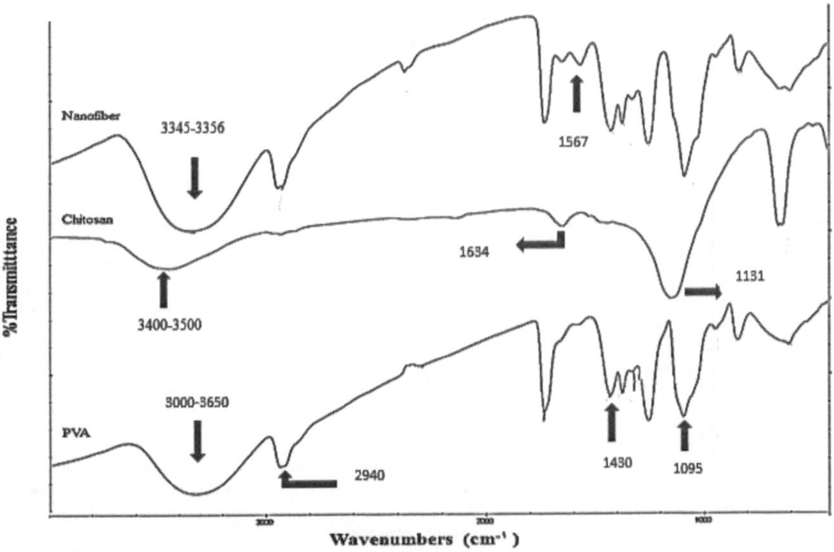

FIGURE 30.4 FTIR spectra of nanofiber CS/PVA: 60/40, PVA and CS powder.

number range of 3345–3356 cm^{-1} that represent to OH stretching and -NH stretching. FTIR Spectroscopic Measurement exhibited the existence of relevant functional groups of both PVA and CS in CS/PVA nanofibers web.

30.3.5 DSC ANALYSIS

DSC thermograms of the electrospun fibers CS/PVA, pure PVA and CS are shown in Figure 30.5. The pure PVA showed a relatively large and sharp endothermic curve with a peak at 200°C. More polysaccharides do not melt but they, because of associations through hydrogen bonding, degrade under heating above a certain temperature. Below degradation temperature of polymer, its thermogram shows a very broad endothermic peak that is associated with the water evaporation. In CS thermogram, a broad peak at approximately 100°C was seen that it was corresponding to the water evaporation process. However, for CS/PVA blend nanofibers, endothermic curve became broad and obtuse, and the peak shifted toward the low temperature. This indicated that the crystalline microstructure of electrospun fibers did not develop well. The reason of this phenomenon is that the

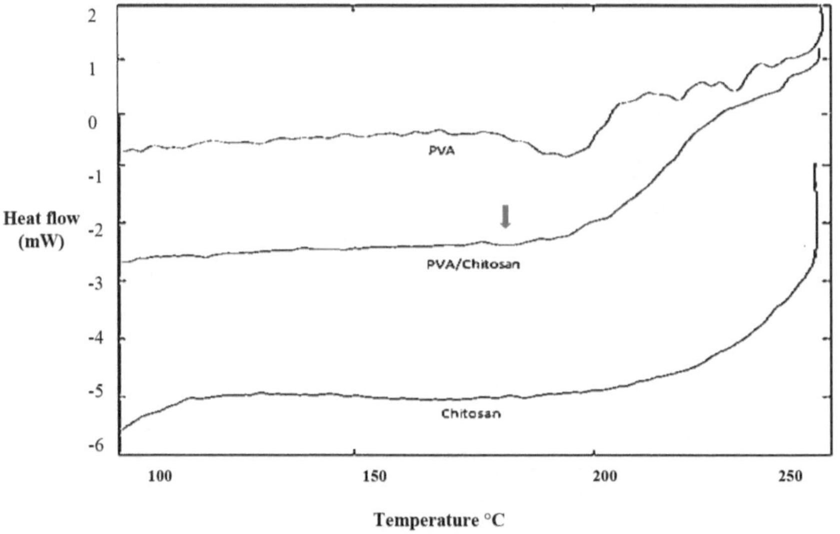

FIGURE 30.5 DSC curves of PVA, CS and PVA/CS electrospun nanofibers

majority of chains are in non-crystalline state due to the rapid solidification process of stretched chains during electrospinning. This demonstrated that CS content in the blend caused to decreasing thermal stability of PVA/CS in comparison with pure PVA. Thus, the filters made of CS/PVA nanofibers are not suitable and applicable for high temperatures.

30.4 CONCLUSIONS

The electrospun nanofibrous web of PVA/CS blends was fabricated. The effects of the blend weight ratio on structure and morphology of the fibers were investigated. The result indicated that the average diameter of the fiber gradually decreased with increasing CS content from 10% to 60%. Above 60% CS, the nanofibers had a lot of beads. In antibacterial test, turbidimetric method was more suitable than agar diffusion method and the web with 60/40 weight ratio was demonstrated most bacteria reduction. FTIR and DSC analysis demonstrated that there were strong intermolecular hydrogen bonds between CS and PVA molecular. It was known the filter made nanofibers CS/PVA can be able to inhibit bacteria growth so can use as antibacterial filter. In DSC analysis, it was known that filters made of CS/PVA nanofibers are not suitable and applicable for high temperatures.

KEYWORDS

- **antibacterial**
- **chitosan**
- **filtration**
- **nanofibers web**

REFERENCES

1. Song, J., Fallgren, H., Morria, M., Ch. Qi, Removal of bacteria and viruses from waters using layered double hydroxide nanocomposites, Sci. Tech. Adv. Mater, 8, 67–70 (2007).

2. Kosmider, K., Scott, J., Polymericnanofibers exhibit an enhanced air filtration performance. Filtr Separat, 39, 20–22 (2002).
3. Bohonak, M., Zydney, L., Compaction and permeability effects with virus filtration membranes, J. Membr. Sci, 254, 71–79 (2005).
4. Tahaikta, M., R. El Habbania, AitHaddoua, A., Acharya, I., Amora, Z., Takya, M., Alamib, A., Boughribab, A., Hafsib, M., Elmidaouia, A., Fluoride removal from groundwater by nanofiltration, Desalination, 212, 46–53 (2007).
5. Kaur, S., Gopal, R., Ng, W. J., Ramakrishna, S., Matsuura, T., Next-generation fibrous media for water treatment, MRS Bulletin 33 (1), 21–26 (2008).
6. Wang, J., Kim, S. C., Pui, D. Y. H., Investigation of the figure of merit for filters with a single nanofiber layer on a substrate, J. Aerosol Sci, 39 (4), 323–334 (2008).
7. Kim, J. S., Reneker, D. H., Polybenzimidazolenanofiber produced by electrospinning, Polymer Engineering & Science, 39 (5), 849–854 (1999).
8. Fang, S. W., Li, C. F., Shih, D. Y. C., Antifungal activity of Chitosan and its preservative effect on low-sugar candied kumquat, J. Food Prot, 57, 136–140 (1994).
9. Ignatova, M., Manolova, N., Markova, N., Rashkov, I., Electrospun non-woven nanofibrous hybrid mats based on Chitosan and PLA for wound-dressing applications, Macromol. Biosci, 9, 102–111 (2009).
10. Pillai, C. K. S., Paul, W., et al., Chitin and Chitosan polymers: Chemistry, solubility and fiber formation, Progress in Polymer Science, 34 (7), 641–678 (2009).
11. Kumar, M., A review of chitin and Chitosan applications, Reactive & Functional Polymers 46 (1), 1–27 (2000).
12. Kumar, M., Muzzarelli, R. A. A., et al., Chitosan chemistry and pharmaceutical perspectives, Chemical Reviews 104 (12), 6017–6084 (2004).
13. Ki Myong Kim, JeongHwa Son, Sung-Koo Kim, Curtis L. Weller andMilford A, Properties of Chitosan Films as a Function of pH and Solvent Type, Hanna Journal of Food Science, 71 (3), 119–124 (2006).
14. Helander, I. M., Nurmiaho-Lassila, E. L., Ahvenainen, R., Rhoades, J., Roller, S., Chitosan disrupts the barrier properties of the outer membrane of Gram-negative bacteria, International Journal of Food Microbiology, 71 (2–3), 235–244 (2001).
15. Devlieghere, F., Vermeulen, A., Chitosan: antimicrobial activity, interactions with food components and applicability as a coating on fruit and vegetables, Debevere, J., Food Microbiol, 21, 703 -714 (2004).
16. Aranaz, I., Mengíbar, M., Harris, R., Paños, I., Miralles, B., Acosta, N., Galed, G., Heras, Á., Functional characterization of chitin and chitosan, Curr. Chem. Biol., 3 (2), 203–230 (2009).
17. Neamnark, A., Rujiravanit, R., Supaphol, P., Electrospinning of hexanoyl Chitosan, Carbohydr. Polym, 66, 298–305 (2006).
18. Duan, B., Dong, C., Yuan, X., Yao, K., Electrospinning of Chitosan solutions in acetic acid with poly (ethylene oxide), J. Biomater. Sci. Polym. Ed, 15 (6), 797–811 (2004).
19. Jia, Y. T., Gong, J., Gu, X. H., Kim, H. Y., Dong, J., Shen, X. Y., Fabrication and characterization of poly (vinyl alcohol)/chitosan blend nanofibers produced by electrospinningmethod, Carbohydr. Polym., 67 (3), 403–409 (2007).

20. Homayoni, H., Ravandi, S. A. H., Valizadeh, M., Electrospinning of Chitosan nanofibers: Processing optimization, Carbohydr. Polym, 77 (3), 656–661 (2009).
21. Geng, X., Kwon, O. H., Jang, J., Electrospinning of Chitosan dissolved in concentrated acetic acid solution, Biomaterials, 26 (27), 5427–5432 (2005).
22. Vrieze, S. D., Westbroek, P., Camp, T. V., Langenhove, L. V., Electrospinning of chitosannanofibrous structures: Feasibility study, J. Mater. Sci, 42, 8029 -8034 (2007).
23. Boonsongrit, Y., Mueller, B. W., Mitrevej, A., Characterization of drug–Chitosan interaction by H NMR, FTIR and isothermal titration calorimetry. Eur. J. Pharm. Biopharm., 69, 388–395 (2008).
24. Mincheva, R., Manolova, N., Sabov, R., Kjurkchiev, G., Rashkov, L., Hydrogels from Chitosan cross linked with polyethylene glycol di acid as bone regeneration materials, E-polymer, 58, 1–11 (2004).

CHAPTER 31

THERMAL CURING OF COMPOSITES BASED EPOXY

LIA WIJAYANTI PRATOMO[1] and MARC J.M. ABADIE[1, 2]

[1]*School of Materials Science and Engineering, Nanyang Technological University, 639798, Singapore*

[2]*Institut Charles Gerhardt de Montpellier – Agrégats, Interfaces et Matériaux pour l'Energie (IGCM AIME UMR CNRS 5253), Université Montpellier 2, Place Bataillon, 34095 Montpellier Cedex 5, France, E-mail: abadie@univ-montp2.fr*

CONTENTS

ABSTRACT

This paper illustrates the benefit of kinetics to optimize a thermal curing formulation. The differential scanning calorimetry (DSC) was used to study the curing behavior of a system based epoxy (Epolam 5015).

Two thermal curing methods have been developed, non-isothermal and isothermal curing. Non-isothermal curing at five different heating rates and isothermal curing at five different temperatures, for various formulation of Epolam 5015 system was conducted. Kinetic parameters such as activation energy (Ea), rate constant (*k*) and enthalpy of reaction were studied. Results obtained showed that the optimum system of Epolam 5015 resin and hardener was 100/30 wt/wt%, where its Ea value is the lowest among other formulations. In addition, autocatalytic and second order kinetic model have proven to give the closest estimation to the non-isothermal and isothermal curing, respectively.

31.1 INTRODUCTION

Epoxy resins are widely used in diverse applications including fiber-reinforced composites, surface coatings, printed circuit boards, rigid foams and adhesives [1]. In all of these applications, a curing process is involved in which the monomeric or oligomeric polyfunctional epoxide is transformed into a cross-linked macromolecular structure. An understanding of the curing reaction and its kinetics, together with the availability of reliable methods of monitoring them, is important in order to obtain consistent products with the optimal desired physical and mechanical properties. The control of the curing process is especially important in the case of structural materials that require a critical design specification [2].

There are three curing methods that have been developed till today. They are thermal, electron beam (e-beam), and UV curing [1]. Thermal curing is the oldest among all these methods. It has been around for many years. The understanding of the thermal-curing process is now mature enough where applications such as large and complex structures are being manufactured and repaired. One of the newer trends in the composites industry is using radiation-curable resin systems with the radiation sources being ultraviolet (UV) or electron beam (EB). These methods of cure use radiation to physically or chemically change organic materials, forming cross-linked polymer networks. However, to date, thermal-curing systems is still proven to dominate the market place.

This chapter aims to investigate the thermal curing behavior of Epolam 5015 epoxy resin and hardener system. It shows the interest through kinetics to determine the optimum formulation of Epolam 5015 resin and hardener system thanks to the determination of kinetic parameters (i.e., activation energy and pre-exponential factor of the epoxy system) and the determination of best-fitted curing model, which corresponds to each thermal curing behavior.

31.1.1 THERMAL CURING

Epoxy resins are unique among thermosetting resins because broadly used in coatings (45%), electronics/electrical insulation (28%), composites (14%), construction (7%), and adhesives (6%) applications [3].

In order to obtain epoxy system for high performance applications, numerous curing processes have been done in the last decade. Thermal curing utilizes the application of heat to produce strong cross-linked network. In thermal curing, epoxy resins will react with a large number of chemical species called curatives or hardeners. Sometimes, it is incorrectly called catalysts. The hardeners are used in the amount that is sufficient to react with each epoxide group, while this reaction can be accelerated by using catalysts [4].

Curing reaction of epoxy and amine is a type of step-growth polymerization or condensation polymerization. This polymerization involves monomers that have at least two functional groups, so that the polymer can grow at either end of chain [5]. It can be monitored using differential scanning calorimetry (DSC). Dynamic DSC measures the heat flow to the sample as a function of time or temperature.

Figure 31.1 displays a typical DSC thermogram of an initially uncured sample submitted to a non-isothermal curing. The first heating corresponds to curing process. It is recorded from 30 to 250°C. The exothermic reaction starts near 40°C (T_{onset}) and ends at about 220°C (T_{end}) for 22 min (T_{cure}). The theoretical overall conversion attained at this point is one. The area underneath the exothermic peak is measured as the total heat of a complete reaction, ΔH_{tot}. The second heating is used to identify whether the system is fully cured. A fully cured system is attained when there is no exotherm peak observed in the second curve. In addition, the shift in the second heating curve around 40°C represents the glass transition temperature of the fully-cured system, T_g.

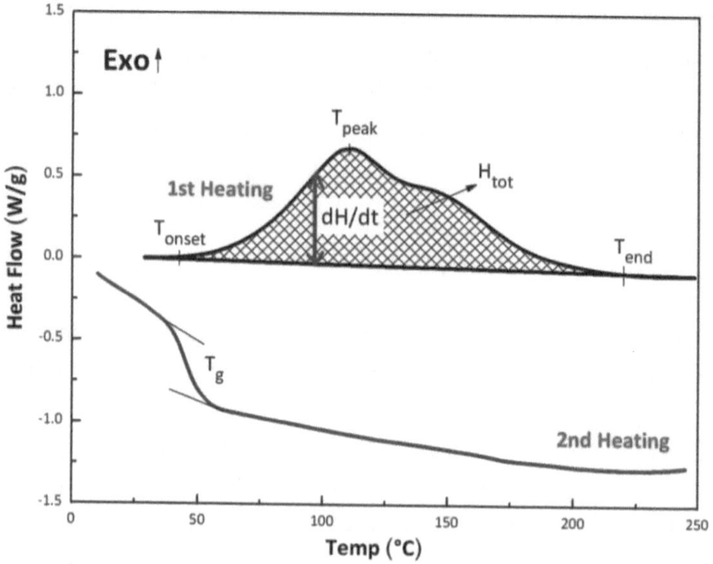

FIGURE 31.1 Typical DSC thermogram of non-isothermal curing.

The kinetic study of an exothermal reaction requires the following fundamental data: degree of cure, conversion rate, and glass transition temperature. All of these data can be determined using experimental results (as explained in Figure 31.1). The following is thorough explanation of each fundamental data.

31.1.1.1 Conversion Rate ($d\alpha/dt$) and Degree of Curing (α)

Thermal curing method continuously observes the conversion rate and degree of cure (conversion) over the entire course of reaction. This method assumes that the measured heat flow (dH/dt) is proportional to the conversion rate ($d\alpha/dt$). The relation can be expressed by the following equation (conversion rate, [7–9]:

$$d\alpha / dt = \frac{dH / dt}{\Delta H_{tot}} \tag{1}$$

The degree of curing was evaluated as fractional conversion (α_t). It is defined as the ratio of the heat recorded up to time t (ΔH_t) to the total heat of the complete reaction (ΔH_{tot}). The fractional conversion (degree of curing [7–9]) was quantitatively calculated as follows:

$$\alpha_t = \frac{\Delta H_t}{\Delta H_{tot}} \tag{2}$$

31.1.1.2 Total Heat of Reaction

The overall reaction enthalpy is an important parameter in Eqs. (1) and (2). For too low heating rates, some initial and final reactions may remain unrecorded because of insufficient instrument sensitivity. As for too high heating rates, the later stages of curing may interfere with thermal decomposition processes [10]. In this study, the total heating rate, ΔH_{tot} was taken by scanning the sample from 0 to 250°C at heating rate of 10°C/min.

31.1.1.3 Baseline of DSC Thermal Curing

The calculation of conversion and conversion rate requires the numerical integration of the partial areas of exothermic reaction peaks and therefore the need to draw a baseline. As it is observed in Figure 31.1, the heat flow (dH/dt) in Eq (1), was taken as the height of exothermic curve from the baseline.

In thermal curing, a linear baseline is drawn between the T_{onset} and T_{end} as illustrated in Figure 31.1 [11] where T_{onset} and T_{end} represent the starting and ending point of curing mechanism, respectively.

31.1.1.4 Glass Transition Temperature

Glass transition temperature or T_g is an essential data in the curing mechanism. As the curing progresses, T_g increases with the increasing molecular mass and/or crosslink density [12]. It usually increases from a value below room temperature for a fresh mixture to a value far beyond room temperature for the cured system [13].

It is important to know the glass transition temperature of the fully cured system, especially in the case of isothermal curing. While curing isothermally, T_g will often rise up to material gradually transforms from a liquid or rubbery state to a glassy state. This process is called vitrification. The reaction under vitrification situation proceeds in mobility-restricted conditions, which greatly reduces the reaction rate before the completion of curing. Consequently, the experimental conversion and conversion rate are much lower than those predicted by the kinetic equation [14].

31.1.2 KINETIC ANALYSIS

All mathematical approaches to describe the curing kinetics of thermosets are based on the following fundamental rate law (rate law equation) [15–17]:

$$\frac{d\alpha}{dt} = kf(\alpha) \tag{3}$$

where α is the degree of curing, $f(\alpha)$ is a function that depends on the reaction mechanism, and k is the temperature dependent rate constant given by Arrhenius equation [15–17]:

$$k = A\exp\left(\frac{-Ea}{RT}\right) \tag{4}$$

where A is the pre-exponential factor or Arrhenius frequency factor (s^{-1}), Ea is the activation energy ($J.mol^{-1}$), R is the gas constant ($8.314\ J.mol^{-1}.K^{-1}$), and T is the absolute temperature (K) of the sample.

To take into account the temperature dependence of rate constant, the rate formula in Eq. (3) is combined with the Arrhenius equation in Eq. (4). As a result, the kinetics of curing reaction (modified kinetic equation) can be described as follows:

$$\frac{d\alpha}{dt} = A\exp\left(\frac{-Ea}{RT}\right)f(\alpha) \tag{5}$$

We have use two thermal curing methods; they are non-isothermal and isothermal curing. Figure 31.2 shows the mechanism of kinetic analysis of the two thermal curing used in this study.

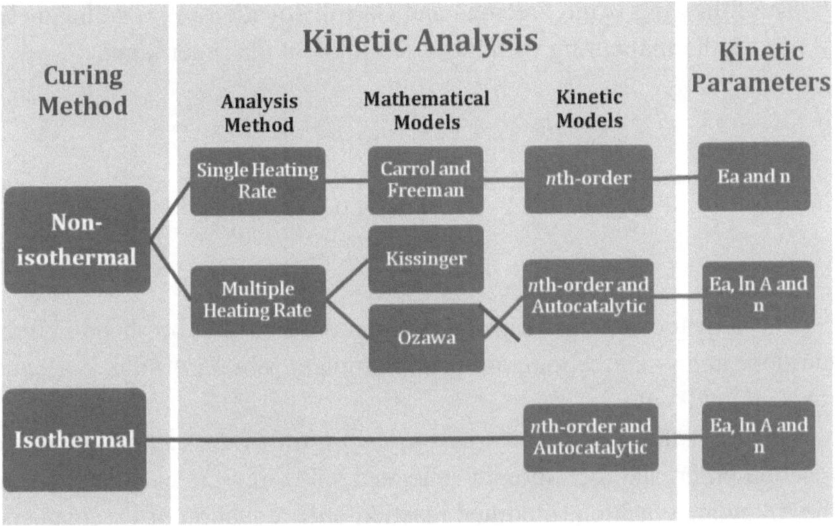

FIGURE 31.2 Mechanism of kinetic analysis.

31.1.2.1 Non-Isothermal Method

Non-isothermal curing is one of thermal curing method where temperature of material is gradually increased to a temperature just below decomposition temperature of the material. Non-isothermal methods that can be used include single heating rate (single HR) and multiple heating rate (Multiple HR) method.

31.1.2.1.1 Single HR Method

The single HR method measures the curing process based on only a single constant heating rate cycle. This method use *Freeman and Carroll* mathematical model and analyze according to the assumption of *n*th-order kinetic, from which the kinetic parameters (Ea and n) are then determined [18]. This fact that it is based on the assumption of *n*th order kinetic probably limits the reliability of the system as many systems are in fact not of the *n*th-order type. However, it can give us the advantages to provide extensive information from only a single dynamic scan.

The following is the Freeman and Carroll equation [19] on single HR of non-isothermal curing with the assumption of *n*th order kinetic:

$$\frac{dln\left(\frac{d\alpha}{dt}\right)}{dln(1-\alpha)} = -\frac{Ea}{R}\left[\frac{d\left(\frac{1}{T}\right)}{dln(1\text{-}\alpha)}\right] + n \tag{6}$$

The activation energy (Ea) and reaction order (n) can be obtained from the slope and y-interception of linear plotting of [dln(dα/dt)/dln(1–α)] and [d(1/T)/dln(1–α)].

However, this equation showed a large error in the estimation of the reaction order, and therefore the amended value of n are calculated using the following equation (modified reaction order equation of Freeman and Carroll method [20]):

$$n = \frac{Ea}{R}\left[\frac{1-\alpha_{max}}{a'_{max}T^2_{max}}\right] \tag{7}$$

31.1.2.1.2 Multiple HR method

Multiple-heating-rate methods are isoconversion methods. It assumes that the activation energy (Ea) is temperature-independent but it is conversion dependent [18, 21]. This assumption makes it equally effective for both the *n*th-order and the autocatalytic kinetic models. This method requires several thermograms, but demands little work as far as calculations are concerned. Two mathematical approaches that have been shown to be effective in multiple heating rate methods are proposed by Kissinger and Ozawa.

Kissinger mathematical approach

Kissinger uses differentiation approach to determine its kinetic parameters. The Kissinger equation is based on the assumption that the value of conversion at the peak of exothermic curing reaction is constant and is independent of heating rate [22–24]. The Kissinger equation at the peak of exothermic reaction for each heating rate can be expressed as follows:

$$\frac{d\ln\left(\frac{\beta}{T_{peak}^{\ 2}}\right)}{d\left(\frac{1}{T_{peak}}\right)} = -\frac{Ea}{R} \tag{8}$$

The above mentioned equation allows a linear plotting of $\ln(\beta/T_{peak}^{\ 2})$ and $1/T_{peak}$ to obtain Ea from its slope.

However, the Kissinger equation needs to be modified, where T_{peak} is substituted with T_i (isoconversion temperature: temperature at every conversion measured) to obtain ln A. The equation is described as follows (Kissinger isoconversion equation [22–24]):

$$\ln\left(\frac{\beta}{T_i^2}\right) = \ln\left(\frac{AR}{g(\alpha)Ea}\right) - \frac{Ea}{RT_i} \tag{9}$$

where, g(α) is the integration function of conversion [22–23], defined as:

$$g(\alpha) = \int_0^{\alpha} \frac{d\alpha}{f(\alpha)} \tag{10}$$

Consequently, the value of ln A can be obtained from the *y*-interception of linear plotting of $\ln(\beta/T_i^2)$ and $1/T_i$.

Ozawa mathematical approach

Ozawa uses integration approach to determine its kinetic parameters. Ozawa equation [22–25] provides a simple relationship between the activation energy, heating rate and isoconversion temperature. Its equation can be expressed as follows:

$$\ln\beta = \ln\frac{AEa}{g(\alpha)R} - 5.330 - 1.052\left(\frac{Ea}{RT}\right) \tag{11}$$

According to Eq. (11), Ea and ln A can be obtained from the slope and y-interception of the linear plotting of lnβ and 1/T. The advantage of this approach is that the activation energy can be measured over the entire course of the curing reaction.

31.1.2.2 Isothermal Method

Isothermal curing is a curing process where the material is set to cure at a constant temperature. Isothermal method does not use any mathematical approach to determine its kinetic parameters; instead it only uses the fundamental rate law and the application of the two kinetic models [22, 26].

Isothermal method uses an exponential logarithmic plot of Arrhenius equation [22, 26]:

$$ln(k) = \frac{-Ea}{R}\left(\frac{1}{T}\right) + ln(A) \quad (12)$$

The linear plotting of ln k and 1/T gives us the value of Ea and ln A from the slope and y-interception, respectively.

31.1.2.3 Kinetic Models

The modeling equations for the cure kinetics of thermosetting materials generally fall under two general categories: *n*th-order and autocatalytic.

31.1.2.3.1 nth-Order Kinetic Model

For thermoset materials that follow *n*th-order reactions [23–24] suggest that the rate of conversion is proportional to the concentration of unreacted material, such that $f(\alpha)=(1-\alpha)^n$ and therefore:

$$\frac{d\alpha}{dt} = k(1\text{-}\alpha)^n \quad (13)$$

where *n* is the reaction order.

The above equation indicates that the maximum conversion rate occurs at around t=0. Also, the reaction rate ($d\alpha/dt$) is dependent only on the amount of unreacted material and assumes that the reaction products do not become involved in the reaction.

In this study, the three *n*th-order kinetic models will be used. They are first, second and third order kinetic models.

31.1.2.3.2 Autocatalytic Kinetic Model

Autocatalyzed curing reactions on the contrary assumes that at least one of the reaction products is also a reactant [27–28], and thus are characterized by an accelerating conversion rate, which typically reaches its maximum between 30 and 40% conversion. The kinetics of autocatalyzed reaction is described by the following equation (conversion rate of autocatalytic reaction [27–28]):

$$\frac{d\alpha}{dt} = k\alpha^{n}(1-\alpha)^{m} \tag{14}$$

where m and n are the reaction orders and $f(\alpha)$ correspond to $\alpha^{n}(1-\alpha)^{m}$.

According to the autocatalytic model, the reaction rate is zero or very small initially and attains a maximum value at some intermediate conversion. The initial rate of autocatalytic reactions may not be necessarily zero, as there is a possibility that reactants can be converted into products via alternative paths. However, we will only focus on zero initial rates.

31.1.2.3.3 List of All Kinetic Model

In this study, the kinetic analysis will use *n*th-order kinetic and autocatalytic model to simulate the curing behavior of epoxy system. Following is the list of kinetic models with their $f(\alpha)$ and $g(\alpha)$, respectively.

TABLE 31.1 Different Types of Curing Kinetic Model

Model	f(α)	g(α)
n^{th} order kinetic		
• $n = 1$	$(1-\alpha)$	$-\ln(1-\alpha)$
• $n = 2$	$(1-\alpha)^{2}$	$-1 + (1-\alpha)^{-1}$
• $n = 3$	$(1-\alpha)^{3}$	$2-^{1}[-1+(1-\alpha)^{-2}]$
Autocatalytic		
• $n + m = 2; n = 1.5$	$\alpha^{05}(1-\alpha)^{1.5}$	$[(1-\alpha)\alpha^{-1}]^{-0.5}(0.5)^{-1}$

31.2 EXPERIMENTAL PART

31.2.1 MATERIALS

A type of epoxy resin and hardener was used in this study. It was Epolam 5015 resin and hardener. The materials were used as received from Axson. Table 31.2 shows the chemical structure of epoxy resin and hardener we have used.

31.2.2 SAMPLE PREPARATION

Three formulations were prepared. All of them use the same resin and hardener. However, the composition of each resin and hardener is different for each formulation as shows in Table 31.3.

Because most amines are reactive at room temperature, therefore, the formulations were mixed just prior to application.

TABLE 31.2 Chemical Structure of Epolam 5015 Resin and Hardener

Product Name	Chemical Structure
Epolam 5015 Resin	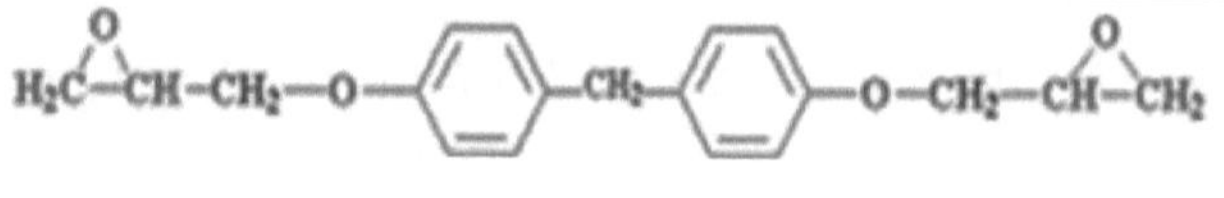
Epolam 5015 Hardener	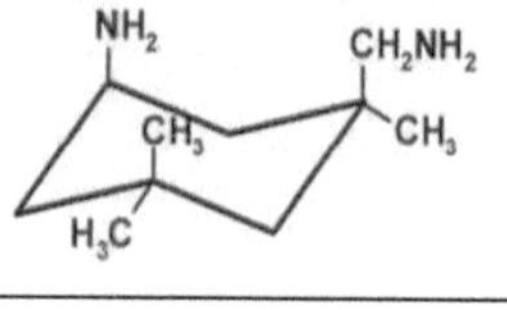

TABLE 31.3 Composition of Various Formulations

Formulation	Epolam 5015 Resin (% by weight)	Epolam 5015 Hardener (% by weight)
1	100	30
2	100	20
3	100	10

31.2.3 EXPERIMENTAL PROCEDURE

Thermal curing of Epolam 5015 system was done using both non-isothermal and isothermal curing method and monitored by DSC. The following were the experimental procedure for both curing method used in this study.

1. Sample was prepared with aluminum hermetic pan and clamped. The amount of sample was in the range of 5 to 7 mg.
2. *Non-isothermal curing* was performed at the heating rate of 5, 7.5, 10, 15 and 20°C/min. *Isothermal curing* was performed at temperature of 120, 125, 130, 135 and 140°C.
3. Curing peak of resulting curing peak was integrated using *integrated peak linear* function in TA Analysis to obtain the fundamental data of exothermic reaction.

31.3 RESULTS AND DISCUSSION

31.3.1 NON-ISOTHERMAL CURING

The non-isothermal curing reaction of Epolam 5015 system was investigated by DSC at five different heating rates as mentioned above for three different formulations.

31.3.1.1 Formulation #1 (100/30 Ratio)

The first formulation is a mixture of 100 wt% and 30 wt% of Epolam 5015 resin and hardener respectively.

In Figure 31.3, the heat flow is plotted as a function of the temperature for five different heating rates. It is seen that all exothermic reactions are recorded from 30 to 250°C, and the curing peak shifts towards the higher temperature with higher heating rates.

The detailed result of curing experiment can be seen in the Table 31.4.

The curing data in Table 31.4 shows that the increase in heating rate leads to decrease of the curing time. In addition, glass transition temperature (T_g) and total heat flow (ΔH_{tot}) remains relatively constant while

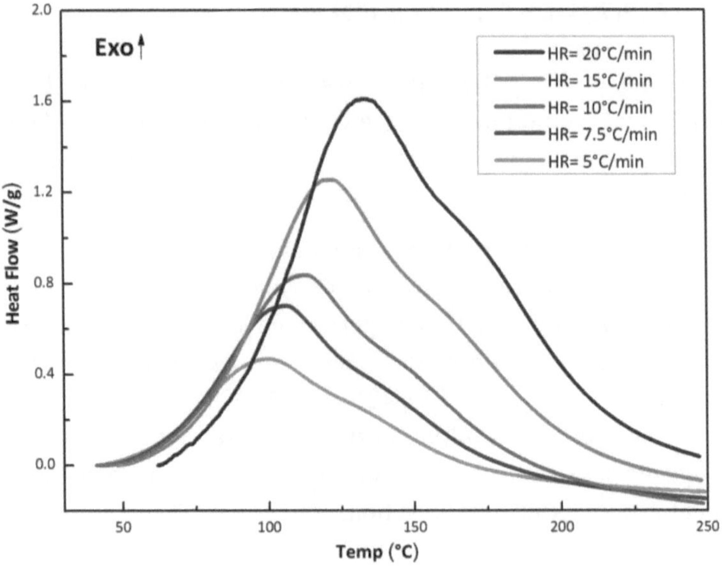

FIGURE 31.3 DSC plot of non-isothermal curing (formulation #1) at different heating rates.

TABLE 31.4 Kinetic Data of Non-Isothermal Curing (Formulation #1)

β (°C/min)	T_{onset} (°C)	T_{peak} (°C)	T_{end} (°C)	T_g (°C)	ΔH_{tot} (J/g)	t_{cure} (min)
5	57.08	100.32	249.13	95.49	408.1 ± 20.41	41.50
7.5	63.05	106.33	249.2	95.41	405.9 ± 20.30	27.40
10	66.22	113.45	248.44	96.10	395.3 ± 19.77	20.75
15	73.82	122.61	247.77	97.33	387.3 ± 19.37	13.30
20	87.43	134.54	247.25	96.02	371.9 ± 18.60	9.25

subjected to different heating rates. The average total heat released for non-isothermal curing was 393.70 ± 14.80 J/g.

The result of curing experiment for formulation #2 and #3 gave the same trend as results for formulation #1.

31.3.1.1.1 Comparison of the Three Formulations

The composition of resin and hardener in each formulation shows a great influence on the curing process. As the ratio of hardener in the formulation decreases, the average total heat released and glass transition temperature of the system also decreases. These relations can be seen in Table 31.5 below.

Figure 31.4 shows that single exothermal signal is observed in formulation #1 and two exothermal peaks appear in formulation #2 and #3. Additionally, the second peak can clearly be observed in formulation #3 where the ratio of hardener is the least. However, the nature of these two peaks still remains unclear as the appearance of two peaks usually occurred

TABLE 31.5 Experimental Data of Non-Isothermal Curing At Three Different Formulations

Formulation	Ratio of resin and hardener	T_g (°C)	ΔH_{tot} (J/g)
#1	100/30	96.07 ± 0.77	393.70 ± 14.80
#2	100/20	44.22 ± 0.62	317.70 ± 5.97
#3	100/10	–4.22 ± 1.06	164.98 ± 3.63

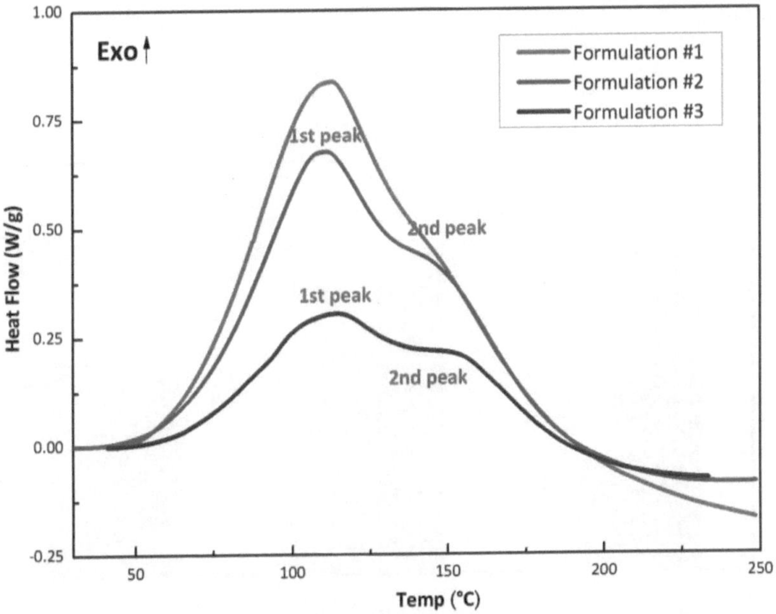

FIGURE 31.4 Comparison of non-isothermal curing at HR = 10°C/min.

when accelerator or catalyst was used, with the second peak corresponds to the decomposition of accelerator used [4].

31.3.1.1.2 Kinetic Models

Determination of kinetic model is essential to determine which method is used to analyze non-isothermal curing.

Figure 31.5 shows the values of conversion correspond to the first exothermic peak of all formulations, which is located at conversion of 30–40%. This information suggested that the curing process undergo autocatalytic reaction. Consequently, single heating rate method will not be used since it is based on the assumption of *n*th-order kinetic. Instead, multiple heating rate method using Kissinger and Ozawa approached will be performed to find kinetic parameters (i.e., Ea and ln A).

The second exothermic peak will be simulated using *n*th-order kinetic as shown in Figure 31.6. At conversion higher than 60%, the conversion rate decreases with increases degree of conversion. This behavior fits the *n*th-order reaction where its conversion rate at the start of reaction is the maximum.

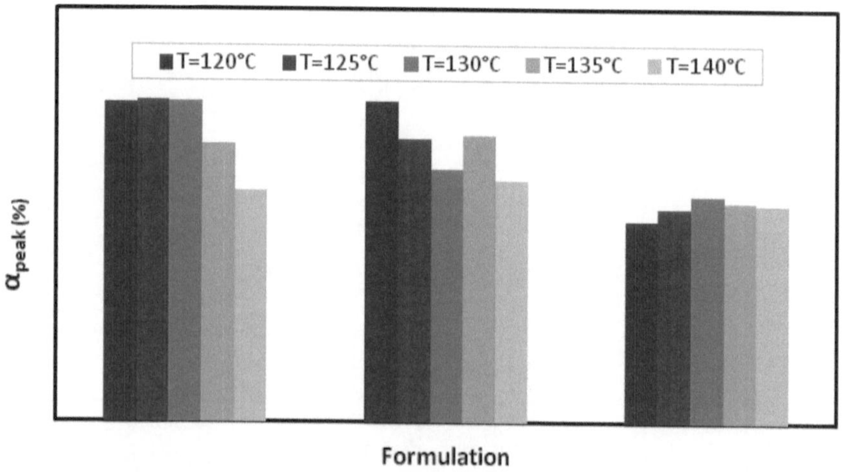

FIGURE 31.5 Conversion profile at peak of non-isothermal curing.

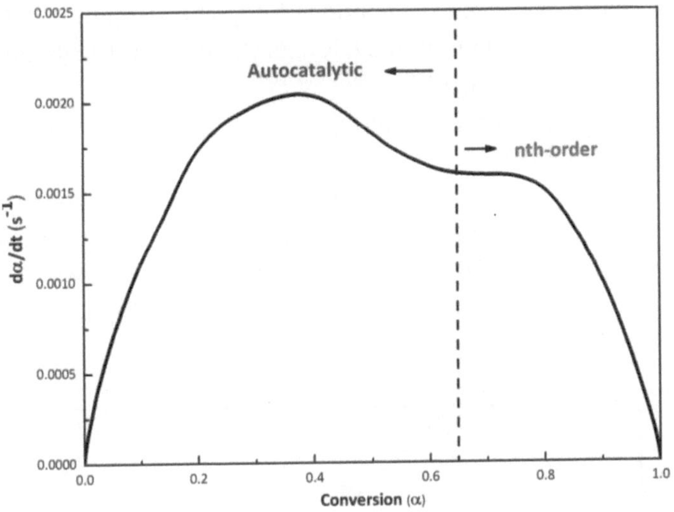

FIGURE 31.6 Plot of conversion rate and conversion at HR = 10°C/min for formulation #3.

31.3.1.1.3 Kinetic Analysis

The non-isothermal kinetics of Epolam 5015 system is investigated by Multiple HR method using Kissinger and Ozawa approaches for all formulations. As previously stated, the first and second peak for all formulations will be based on the assumption of autocatalytic and *n*th-order reaction, respectively.

(a) Kissinger method

Kissinger assumed that the value of conversion at the peak of exothermic curing reaction is constant and is independent of heating rate. Figure 31.6 shows that the conversion values at the first peak were around 30–40% at five different heating rates for three different formulations. In addition, the second peaks were found to be around 60% for formulation #2 and #3. These evidences confirm that the assumption made was valid for the system tested.

(i) First exothermic peak (Formulation #1)

This section only explains the first exothermic peak for formulation #1, as the kinetic parameters for formulation #2 and #3 can be found using the same method.

On the basis of Kissinger equation in Eq. (8), the plot of $\ln(\beta/T_{peak}^{2})$ and $1/T_{peak}$ obtained from multiple HR data should be a straight line whose slope allow the evaluation of Ea.

According to the linear plot in Figure 31.7, activation energy was calculated, yielding a value of 44.32 ± 0.54 kJ/mol.

Moreover, the value of ln A is obtained using the combination of autocatalytic model and kinetic value of ln(g(α)/A) from y-interception of the isoconversion plot of $\ln(\beta/T_i^{2})$ and $1/T_i$ shown in Figure 31.8 and Table 31.6.

The isoconversion value of Ea and ln A can be seen in the Table 31.6.

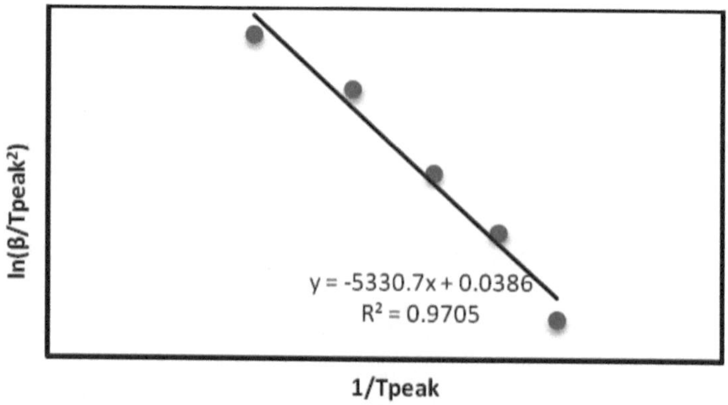

FIGURE 31.7 Kissinger plot at the peak of non-isothermal curing.

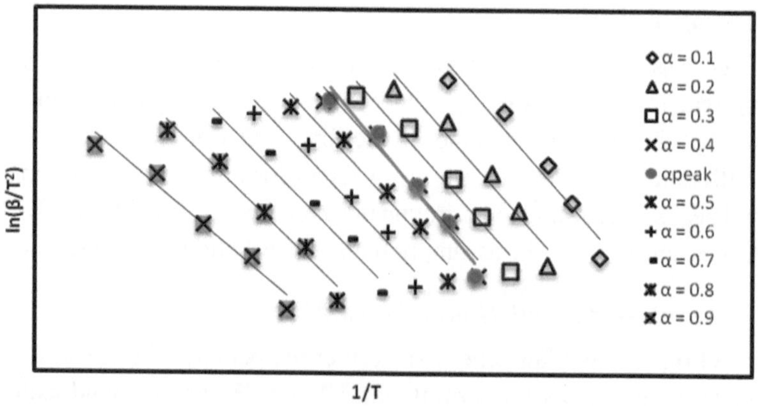

FIGURE 31.8 Kissinger isoconversion plot.

TABLE 31.6 Kissinger Isoconversion Data

α	Ea (kJ/mol)	ln(g(α)/A)	ln A	R^2
0.1	42.8	–9.24	8.83	0.9447
0.2	41.83	–8.41	8.41	0.9556
0.3	41.42	–7.94	8.21	0.9598
0.4	40.86	–7.47	7.96	0.9603
0.5	40.02	–6.91	7.6	0.9606
0.6	38.87	–6.24	7.14	0.9629
0.7	37.67	–5.53	6.65	0.9683
0.8	36.21	–4.71	6.09	0.9738
0.9	31.3	–2.75	4.54	0.9677

The average value of Ea calculated using isoconversion Kissinger method was 39.00 ± 3.56 kJ/mol. This value was close to the value measured from the peak of exotherm reaction. The average value of ln A was 7.27 ± 1.34.

(ii) Second exothermic peak (Formulation #2 and #3)

The kinetic parameters of second exothermic peak can be obtained using the same Kissinger method for the first peak, where the second peak for each heating rates used were around 60% conversion. In addition, the isoconversion data was taken between 65–95% conversion.

The kinetic parameters (Ea and ln A) for both first and second exothermic peak for formulation #2 and #3 can be found in Table 31.7.

As it is observed in Table 31.7, the values of second Ea (Ea of second peak) are lower compare to the values of first Ea (Ea of first peak).

(b) Ozawa method

According to Ozawa equation in Eq. (11), the activation energy can be measured over the entire course of the curing reaction. This relation makes Ozawa method is one of isoconversion method where finding the value of Ea for any conversion is required.

TABLE 31.7 Kissinger's Kinetic Parameters (Formulation #2 and #3)

Model	Ea (kJ/mol)		ln A	
	Formulation #2	**Formulation #3**	**Formulation #2**	**Formulation #3**
First peak				
Autocatalytic	50.74 ± 0.46	54.74 ± 0.49	9.92 ± 0.76	11.94 ± 1.01
Second peak				
First order kinetic			8.00 ± 0.59	9.98 ± 0.29
Second order kinetic	44.09 ± 0.42	52.32 ± 0.52	9.04 ± 0.15	10.94 ± 0.29
Third order kinetic			10.30 ± 0.46	12.12 ± 0.86

(i) First exothermic peak (Formulation #1)

This section only explains the first exothermic peak for formulation #1. The kinetic parameters for formulation #2 and #3 can be found using the same method.

Figure 31.9 shows that the plot of ln β against 1/Ti enables the calculation of the activation energy at any conversion by the Ozawa method.

The value of ln A is obtained using similar method as in Kissinger method. The isoconversion result of Ea and ln A were shown in the Table 31.8.

The regression coefficient shown in Table 31.8 proved that the linear relationship between ln(β) and 1/T are in a good term. The average value of Ea and ln A calculated using isoconversion Ozawa method is 42.47 ± 3.74 kJ/mol and 9.32 ± 1.26, respectively.

The value of Ea obtained in this method is in an agreement with the value obtained using Kissinger method.

(ii) Second exothermic peak (Formulation #2 and #3)

The kinetic parameters of second exothermic peak can be obtained using the same Ozawa method for the first peak, where the data can be taken from conversion above 60%. The kinetic parameters (Ea and ln A) for both first and second exothermic peak for formulation #2 and #3 can be found in Table 31.9.

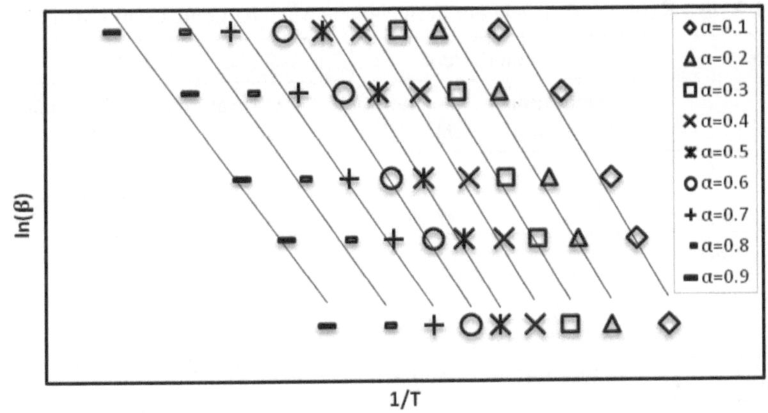

FIGURE 31.9 Ozawa isoconversion plot.

TABLE 31.8 Ozawa Isoconversion Data

α	Ea (kJ/mol)	ln(g(α)/A)	ln A	R^2
0.1	46.44	–11.19	10.78	0.9576
0.2	45.69	–10.47	10.47	0.9666
0.3	45.43	–10.05	10.32	0.9701
0.4	45.02	–9.64	10.13	0.9707
0.5	44.19	–9.07	9.77	0.9839
0.6	41.97	–8.17	9.07	0.9818
0.7	38.79	–6.96	8.07	0.9788
0.8	38.17	–6.45	7.84	0.9780
0.9	36.53	–5.62	7.41	0.9675

Table 31.9 shows that the values of second Ea are similar to the value first Ea. In addition, the kinetic parameters obtained using this method is relatively higher than the result in Kissinger method.

31.3.1.1.4 Simulation of Non-Isothermal Curing

Kissinger and Ozawa method have provided the result of kinetic parameters for non-isothermal curing for all different formulations as shown in Table 31.10.

TABLE 31.9 Kinetic Parameters For Formulation #2 and #3 Using Ozawa Method

Model	Ea (kJ/mol)		ln A	
	Formulation #2	Formulation #3	Formulation #2	Formulation #3
First peak				
Autocatalytic	50.99 ± 1.20	57.60 ± 1.07	11.85 ± 0.59	13.65 ± 0.83
Second peak				
First order kinetic			10.15 ± 0.47	11.87 ± 0.22
Second order kinetic	49.49 ± 1.47	56.48 ± 0.34	11.18 ± 0.04	12.83 ± 0.34
Third order kinetic			12.44 ± 0.58	14.01 ± 0.92

Table 31.10 only shows the kinetic parameters of the best-fitted model for non-isothermal curing.

As previously stated, the first and second exothermic peak is simulated using autocatalytic and *n*th-order kinetic model. In addition, the curing data of second exothermic peak are found to fit the first order kinetic model. The simulation of these two models can be found in the following figures.

Figures 31.10 and 31.11 displays the simulation of non-isothermal curing using Kissinger and Ozawa kinetic parameters represented by different heating rate of each formulation for the purpose to see it clearly. The autocatalytic model using both methods shows an accurate reading for formulation #1 and #2. However, the simulation using Ozawa method for formulation #3 shows a much higher estimation than the curing experimental data. This phenomenon could be due to a higher value of Ea estimated by Ozawa method as shown in Table 31.9.

The simulation for second exothermic peak using first order kinetic model have been proven to give the same trend as shown in the experimental curing data. However, this model have not yet given an accurate reading of curing behavior, as the error can clearly be seen in Figures 31.10 and 31.11.

Overall, Kissinger gives a better simulation for non-isothermal curing compare to Ozawa method, as their estimation is much closer to the experimental values for all formulations used. The detailed simulation for formulation #1 can be found in Appendix A1–3.

TABLE 31.10 Kinetic Parameters of Non-Isothermal Curing

Model	Ea (kJ/mol)		ln A	
	First peak (Autocatalytic)	Second peak (1st order)	First peak (Autocatalytic)	Second peak (1st order)
Formulation #1				
Kissinger	44.32 ± 0.54	-	7.27 ± 1.34	-
Ozawa	42.47 ± 3.74	-	9.32 ± 1.26	-
Formulation #2				
Kissinger	50.74 ± 0.46	44.09 ± 0.42	9.92 ± 0.76	8.00 ± 0.59
Ozawa	50.99 ± 1.20	49.49 ± 1.47	11.85 ± 0.59	10.15 ± 0.47
Formulation #3				
Kissinger	54.74 ± 0.49	52.32 ± 0.52	11.94 ± 1.01	9.98 ± 0.29
Ozawa	57.60 ± 1.07	56.48 ± 0.34	13.65 ± 0.83	11.87 ± 0.22

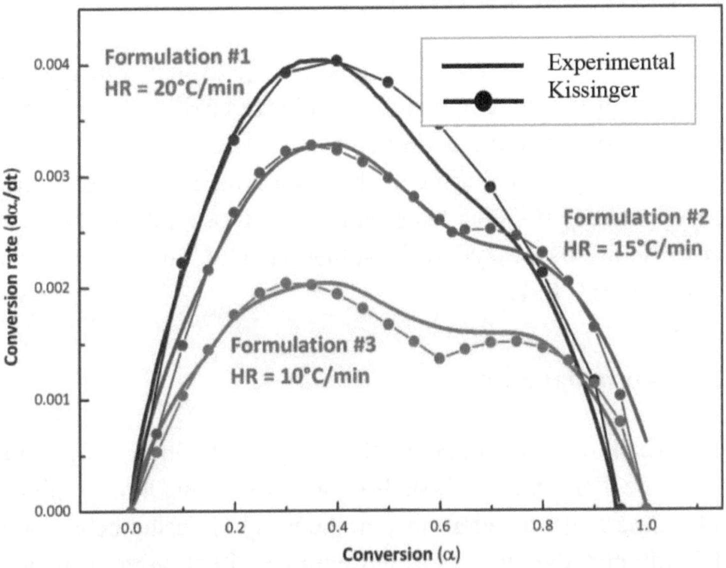

FIGURE 31.10 Simulation of non-isothermal curing using Kissinger kinetic parameters.

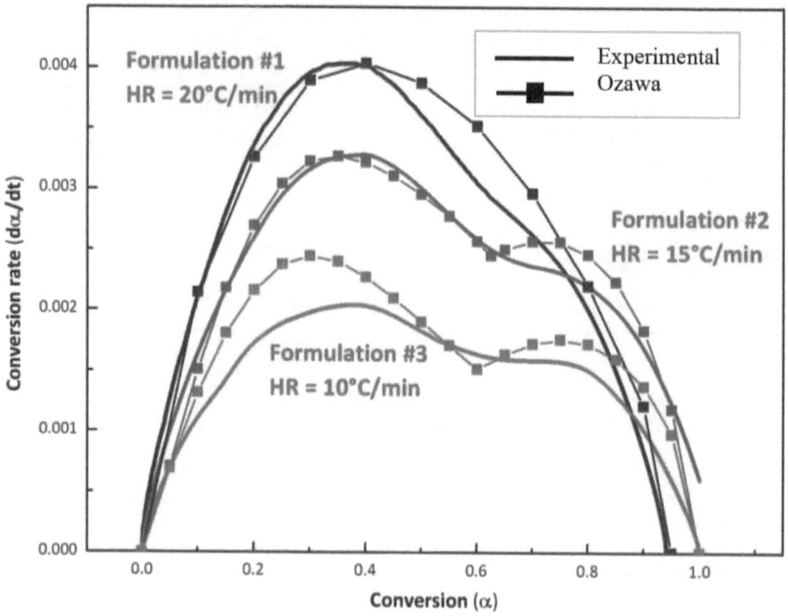

FIGURE 31.11 Simulation of non-isothermal curing using Ozawa kinetic parameters.

31.3.2 ISOTHERMAL CURING

The isothermal curing reaction of Epolam 5015 system was investigated by DSC at five different temperatures and for three different formulations. At higher temperatures, the curing time was so fast that the heat released was unrecorded by the calorimeter. Therefore, no experiments were conducted isothermally at higher temperature than those indicated above. This type of difficulty associated with isothermal DSC and not encountered with dynamic DSC tests.

31.3.2.1 Formulation #1

Figure 31.12 shows that single exothermic peak is observed for this formulation. The DSC plot of heat flow against curing time is limited to 5 minutes time to get a better view on the trend of curing behavior. In the end of 5 minutes, the curing has not completed yet as the signals is not leveled off to the baseline.

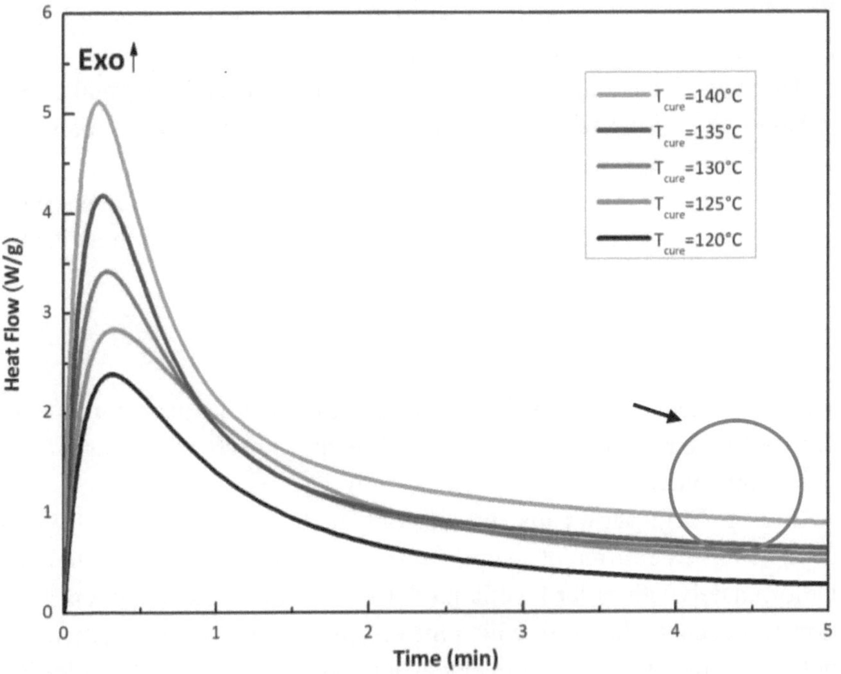

FIGURE 31.12 DSC plot of isothermal curing (formulation #1) at different temperatures.

The detailed result of curing experiment can be seen in the Table 31.11.

Table 31.11 shows that the total heat flow (ΔH_{tot}) remains relatively constant while subjected to different curing temperatures. The average total heat released for isothermal curing is 383.96 ± 13.81 J/g. This value is comparable to the value obtained using non-isothermal curing.

TABLE 31.11 Kinetics Data of Isothermal Curing (Formulation #1)

T_{cure} (°C)	T_g (°C)	ΔH_{tot} (J/g)	$t_{\alpha=0.9}$ (min)
120	85.33	374.81 ± 18.74	35.50
125	86.69	390.18 ± 19.51	11.53
130	88.12	395.95 ± 19.80	17.17
135	90.28	394.53 ± 19.73	26.65
140	93.45	364.35 ± 18.22	23.76

In addition, the glass transition temperature (T_g) shows increasing behavior as the curing temperature increased.

However, it can be seen that the value of T_g for each isothermal curing is below the curing temperature, which means vitrification process was avoided during the curing process.

The isothermal curing results for formulation #2 and #3 can be seen in appendix A2–1, as they gave the same trend as the results in formulation #1.

31.3.2.1.1 Kinetic Models

The experimental result in Figure 31.12 shows that the peak maxima are detected at the early curing time, which is the expected behavior of *n*th-order reaction.

Figure 31.13 shows that the peaks of exotherms are located within 0–12.5% at five different temperatures for three different formulations. Unfortunately, *n*th-order kinetic model assumed that the maximum reaction rate occurs when t=0, which means, in this case, the simulation will only be applicable from conversion around 10% onwards. In addition, three models of *n*th-order kinetics will be used, they are first, second and third order.

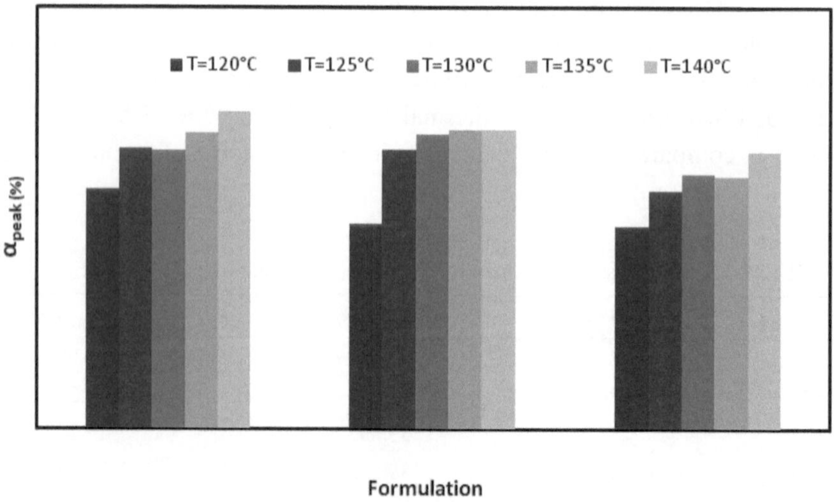

FIGURE 31.13 Conversion profile at peak of isothermal curing.

31.3.2.1.2 *Kinetic Analysis*

The kinetic analysis of isothermal curing will be performed using fundamental rate law and *n*th-order kinetic model. In the end of each analysis, kinetic parameters (Ea, ln A and n) will be obtained and used to simulate the curing process.

We present the best-fitted model of formulation #1, which is the second order kinetic model. The same method was used to the others kinetic models

(a) Second Order Kinetic Model

Arrhenius equation in Eq. (4) proposed kinetic parameters can be obtained using the linear plotting of ln k vs. 1/T. Firstly, the value of k can be obtained from the slope of the linear relation between conversion rate and f(α) as stated in fundamental rate law shown in Figure 31.14.

Table 31.12 shows the value of k for every isothermal temperature obtained from Figure 31.16. It can be seen that the linear relation of the conversion rate and f(α) is in a good term as specified in fundamental rate law, which is indicated by its regression coefficient for more than 98% for all cases. The results also clearly show that the rate constant increases as isothermal temperature increases.

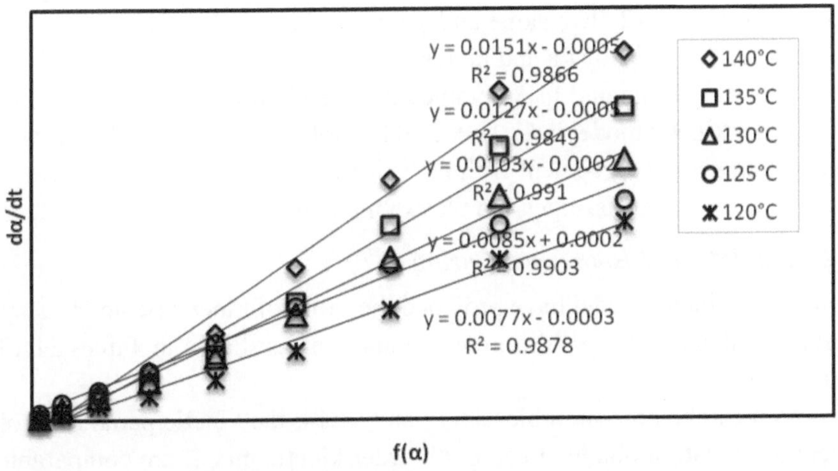

FIGURE 31.14 Second order kinetic plot of conversion rate and f(α).

TABLE 31.12 Rate Constant Value At Five Different Isothermal Temperatures

T(°C)	k (s^{-1})	R^2
120	0.0077	0.9878
125	0.0085	0.9903
130	0.0103	0.9910
135	0.0127	0.9848
140	0.0151	0.9846

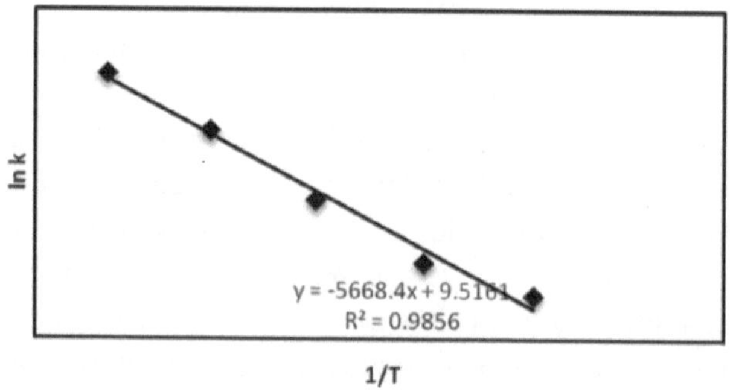

FIGURE 31.15 Second order kinetic plot of ln k and 1/T.

Figure 31.15 below display a linear plotting of ln k and 1/T using the data in Table 31.12. The slope and y-interception of this plot allow us to determine the value of Ea and ln, respectively.

The value of Ea and ln A obtained from isothermal method using second order kinetic model is 47.16 ± 0.40 kJ/mol and 9.52 ± 0.98. These values are in a good agreement with the value obtained from non-isothermal curing using both Kissinger and Ozawa method.

(b) Simulation of Isothermal Curing

*n*th-order kinetic model have provided the result of kinetic parameters for isothermal curing for all different formulations as shown in Tables 31.13 and 31.14.

As it can be seen in Tables 31.13 and 31.14, the kinetic parameters of each formulation obtained using *n*th-order kinetic model are comparable to those results in non-isothermal curing.

TABLE 31.13 Ea of Isothermal Curing Using *n*th-Order Kinetic Model

Model	Ea (kJ/mol)		
	Formulation #1	**Formulation #2**	**Formulation #3**
First order kinetic	47.50 ± 0.27	51.69 ± 0.80	53.68 ± 0.40
Second order kinetic	47.16 ± 0.40	56.19 ± 1.05	54.04 ± 0.40
Third order kinetic	46.87 ± 0.52	57.97 ± 1.15	54.30 ± 0.43

TABLE 31.14 ln A of Isothermal Curing Using *n*th-Order Kinetic Model

Model	ln A		
	Formulation #1	**Formulation #2**	**Formulation #3**
First order kinetic	9.61 ± 0.67	10.73 ± 1.98	11.24 ± 1.00
Second order kinetic	9.52 ± 0.98	12.06 ± 2.61	11.37 ± 1.00
Third order kinetic	9.52 ± 1.28	12.67 ± 2.85	11.54 ± 1.06

The simulation of formulation #1 with T_{cure} = 130°C using all three *n*th-order kinetic model can be seen in Figure 31.16. The Figure 31.16 shows that the first order kinetic model failed to simulate isothermal curing as it

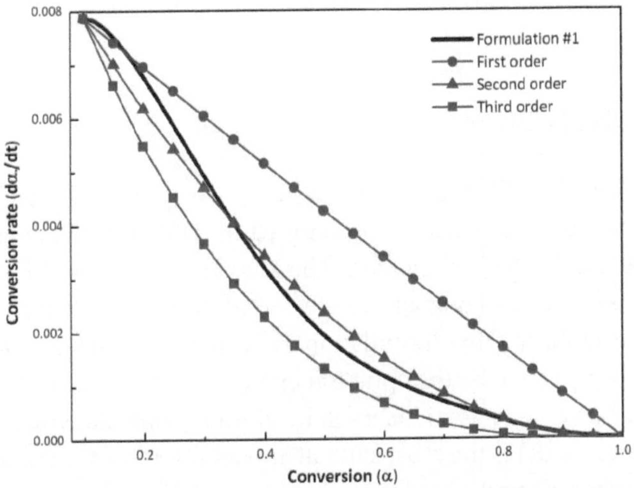

FIGURE 31.16 Simulation of isothermal curing at 130°C using *n*th-order kinetic model.

gives a linear relation. On the other hand, second and third order provides non-linear correlation, however, the second order proves to give the best fitted model for isothermal curing as it offers much closer estimation to the experimental result.

The simulation of each formulation using second order kinetic model can be seen in the Figure 31.17.

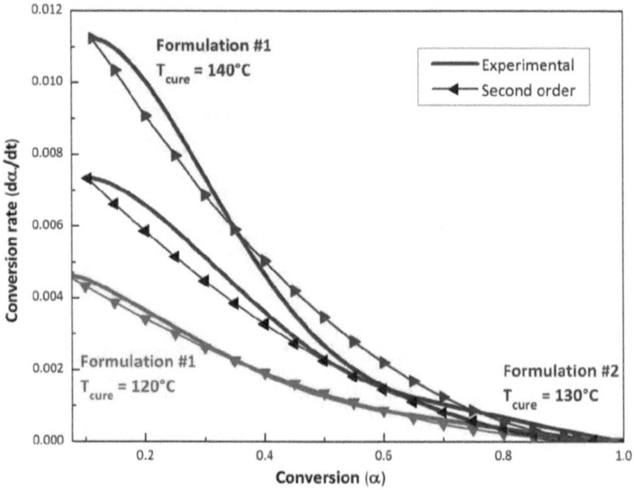

FIGURE 31.17 Simulation of isothermal curing for all formulations.

31.4 CONCLUSION

The cure kinetics of Epolam 5015 system using three different formulations of resin and hardener was examined by DSC technique using non-isothermal and isothermal method. The kinetic parameters obtained using both methods are in a good agreement for all three formulations.

It was established that formulation #1, which have the least amount of hardener, is proven to be the optimum epoxy system of Epolam 5015 with the lowest Ea value of 45–47 kJ/mol. In addition, autocatalytic and second order kinetic model is the most suitable model for non-isothermal and isothermal curing, respectively for all three formulations.

TABLE 31.15 Activation Energy of Epolam 5015 System For All Three Formulations

Model	Ea (kJ/mol)		
	Formulation #1	**Formulation #2**	**Formulation #3**
Non-isothermal (Autocatalytic)			
Kissinger	44.32 ± 0.54	50.74 ± 0.46	54.74 ± 0.49
Ozawa	42.47 ± 3.74	50.99 ± 1.20	57.60 ± 1.07
Isothermal (2nd order kinetic)	47.16 ± 0.40	56.19 ± 1.05	54.04 ± 0.40

KEYWORDS

- **acrylates**
- **kinetics**
- **thermal curing**
- **thermoset**

REFERENCES

1. Lopata, V. J., D. R. S., *Electron beam processing for composite manufacturing and repair*. 2003. p. 32–42.
2. Li Shuyan, Lemmetyinen, E. V., helge. Journal of Applied Polymer Science, 81(6), 1474, 2000.
3. Sbirrazzuoli, N., Alice Mititelu, S. V., Sladic C, Vincent, L. Macromolecular Chemistry and Physics, 204(15), 1815, 2003.
4. Atarsia, A., R. B., Journal of Polymer Engineering and Science, 40(3), 607, 2000.
5. Foundation, T. C. H. *Step growth polymerization.* Other Science from Wallace Carothers 2000.
6. *Making epoxy resins*. Macrogalerria 1995–2005.
7. *Principles and Applications of Thermal Analysis*, P. Gabbott, Editor. 1988, Blackwell Publishing.
8. Ghaemy, M., Rostami, A. A., Omrani, A. Polym Intern., 55(3), 279–284, 2006.
9. Barghamadi, M., Ghaemy, M., and Alizadeh, R. Iranian Polymer Journal, 18(6), 431, 2009.
10. Van Assche, G., Van Mele, S. S., B. Thermochimica Acta, 388(1–2), 327, 2002.

11. Sun, L., Peng S-S., Sterling Arthur, M., Negulescu Ioan, I., Stubblefield Michael, A. J., of Applied Polymer Science, 83, 1074, 2002.
12. Mina, B. G., SStachurski, Z. H., Hodgkin, J. H. Polymer, 34(23), 4908, 1993.
13. Zhang, J., Fox, B., Guo, Q. J. of Applied Polymer Sci., 107(4), 2231, 2007.
14. Lei Zhao, Xiao Hu. Polymer, 48(20), 6125, 2007.
15. *Thermal analysis of polymers: Fundamentals and Applications*, Editor: Menczel Joseph, D., Bruce Prime, R., 2008, A John Wiley & Sons, Inc.: Canada.
16. *Thermal Analysis: Fundamentals and Applications to Polymer Science*. Editor: Hatakeyama, T., Quin, F. X. 1999, John Wiley & Sons, Inc.
17. *Modulated-Temperature Differential Scanning Calorimetry: Theoretical and Practical Applications in Polymer Characterisation*, D. J. H. Mike Reading, Editor. 2006: The Netherlands.
18. Shi Zixing, Yu, D., Wang Yizhong, Xu Riwei. J., Applied Polymer Sci., 88(1), 194, 2003.
19. Segal, E., Fatu, D. Journal of Thermal Analysis, 9(1), 65, 1976.
20. Jerez, A. J. of Thermal Analysis 26(2), 315, 1983.
21. Li Chao-Rui, Tang Tong, B. J., of Materials Science, 34(14), 3467, 1999.
22. Perrin, F. X., Nguen, T. M. H., Vernet, J. L. J. Macromolecular Chem., and Physics, 208(7), 718, 2007.
23. Boey, F. Y. C., Qiang, W. Polymer, 41(6), 2081, 2000.
24. Xiong, Y., Boey, F. Y. C., Rath, S. K. Journal of App. Polymer Sci., 90(8), 2229, 2003.
25. Ozawa, T. J. of Thermal Analysis, 7(3), 601, 1975.
26. Luo Zh., Wei, L., Li, W., Lilu, F., Zhao, T. J. Applied Polymer Sci., 109(1). 525, 2008.
27. Ramírez, C., Rico, M., López, J., Montero, B., Montes, R. J. Applied Polymer Sci., 103(3), 1759, 2007.
28. Keenan, M. R. J. of Applied Polymer Science, 33(5), 1725, 2003.

INDEX

C

D

F

G

H

Q

R

S

U

V

W

X

Y

Z